Laser Fabrication and Machining of Materials

Narendra B. Dahotre • Sandip P. Harimkar

Laser Fabrication and Machining of Materials

 Springer

Narendra B. Dahotre
Professor of Materials Science
   and Engineering
326 Dougherty Hall
Department of Materials Science
   and Engineering
The University of Tennessee
Knoxville, TN 37996

Sandip P. Harimkar
420, Dougherty Hall,
Department of Materials Science
   and Engineering
The University of Tennessee
Knoxville, TN 37996

ISBN 978-0-387-72343-3   e-ISBN 978-0-387-72344-0

Library of Congress Control Number: 2007932417

Printed on acid-free paper.

9 8 7 6 5 4 3 2 1

springer.com

Preface

This book covers the fundamental principles and physical phenomena governing laser-based fabrication/machining processes and their existing and potential applications. It provides a link between advanced materials and advanced processing/ manufacturing techniques. It connects physical science and engineering aspects together. Laser machining is an emerging area in various applications ranging from bulk machining in metal forming to micromachining and microstructuring in electronics and biomedical applications. The uniqueness of lasers lies in flexible manufacturing using lasers as assist in conventional machining techniques that has emerged into new fields like laser-assisted mechanical machining (LAM), laser-assisted chemical machining (LCM), laser-assisted etching, etc. The book is intended to give a comprehensive overview of the principles and applications of lasers applied to material removal as well as forming processes in manufacturing, micromachining, and medical applications. In addition, the emerging field of LAM and machining using intersecting beams is introduced. These machining techniques have the potential for direct three-dimensional machining and microstructuring for a number of applications.

The content of this book is divided into four parts comprising 14 chapters. Part I deals exclusively with the basics of lasers used in materials processing and the principles of laser materials processing including physical processes during laser–material interactions. This part is intended to emphasize a dire need for understanding the physical principles of the complex tool (lasers) and processes during its use. Such understanding will allow better insight into the use of lasers in various processes described in the following parts of the book and possibly trigger thoughts for the development of new uses of lasers. Part II begins with an overview of conventional manufacturing processes and their configuration for laser-based manufacturing. The remaining chapters in Part II are mostly based on laser material removal processes such as drilling, cutting, and machining. As all laser-based processes are nonphysical (nonmechanical) and mostly thermal, the thermal principles governing material removal under various conditions are explained in these chapters. Part III includes chapters on laser processes mostly based on material addition such as rapid prototyping and welding along with the governing thermodynamic principles. Part IV dwells upon new and novel uses of lasers. Although use of lasers is new in medicine, it is rapidly making headways in the field of biomedicine. As mentioned earlier, Part I provides sufficient basics of lasers and laser processes to spawn thoughts in one's mind for new applications of lasers. Examples of such new applications are described in chapters on laser interference processing, laser shock processing, and laser dressing of grinding wheels. It is hoped that the book will provide further impetus to many more readers to take lasers into new frontiers of applications.

Thus, unlike many other books on similar topics, this book deals with the subject of laser-based fabrication and machining in a comprehensive manner. The subject matter in the proposed book extends over a wide range of disciplines that include, but are not limited to, mechanical, electrical/electronics, materials, and manufacturing. Such interdisciplinary discussion of the topic is a need of the time for many undergraduate and graduate academic programs and research activities that themselves are increasingly wielding interdisciplinary flavor. Furthermore, the scope of the subject matter ranging over macro-, to micro-, to nanolevels of processing/manufacturing is very important to educate and prepare individuals for the next generation manufacturing. In addition, the integration of fundamental principles and physical phenomena governing laser-based fabrication/machining processes and their existing and potential applications have widened the scope of the book across the sectors of academia, research, and industry.

Knoxville, TN   Narendra B. Dahotre
July 12, 2007   Sandip P. Harimkar

Acknowledgments

We are indebted to many individuals, publishers, and organizations for granting us the permission to reproduce copyrighted material in the present work. We also express our sincere appreciation to Jennifer Mirski, Gregory Franklin, and Alexander Greene of Springer Science+Business Media LLC for their guidance and assistance in producing this work. We also appreciate the assistance of Anoop Samant and Sameer Paital in enhancing the figures. Finally, we thank our family, friends, and teachers for their continuous encouragement and support.

Contents

Part I   Fundamentals of Laser Processing. **1**

1   Basics of Lasers **3**
   1.1   Introduction.   3
   1.2   Nature of Electromagnetic Radiation   3
   1.3   Laser Operation Mechanism.   6
      1.3.1   Population Inversion   6
      1.3.2   Stimulated Emission   8
      1.3.3   Amplification   9
   1.4   Properties of Laser Radiation   13
      1.4.1   Monochromaticity.   13
      1.4.2   Collimation   15
      1.4.3   Beam Coherence.   16
      1.4.4   Brightness or Radiance.   17
      1.4.5   Focal Spot Size   17
      1.4.6   Transverse Modes   18
      1.4.7   Temporal Modes.   20
      1.4.8   Frequency Multiplication   22
   1.5   Types of Industrial Lasers   23
      1.5.1   Solid-State Lasers.   24
      1.5.2   Gas Lasers.   25
      1.5.3   Semiconductor Lasers.   27
      1.5.4   Liquid Dye Lasers.   31

2   Laser Materials Interactions. **34**
   2.1   Introduction.   34
   2.2   Absorption of Laser Radiation   34
   2.3   Thermal Effects   40
      2.3.1   Heating   40
      2.3.2   Melting   43
      2.3.3   Vaporization   45
      2.3.4   Important Considerations for Thermal Analysis   50

2.4   Vapor Expansion and Recoil Pressures   55
2.5   Plasma Formation...   59
2.6   Ablation   61

Part II   Laser Machining.....................................   67

3   Manufacturing Processes: An Overview   69
3.1   Introduction..   69
3.2   Manufacturing Processes   69
       3.2.1   Casting Processes   70
       3.2.2   Forming Processes   71
       3.2.3   Joining Processes   74
       3.2.4   Machining Processes   75
3.3   Lasers in Manufacturing................................   79
       3.3.1   Laser Casting   79
       3.3.2   Laser Forming/Shaping.........................   80
       3.3.3   Laser Joining.................................   83
       3.3.4   Laser Machining...............................   84
3.4   Selection of Manufacturing Processes   85
       3.4.1   Properties of Materials   85
       3.4.2   Geometrical Complexity of Product..............   88
       3.4.3   Quality Parameters   90
       3.4.4   Manufacturing Economics   92

4   Laser Drilling   97
4.1   Introduction..   97
4.2   Laser Drilling Approaches   98
4.3   Melt Expulsion During Laser Drilling   99
4.4   Analysis of Laser Drilling Process........................   103
4.5   Quality Aspects...   117
4.6   Practical Considerations   124
       4.6.1   Effect of Laser Parameters   125
       4.6.2   Effect of Focusing Conditions   128
       4.6.3   Effect of Assist Gas Type, Gas Pressure,
                and Nozzle Design   129
4.7   Laser Drilling Applications   134
       4.7.1   Drilling of Cooling Holes.......................   134
       4.7.2   Drilling of Diamonds   136
       4.7.3   Microdrilling.................................   136
4.8   Advances in Laser Drilling.............................   137

5   Laser Cutting   144
5.1   Introduction..   144
5.2   Laser Cutting Approaches   145
       5.2.1   Evaporative Laser Cutting   146

5.2.2   Laser Fusion Cutting. 152
5.2.3   Reactive or Oxygen-Assisted Laser Cutting. 158
5.2.4   Controlled Fracture Technique 167
5.3   Quality Aspects. 171
    5.3.1   Striations. 171
    5.3.2   Dross. 176
    5.3.3   Heat-Affected Zone 179
5.4   Practical Considerations. 183
    5.4.1   Effect of Laser Type 183
    5.4.2   Effect of Laser Power 183
    5.4.3   Effect of Optical System. 186
    5.4.4   Effect of Nozzle Parameters 187
    5.4.5   Effect of Assist Gas Type 190
5.5   Laser Cutting of Various Materials 193
    5.5.1   Metallic Materials. 193
    5.5.2   Polymers. 197
    5.5.3   Ceramics and Glasses 198
5.6   Laser-Cutting Applications. 200
5.7   Advances in Laser Cutting 200
    5.7.1   Laser Cutting Assisted by Additional Energy
           Sources 200
    5.7.2   Underwater Cutting 201
    5.7.3   Laser Cutting of Composites and Laminates 202

6   **Three-Dimensional Laser Machining. 207**
6.1   Introduction. 207
6.2   Laser-Assisted Machining 207
    6.2.1   LAM Process 208
    6.2.2   Analysis of LAM Process. 209
    6.2.3   LAM Process Results 215
6.3   Laser Machining. 225
    6.3.1   Machining Using Single Laser Beam. 225
    6.3.2   Machining Using Intersecting Laser Beams. 233
6.4   Applications of Three-Dimensional Laser Machining. 243

7   **Laser Micromachining 247**
7.1   Introduction. 247
7.2   Laser Micromachining Mechanisms 247
    7.2.1   Laser Ablation 248
    7.2.2   Laser-Assisted Chemical Etching. 260
7.3   Laser Micromachining Techniques 265
    7.3.1   Direct Writing Technique. 265
    7.3.2   Mask Projection Technique 267
    7.3.3   Interference Technique 268
    7.3.4   Combined Techniques. 270

7.4   Laser Micromachining Applications   271
        7.4.1   Microvia Drilling   271
        7.4.2   Drilling of Inkjet Nozzle Holes   272
        7.4.3   Resistor Trimming   273
        7.4.4   Laser Scribing and Dicing   276
        7.4.5   Laser Marking and Engraving   277
        7.4.6   Biomedical Applications   280
        7.4.7   Thin Film Applications.   282
        7.4.8   Fuel Injector Drilling   283
        7.4.9   Stripping of Wire Insulation.   285

Part III   Laser Fabrication   289

8   Laser Forming   291
     8.1   Introduction.   291
     8.2   Laser Forming Processes   292
     8.2.1   Bending or Temperature Gradient Mechanism   294
     8.2.2   Buckling Mechanism.   296
     8.2.3   Upsetting Mechanism   299
     8.3 Analysis of Laser Forming Processes   299
     8.3.1   Temperature Gradient Mechanism   300
     8.3.2   Buckling Mechanism.   308
     8.4   Practical Considerations.   312
     8.4.1   Processing Parameters.   312
     8.4.2   Bending Rate and Edge Effects.   323
     8.4.3   Laser Forming Strategies and Control.   329
             8.4.4   Microstructure and Properties of Laser-Formed
                      Components   337
     8.5   Laser Forming Applications.   341
             8.5.1   Correction of Bending Angles   341
             8.5.2   Laser Forming of Complex Shapes   341
             8.5.3   Rapid Prototyping.   343
             8.5.4   Flexible Straightening of Car Body Shells.   343
             8.5.5   Microfabrication   344
             8.5.6   Laser Forming in Space   344
     8.6   Advances in Laser Forming   344
             8.6.1   Laser Forming of Tubes   344
             8.6.2   Laser Forming with Preload.   347
             8.6.3   Laser Forming with Two Beams.   348
             8.6.4   Laser Forming of Composite Materials   349

9   Laser-Based Rapid Prototyping Processes   353
     9.1   Introduction.   353
     9.2   Basics of Rapid Prototyping Processes   353
     9.3   Classification of Rapid Prototyping Processes.   355

9.3.1 Liquid-Based RP Processes.......................... 356
9.3.2 Powder-Based RP Processes........................ 357
9.3.3 Solid-Based RP Processes.......................... 358
9.3.4 Concept Modelers................................ 359
9.4 Laser-Based Rapid Prototyping Processes.................. 359
9.4.1 Stereolithography 359
9.4.2 Selective Laser Sintering.......................... 379
9.4.3 Laminated Object Manufacturing (LOM)............. 397
9.4.4 Laser Engineered Net Shaping (LENS) 402
9.5 Applications of Rapid Prototyping Processes................ 405
9.5.1 Applications in Rapid Prototyping 406
9.5.2 Applications in Rapid Tooling 407
9.5.3 Applications in Rapid Manufacturing................ 407
9.5.4 Medical Applications 407

10 Laser Welding... 412
10.1 Introduction... 412
10.2 Laser Welding Process 412
10.3 Analysis of Laser Welding Process 415
10.4 Quality Aspects...................................... 427
10.4.1 Porosity....................................... 427
10.4.2 Cracking 429
10.4.3 Heat-Affected Zone.............................. 431
10.4.4 Mechanical Properties of Laser Welds 432
10.5 Practical Considerations 433
10.5.1 Effect of Laser Parameters 433
10.5.2 Effect of Focusing Conditions..................... 435
10.5.3 Effect of Shielding Gas........................... 435
10.5.4 Effect of Welding Speed 437
10.5.5 Joint Configurations 437
10.6 Laser Welding of Various Materials...................... 438
10.6.1 Metallic Materials............................... 438
10.6.2 Ceramics 439
10.6.3 Polymers 440
10.6.4 Composites 440
10.6.5 Dissimilar Materials 442
10.7 Advances in Laser Welding 443
10.7.1 Arc-Augmented Laser Welding 443
10.7.2 Multibeam/Dual-Beam Laser Welding.............. 443
10.7.3 Laser Welding of Tailor-Welded Blanks............. 444

Part IV Special Topics in Laser Processing 449

11 Laser Interference Processing. 451
  11.1 Introduction... 451
  11.2 Theory of Interference 451
  11.3 Interferometry for Surface Processing of Materials 453
    11.3.1 Laser and Materials Aspects 454
    11.3.2 Interferometer Design Aspects 456
  11.4 Applications of Laser Interference Processing. 459
    11.4.1 Crystallization and Structuring of
         Semiconductor Films 459
    11.4.2 Structuring of Monolayer and Multilayer
         Metallic Films. 464
    11.4.3 Structuring of Biomaterials. 472

12 Laser Shock Processing 477
  12.1 Introduction... 477
  12.2 Fundamentals of Laser Shock Processing 477
  12.3 Analysis of Laser Shock Processing Process 480
  12.4 Processing Parameters. 484
    12.4.1 Effect of Laser Intensity 484
    12.4.2 Effect of Pulse Duration 484
    12.4.3 Effect of Temporal Pulse Shape 486
    12.4.4 Effect of Laser Wavelength. 486
    12.4.5 Effect of Spatial Energy Distribution 488
    12.4.6 Effect of Coating Material. 489
  12.5 Mechanical Effects During Laser Shock Processing 491
  12.6 Microstructure Modification During Laser Shock
       Processing.. 493
  12.7 Applications of Laser Shock Processing 496

13 Laser Dressing of Grinding Wheels 499
  13.1 Introduction... 499
  13.2 Grinding Process and Need of Wheel Dressing 499
  13.3 Laser-Based Wheel Dressing Techniques. 502
    13.3.1 Laser-Assisted Dressing 502
    13.3.2 Laser Dressing. 506
    13.3.3 Recent Approaches of Laser Dressing 511

14 Lasers Processing in Medicine and Surgery 522
  14.1 Introduction... 522
  14.2 Laser–Tissue Interactions 522
    14.2.1 Photothermal Interactions 526
    14.2.2 Photochemical Interactions (Photoablation) 529
    14.2.3 Photodisruptive Interactions 532

14.3 Laser Applications in Medicine and Surgery 534
 14.3.1 Lasers in Ophthalmology 534
 14.3.2 Lasers in Dermatology 540
 14.3.3 Lasers in Otolaryngology–Head and Neck
  Surgery 545
 14.3.4 Lasers in Angioplasty 545
 14.3.5 Lasers in Osteotomy 547
 14.3.6 Lasers in Dentistry 547

Index **555**

Part I
Fundamentals of Laser Processing

Chapter 1
Basics of Lasers

1.1 Introduction

Laser is an acronym for Light Amplification by Stimulated Emission of Radiation. The world's first laser was demonstrated by Maiman using a ruby crystal (Maiman 1960). It is essentially a coherent, convergent, and monochromatic beam of electromagnetic radiation with wavelength ranging from ultraviolet to infrared. Lasers have now found applications in almost every field of engineering, medicine, electronics, etc., where one or more properties of the laser radiation are important (Ion 2005; Dausinger et al. 2004; Wolbarsht 1991; Steen 1991). In order to realize the applicability and capability of a laser radiation in any application, it is necessary to understand the basic operation mechanism and properties of laser radiation. These aspects of laser radiations along with the important industrial laser types are briefly discussed in this chapter.

1.2 Nature of Electromagnetic Radiation

Electromagnetic radiations consist of propagating waves associated with the oscillating electric field (E) and magnetic field (H). These components oscillate at right angles to each other and also to the direction of propagation of wave. Since the magnetic field vector is perpendicular to the electric field vector, the description of the propagation of the wave generally considers the oscillation of the electric field vector only. When the oscillations of the electric field vector are in particular order, the light is said to be polarized. In a plane polarized light the electric vector oscillates in a single plane as the wave travels. This is illustrated in Fig. 1.1 for a wave propagating in x-direction while the electric vector is oscillating in x–y plane. In contrast, the electric vectors in the completely unpolarized light can assume any possible directions (i.e., electric vector oscillating randomly in more than one plane). For a plane-polarized wave shown in Fig. 1.1, the electric vector oscillating in y-direction varies with space and time.

N.B. Dahotre and S.P. Harimkar, *Laser Fabrication and Machining of Materials.*
© Springer 2008

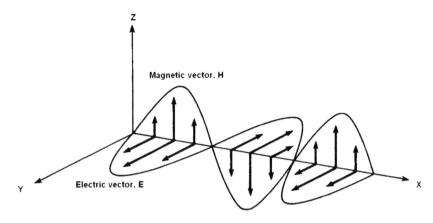

Fig. 1.1 Schematic of the oscillations of electric (E) and magnetic (H) field vectors associated with plane electromagnetic wave

For sinusoidal variation, this can be expressed as (Cullity 1978; Lorrain and Corson 1970):

$$\mathbf{E} = A \sin 2\pi \left(\frac{x}{\lambda} - vt \right), \tag{1.1}$$

where A is the amplitude, λ is the wavelength, and v is the frequency of the wave. The wavelength and frequency of all the electromagnetic waves exhibit a simple relationship:

$$c = n\lambda, \tag{1.2}$$

where c is the velocity of light (2.9979×10^8 m/s in vacuum). The strength of the electromagnetic radiation is often described in terms of intensity of radiation. Intensity is defined as the energy per unit area perpendicular to the direction of motion of the wave and is proportional to the square of amplitude of the wave. Based on the wavelength (or frequency/energy), the electromagnetic spectrum can be divided into various regions. Figure 1.2 presents the entire electromagnetic spectrum consisting of radio waves, microwaves, infrared radiation, visible light, ultraviolet radiation, x-ray, and gamma ray radiation in the order of decreasing wavelength. The description of the exact dividing wavelength between two adjacent regions in electromagnetic spectrum is difficult and hence the wavelengths of various regions are often expressed in approximate ranges.

The wave nature of the electromagnetic radiation (classical theory) discussed so far fails to explain some of the phenomena such as photoelectric effect. To explain such phenomena, quantum theory of electromagnetic radiation has been

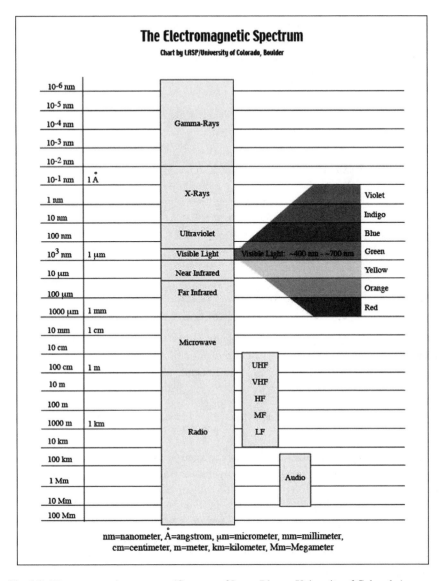

Fig. 1.2 Electromagnetic spectrum. (Courtesy of Laura Bloom, University of Colorado.)

proposed. In quantum theory, the electromagnetic radiation is considered as a stream of particles which are called photons. Each photon is associated with an amount of energy proportional to its frequency. This photon energy is expressed as:

$$E = hn \tag{1.3}$$

where h is the Planck constant (6.63×10^{-34} J/s). Thus, electromagnetic radiation is often said to have dual nature: sometimes wave nature and sometimes particle nature. The wave and the quantum theories of electromagnetic radiation are often regarded as complementary rather than conflicting (Wilson and Hawkes 1987).

1.3 Laser Operation Mechanism

Stimulated emission, the underlying concept of laser operation, was first introduced by Einstein in 1917 in one of his three papers on the quantum theory of radiation (Einstein 1917). Almost half a century later, in 1960, T.H. Maiman came up with the first working ruby laser. The three processes required to produce the high-energy laser beam are population inversion, stimulated emission, and amplification.

1.3.1 Population Inversion

Population inversion is a necessary condition for stimulated emission. Without population inversion, there will be net absorption of emission instead of stimulated emission. For a material in thermal equilibrium, the distribution of electrons in various energy states is given by the Boltzmann distribution law:

$$N_2 = N_1 \exp\left[-\left(E_2 - E_1\right)/kT\right] \tag{1.4}$$

where N_1 and N_2 are the electron densities in states 1 and 2 with energies E_1 and E_2, respectively. T and k are the absolute temperature and Boltzmann constant, respectively. According to the Boltzmann law, the higher energy states are the least populated and the population of electrons in the higher energy states decreases exponentially with energy (Fig. 1.3a). Population inversion corresponds to a non-equilibrium distribution of electrons such that the higher energy states have a larger number of electrons than the lower energy states (Fig. 1.3b). The process of achieving the population inversion by exciting the electrons to the higher energy states is referred to as pumping (Svelto and Hanna 1989). The population inversion explained here for the two-level energy systems is only for the introduction of the concept. In actual practice, it is impossible to achieve the population inversion in two-level energy systems. Population inversion in most of the lasers generally involves three- or four-level energy levels (Fig. 1.4). For a three-level energy system, electrons are first pumped from energy level E_0 to E_2 by the absorption of radiation (of frequency, $v = (E_2 - E_0)/h$) from a pumping source. The lifetime of the electrons in the higher energy level E_2 is generally very short and the electrons from the energy level E_2 rapidly decay into metastable energy level E_1 without any radiation (radiationless decay). Thus, the net population inversion is achieved between

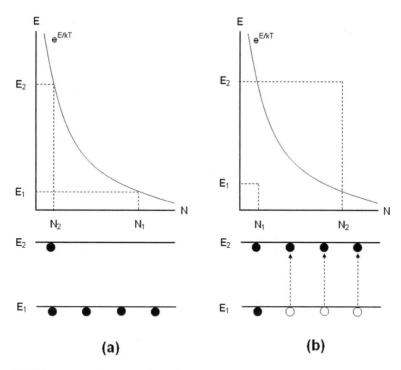

Fig. 1.3 Schematic of the population of electrons in two-level energy systems: **(a)** thermal equilibrium and **(b)** population inversion

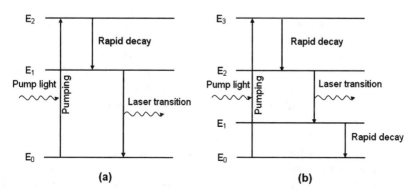

Fig. 1.4 Schematic of the population inversion in **(a)** three-level and **(b)** four-level energy laser systems

the energy level E_1 and E_0, which is responsible for the subsequent emission of laser radiation (Fig. 1.4a). Similar mechanisms cause the population inversion between the energy levels E_2 and E_1 in four-level energy laser systems (Fig. 1.4b) (Ready 1997).

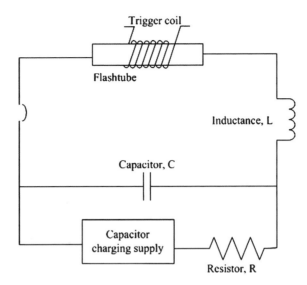

Fig. 1.5 Typical circuitry for the operation of flashlamp

   In general, population inversion is achieved by optical pumping and electrical pumping. In optical pumping, gas-filled flashlamps are most popular. Flashlamps are essentially glass or quartz tubes filled with gases such as xenon and krypton. Some
wavelength of the flash (emission spectrum of flashlamp) matches with the absorption
characteristics of the active laser medium facilitating population inversion. This is used
in solid-state lasers like ruby and Nd:YAG (yttrium–aluminum–garnet). Typical circuit
for the flashlamp operation is shown in Fig. 1.5. Recently, significant interests have
been focused towards using diode lasers of suitable wavelength for pumping the solid-
state lasers. This led to the development of diode-pumped solid-state lasers (DPSS).
The use of diode lasers offers significant advantages over conventional flashlamps such
as better match between the output spectrum of the pumping laser and absorption characteristic of laser medium, increased efficiency, and compact and lighter systems.
Electrical pumping, used in gas lasers, is achieved by passing a high-voltage electric
current directly through the mixture of active gas medium. The collision of discharge
electrons of sufficient kinetic energy excites one of the gases to high energy levels,
which subsequently transfer its excitation energy to the second gas through collision,
achieving the population inversion. There is minimum population inversion, referred to
as threshold condition, required for lasing action (Milloni and Eberly 1988).

1.3.2 Stimulated Emission

Stimulated emission results when the incoming photon of frequency v, such that
$hv = (E_2 - E_1)/h$, interacts with the excited atom of active laser medium with population inversion between the states 1 and 2 with energies E_1 and E_2, respectively.

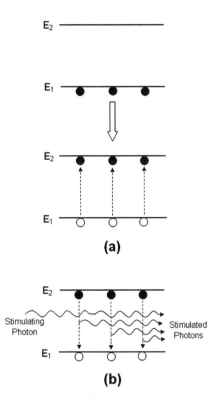

Fig. 1.6 Schematic diagram of laser operation (**a**) pumping and (**b**) stimulated emission

Thus, the incoming photon (stimulating photon) triggers the emission of radiation by bringing the atom to the lower energy state (Fig. 1.6). The resulting radiations have the same frequency, direction of travel, and phase as that of the incoming photon, giving rise to a stream of photons (Haken 1983).

1.3.3 Amplification

Since the stimulated photons are in the same phase and state of polarization, they add constructively to the incoming photon resulting in an increase in its amplitude. Thus, the amplification of the light can be achieved by stimulated emission of radiation. Amplification of laser light is accomplished in a resonant cavity consisting of a set of well-aligned highly reflecting mirrors at the ends, perpendicular to the cavity axis. The active laser material is placed in between the mirrors. Usually, one of the mirrors is fully reflective with reflectivity close to 100%, whereas the other mirror has some transmission to allow the laser output to emerge (Thyagarajan and Ghatak 1981). Figure 1.7 presents the schematic of the amplification process in the

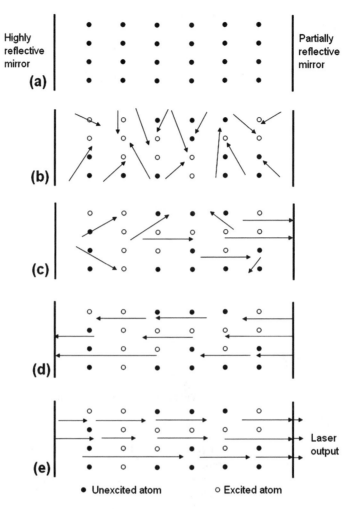

Fig. 1.7 Schematic of amplification stages during operation: (**a**) initial unexcited state (laser off), (**b**) optical pumping resulting in excited state, (**c**) initiation of stimulated emission, (**d**) amplification by stimulated emission, and (**e**) continued amplification due to repeated reflection from the end mirrors resulting in subsequent laser output from one end of the mirror

resonator with flat mirrors at the ends and the active laser material in between the mirrors. When the laser is off, the optical cavity contains all the laser material in its initial unexcited state (Fig. 1.7a). The excitation of the atoms (population inversion) is soon achieved by optical pumping (Fig. 1.7b), followed by initiation of stimulated emission (Fig. 1.7c). The intensity of the stimulated radiation is increased as it travels to the end of the mirrors. Further amplification is accomplished by reflecting the photons into the active medium (Fig. 1.7d). The photons travel the long path back and forth through the lasing medium stimulating more and more emissions resulting in a high-intensity laser beam output from one of the mirrors (Fig. 1.7e) (Chryssolouris 1991).

To initiate and sustain the laser oscillations, the gain within the resonant cavity must be high enough to overcome various losses. This condition is defined by a threshold gain coefficient, k_{th}. If the beam irradiance increases from I_0 to I_1 while traveling from one end of the cavity to the other (cavity length, L), then, in the case of no cavity losses, the irradiance I_1 can be expressed as (Wilson and Hawkes 1987):

$$I_1(L) = I_0 e^{kL} \tag{1.5}$$

where k is defined as the gain coefficient. If we define a single loss coefficient γ for various losses such as diffraction losses from the edges of the mirrors, absorption and scattering by the mirrors, absorption and scattering in the laser medium, etc., then the effective gain coefficient becomes $k - \gamma$ and the irradiance I_1 can be expressed as:

$$I_1(L) = I_0 e^{(k-\gamma)L} \tag{1.6}$$

If r_1 and r_2 are the reflectances of mirrors 1 and 2, respectively, the irradiance after reflection at mirror 1 becomes $r_1 I_0 e^{(k-\gamma)L}$. Then, the final irradiance after the round trip (including reflection from mirror 2) is given by:

$$I_1 = r_1 r_2 I_0 e^{2(k-\gamma)L} \tag{1.7}$$

The round trip gain defined as the ratio of final irradiance to the initial irradiance is given by:

$$G = \frac{I_1}{I_0} = \frac{r_1 r_2 I_0 e^{2(k-\gamma)L}}{I_0} = r_1 r_2 e^{2(k-\gamma)L}. \tag{1.8}$$

To initiate and sustain the laser oscillation, the gain in the resonant cavity must be equal to, or greater than, unity. If the gain is less than unity, the losses in cavity will cause the cessation of oscillations. The threshold condition, thus, can be expressed as:

$$r_1 r_2 e^{2(k_{th}-\gamma)L} = 1, \tag{1.9}$$

where k_{th} is defined as threshold gain coefficient and is given by (Wilson and Hawkes 1987):

$$k_{th} = \gamma + \frac{1}{2L} \ln\left(\frac{1}{r_1 r_2}\right). \tag{1.10}$$

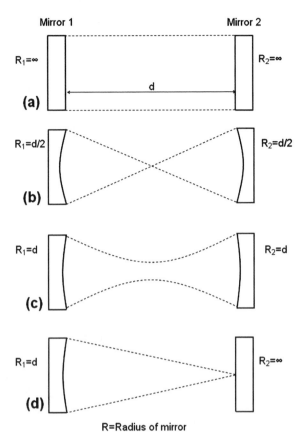

Fig. 1.8 Various mirror configurations for resonant cavities: (**a**) plane-parallel, (**b**) spherical, (**c**) confocal, and (**d**) hemispherical

The preceding discussion on the amplification of stimulated emission assumed that the mirrors of the resonant cavity are flat (plane parallel). However, there are various other configurations which offer significant advantages over the flat mirrors (Haken 1983). Various possible configurations of the resonant cavity mirrors are presented in Fig. 1.8. The important considerations during the designing of the mirrors are the extent of mode volume and the stability of the cavity. The mode volume can be defined as the fraction of the excited laser medium with which the light interacts while oscillating to and fro in the resonant cavity. The extent of mode volume is indicated by the area enclosed by the dashed lines in Fig. 1.8. The stability of the cavity is related with the ability to retain the light rays within the cavity after several reflections between the mirrors. The plane parallel mirrors have the maximum mode volume. However, slight misalignment may cause the light rays to move off the mirrors after few reflections. Thus, the plane parallel mirrors have high mode volume but relatively low stability. Several combinations of spherical mirrors offer very good stability. However, spherical mirrors are associated with small mode volumes.

The stability of the resonant cavity is determined by the radii of curvatures of the end mirrors and the length of cavity. Based on the ray transfer matrix analysis, the condition of the stability can be expressed as (Kogelnik and Li 1966):

$$0 < \left(1 - \frac{d}{R_1}\right)\left(1 - \frac{d}{R_2}\right) < 1, \tag{1.11}$$

where d is the length of the cavity, and R_1 and R_2 are the radii of curvatures of two mirrors. The condition of stability can be expressed graphically by plotting parameters g_1 and g_2 as coordinate axes (Fig. 1.9). These parameters are defined as:

$$g_1 = 1 - \frac{d}{R_1}, \tag{1.12}$$

$$g_2 = 1 - \frac{d}{R_2}.$$

The unstable areas are represented by points lying in the shaded area indicated in the graph (Fig. 1.9) (Kogelnik and Li 1966).

1.4 Properties of Laser Radiation

The laser light is characterized by a number of interesting properties. Various applications of lasers exploit specific combinations of the laser properties. This section briefly explains the most important properties of laser light.

1.4.1 Monochromaticity

Monochromaticity is the most important property of laser beam and is measured in terms of spectral line width. The laser output consists of very closely spaced, discrete, and narrow spectral lines, which satisfies the resonance condition given by:

$$d = \frac{n\lambda}{2}, \tag{1.13}$$

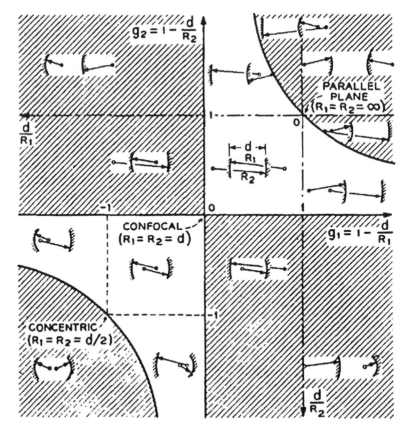

Fig. 1.9 Graphical representation of stability condition for resonant cavities with two mirrors of varying radii of curvatures. (Reprinted from Kogelnik and Li 1966. With permission. Copyright Optical Society of America.)

where d is the cavity length, n is an integer, and λ is the wavelength. These discrete lines, called laser modes or cavity modes, spread over a range of frequencies separated by $c/2d$, where, c is speed of light (Fig. 1.10). The frequencies within this range that are amplified by stimulated emission depend mainly on the losses in the cavity and the gain characteristics. The modes in the laser output correspond to those having gain greater than loss. The number of modes in laser output equals the range of frequencies divided by spacing between the modes. The spectral width of the CO_2 laser ($\lambda = 10.6\,\mu m$) is around 3 GHz. Hence, for a cavity of 0.5 m, the calculated number of modes is around 10 [(3 $\times 10^9)/(3 \times 10^8/2 \times 0.5) = 10$] (Wilson and Hawkes 1987). The number of axial modes may exceed hundreds of modes (e.g., in Nd:glass laser). Monochromaticity is due to narrow spectral widths of individual modes. A laser can be constructed to operate in only one longitudinal mode to give better monochromaticity (Ready 1997).

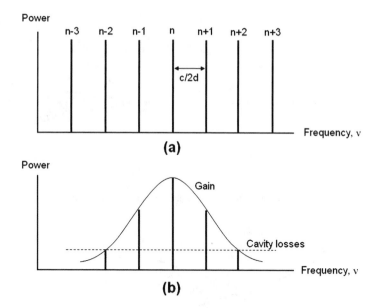

Fig. 1.10 Frequency spectrum in laser output: (**a**) laser cavity modes, and (**b**) axial modes in laser output

1.4.2 Collimation

Collimation of the laser radiation is related with the directional nature of the beam. Highly directional beams are said be highly collimated beams, which can be focused on a very small area even at longer distances. Hence, energy can be efficiently collected on a small area without much loss in the beam intensity. The degree of collimation is directly related with the beam divergence angles (Duley 1983). For a confocal laser cavity of length d, the divergence angle can be expressed as (Thyagarajan and Ghatak 1981):

$$\theta = \sin^{-1}\left(\frac{\lambda}{\pi w_0}\right) \approx \left(\frac{\lambda}{\pi w_0}\right), \tag{1.14}$$

where w_0 is the smallest value of sideways spread of the beam and often called beam waist (Fig. 1.11). The smallest value of sideways spread, w_0, is expressed as:

$$w_0 = \left(\frac{\lambda d}{2\pi}\right)^{1/2}. \tag{1.15}$$

Ideally, the divergence angle should be zero for highest collimation. However, this is impossible due to physical limits set by the diffraction phenomenon. The

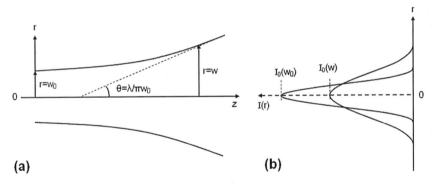

Fig. 1.11 Characteristics of Gaussian beam: **(a)** beam divergence, and **(b)** intensity distribution

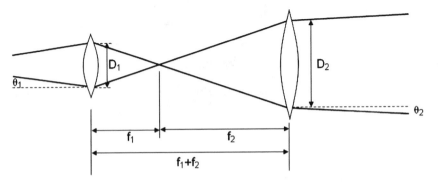

Fig. 1.12 Reverse telescope arrangement for improving the collimation (i.e., reducing the divergence) of the beam

laser beams are characterized by very small divergence angle. The beam divergence angles for most of the lasers (except semiconductor lasers) range from 0.2 to 10 milliradians. Collimation of the laser radiation can be improved by using additional optics such as reverse telescope (Fig. 1.12). With the eyepiece and objective lenses of focal lengths f_1 and f_2, respectively, the beam divergence is decreased by the factor f_1/f_2 and the beam width is enlarged by the factor f_2/f_1.

1.4.3 Beam Coherence

Coherence is the degree of orderliness of waves and is specified in terms of mutual coherence function, $\gamma_{12}(\tau)$, which is a measure of the correlation between the light wave at two points P_1 and P_2 at different times t and $t + \tau$. The absolute value of $\gamma_{12}(\tau)$ lies between 0 and 1, corresponding to completely incoherent beam and coherent beam, respectively (Ready 1997). The two components of beam coherence are spatial and temporal coherence (Haken 1983). Spatial coherence

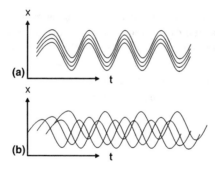

Fig. 1.13 Schematic illustrating the concept of coherence: (**a**) spatially and temporally coherent light, and (**b**) spatially and temporally incoherent light

correlates the phases at different points in space at a single moment in time, whereas temporal coherence correlates the phases at a single point in space over a period of time. Figure 1.13 illustrates the concept of temporal and spatial coherence. The two important quantities related to temporal coherence are coherence time and coherence length. Coherence properties can be improved by operating the laser in a single longitudinal and transverse mode (Rubahn 1999). Coherence of laser beam is of particular interest in the applications such as interferometry and holography.

1.4.4 Brightness or Radiance

Brightness or radiance is defined as the amount of power emitted per unit area per unit solid angle. Laser beams are emitted into very small divergence angles in the range of 10^{-6} steradians, hence it can be focused on a very small area ensuring the correspondingly high brightness of laser beams. Brightness of the laser beam is a very important factor in materials processing and determines the intensity (power density) or fluence (energy density) of the laser beam. The brightness of the source cannot be increased by the optical system; however, high brightness characteristics are influenced by operating the lasers in Gaussian mode with minimum divergence angle and high output power (Ready 1997).

1.4.5 Focal Spot Size

The spot radius is the distance from the axis of the beam to the point at which the intensity drops to $1/e^2$ from its value at the center of the beam (Duley 1983). Focal spot size determines the irradiance, which is of prime importance in materials processing; for example, the dominant mechanism of material removal during laser machining such as surface melting or evaporation and the consequent rate of material removal directly depends on the irradiance at the surface. The maximum irradiance corresponds to the minimum diameter of spot. However, it is not possible to focus the beam to an infinitesimal point and there is always a minimum spot size determined by diffraction limit. For a Gaussian laser beam, the diffraction limited minimum spot radius, r_s, is approximately given by:

$$r_s = \lambda F, \tag{1.16}$$

where F denotes number of the lens. Since it is impractical to work with F-numbers much less than unity, the minimum spot radius is approximately equal to the wavelength of the radiation. The diffraction-limited spot size is set by the principles of optics. The performance of the focusing optics may be further degraded by lens aberrations. In general, best focusing conditions are primarily obtained by using Gaussian beam profile and lenses with minimum aberrations.

1.4.6 Transverse Modes

The cross sections of laser beams exhibit certain distinct spatial profiles termed as transverse modes and are represented as the transverse electromagnetic mode, TEM_{mn}, where m and n are small integers representing the number of nodes in direction orthogonal to the direction of propagation of beam. The various transverse modes are shown in Fig. 1.14 (Kogelnik and Li 1966). The fundamental mode TEM_{00} has Gaussian spatial distribution and is the most commonly used mode in laser machining applications. The intensity distribution in the Gaussian beam can be expressed as:

$$I(r) = I_0 \exp\left[\frac{-2r^2}{w^2}\right], \tag{1.17}$$

where r is the radius of the beam, I_0 is the intensity of the beam at $r = 0$, and w is the radius of the beam at which $I = I_0 e^{-2}$ (Fig. 1.11b). The quality of any beam is often expressed in terms of the factor M^2, which compares the divergence of any given beam with that of a pure Gaussian beam ($M^2 = 1$). For higher-order modes, the value of M^2 is greater than unity. The laser can be made to operate in single Gaussian mode while discarding the higher-order modes by inserting a circular aperture within the cavity; however, this is sometimes accomplished at the expense of some output power.

Sometimes, it is desired to have the uniform distribution of intensity in the beam. This can be achieved by beam shaping optics. Figure 1.15 shows the sche-

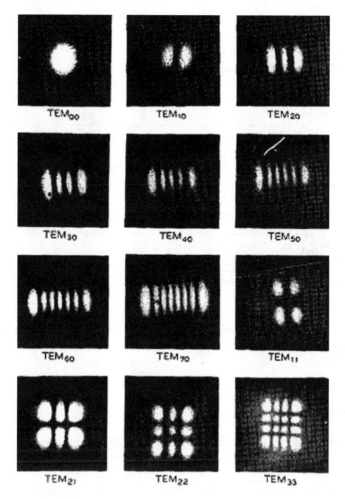

Fig. 1.14 Spatial modes of laser operation. (Reprinted from Kogelnik and Li 1966. With permission. Copyright Optical Society of America.)

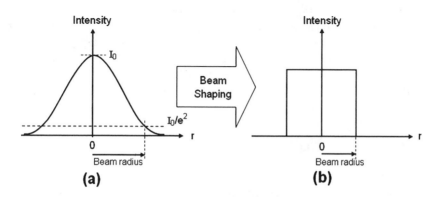

Fig. 1.15 Beam shaping: (**a**) Gaussian beam, and (**b**) shaped "top-hat" beam

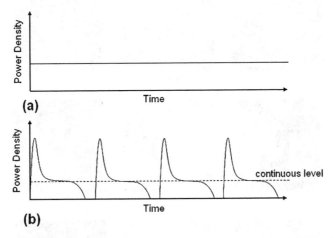

Fig. 1.16 Temporal modes of laser operation: **(a)** continuous wave mode, and **(b)** pulsed mode

matic of one such beam shaping operation where the Gaussian beam is converted into "top-hat" profile with uniform intensity distribution.

1.4.7   Temporal Modes

The output of laser can either be continuous, constant amplitude, known as continuous wave (CW) mode, or periodic, known as pulsed beam mode (Fig. 1.16). In continuous beam operation, constant laser energy is discharged uninterruptedly for a long time. In pulsed mode of operation, the pumped energy is stored until a threshold is reached. Once the threshold is reached, the stored energy is rapidly discharged into short duration pulses of high energy density. In general, most of the gas lasers and some of the solid-state lasers (Nd:YAG, dye, semiconductor laser) are operated in continuous mode. Solid-state lasers such as ruby Nd:glass lasers are primarily operated in pulsed mode. One of the important parameters in the pulsed laser operation is the pulse repetition rate. Pulse repetition rate is defined as the number of pulses emitted per unit time. For pulsed lasers the pulsing may be carried out in various ways: normal pulsing, Q-switching, and mode locking (Ready 1997).

1.4.7.1   Normal Pulsing

In normal pulse operation, the laser pulse duration is primarily controlled by changing parameters of the flashlamp. This is generally achieved by varying the inductance and the capacitance in the circuitry of the flashlamp. No

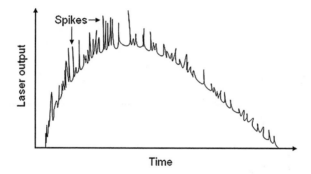

Fig. 1.17 Shape of the normal pulse with a number of spikes of nonuniform amplitudes and durations (typically in the range of microseconds)

intentional attempts are made to change the properties of the resonator during normal pulsed operation and the laser output is allowed to emerge at its natural rate. The typical pulse durations for the normal pulses are of the order of microseconds to milliseconds. The typical shape of the normal pulse is shown in Fig. 1.16b. As indicated in the figure, each pulse is characterized by an initial spike with peak power around two or three times the average power during the pulse. After this initial spike, the power drops down. The pulse (laser output) shown in Fig. 1.16b is approximated by a smooth curve. In actual practice, the normal pulse shape may exhibit the complex shape characterized by spikes of microsecond duration (Fig. 1.17). These microsecond spikes with nonuniform amplitude and spacing are often referred to as relaxation oscillations (Ready 1997).

1.4.7.2 Q-Switching

The properties of the output laser pulse can be modified by changing the properties of the resonant cavity. In Q-switching, short and intense pulse of laser radiation is obtained by changing the Q value of the cavity. Q value of the cavity is the measure of ability of the cavity to store the radiant energy. When the Q value is high, energy will be efficiently stored in the cavity without significant laser radiation. If the Q value of the cavity is lowered, the stored energy will emerge as short and intense pulse of laser beam. Thus, Q-switching involves the "switching" of Q values of the resonant cavity leading to the emergence of short and intense pulse (high peak power) of laser radiation. Various methods of Q switching are: rotating mirror method, electro-optic Q-switching, acousto-optic Q-switching, and passive Q-switching. Pulse repetition rates as high as a few hundred kilo hertz can be obtained in Q-switched laser operation (Ready 1997).

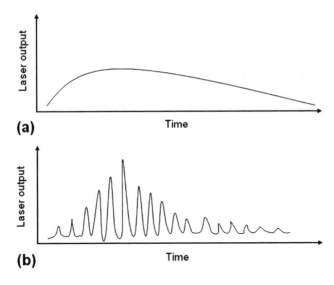

Fig. 1.18 Schematic of the pulse characteristic in the (**a**) Q-switched laser operation and (**b**) Q-switched with mode locking operation

1.4.7.3 Mode Locking

As explained in Section 1.4.7.2, Q-switching produces short and intense pulse after each pumping pulse in the laser output. In mode locking operation, a train of extremely short and equally spaced pulses is produced (Fig. 1.18). The mode locking is due to interaction between the longitudinal modes and results in oscillatory behavior of the laser output. It can be achieved by modulating the loss or gain of the laser cavity at a frequency equal to the intermode frequency separation ($\Delta v = c/2d$). This makes the longitudinal modes maintain fixed phase relationship resulting in mode locking. Typical pulse repetition rates in mode-locked laser operation are in the range of megahertz to gigahertz (Ready 1997).

1.4.8 Frequency Multiplication

The frequency of the laser beam can be multiplied by using frequency multiplier materials. The frequency multiplier materials are characterized by the beam's non-linear response to the electric field. The frequency of the laser radiation can be doubled (second harmonic generation) or tripled (third harmonic generation) without adversely affecting the basic properties of laser radiation. Common frequency multiplier materials include potassium dihydrogen phosphate (KDP), and lithium niobate (Wilson and Hawkes 1987).

1.5   Types of Industrial Lasers

Since the development of the first ruby laser in 1960, the laser action has been demonstrated in hundreds of materials. However, the range and variety of active materials for commercial lasers are still limited. Lasers are generally classified into four main types depending on the physical nature of the active medium used: solid-state lasers, gas lasers, semiconductor lasers, and dye lasers. Table 1.1 gives the list of important lasers in each category. It is beyond the scope of this book to review the principles of operation of all these lasers. Extensive discussion of the principles of laser operation in various laser systems can be found in various standard texts. Here only few laser systems typical of each class are explained.

Table 1.1 Typical wavelengths of various types of lasers

Laser type	Wavelength (nm)
Solid-state lasers	
Nd:YAG	1,064
Ruby	694
Nd:glass	1,062
Alexandrite	700–820
Ti-sapphire	700–1,100
Er:YAG	2,940
Nd:YLF	1,047
Gas lasers	
HeNe	632.8
Argon	488, 514.5
Krypton	520–676
HeCd	441.5, 325
CO_2	10,600
ArF	191
KrF	249
XeCl	308
XeF	351
Copper vapor	510.6, 578.2
Gold vapor	628
Semiconductor lasers	
InGaAs	980
AlGaInP	630–680
InGaAsP	1,150–1,650
AlGaAs	780–880
Liquid dye lasers	
Rhodamine 6G	570–640
Coumarin 102	460–515
Stilbene	403–428

1.5.1 Solid-State Lasers

In solid-state lasers, active medium consists of a small percentage of impurity ions doped in a solid host material. The first practical solid-state laser was the ruby laser developed by Maimam in 1960. Large numbers of lasers such as Nd:YAG, Nd: glass, alexandrite, and Ti:sapphire are now available in this class. Among these, Nd:YAG laser is the most commonly used one in the laser machining applications. Hence, operating principles of Nd:YAG laser are explained here.

1.5.1.1 Nd:YAG Laser

Nd:YAG laser consists of crystalline YAG with a chemical formula $Y_3Al_5O_{12}$ as a host material. The Nd^{3+} ions substitute yttrium ion sites in the lattice with a maximum doping level of around 2%. This is a typical four-level energy laser system earlier illustrated in Fig. 1.4b. Such systems offer significant advantages such as ease of achieving population inversions. Hence, simple designs of flashlamps with modest amount of pumping energy are sufficient to achieve the efficient population inversions. The energy levels involved in the population inversion and the laser transitions are shown in Fig. 1.19. Laser transitions takes place between

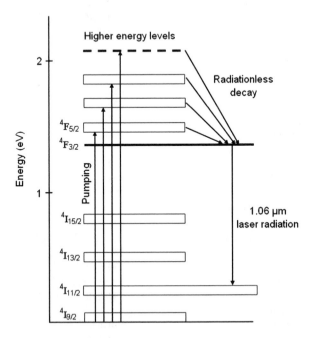

Fig. 1.19 Schematic of the energy levels of neodymium ion showing the levels involved in population inversion (pumping) and laser transitions

the $^4F_{3/2}$ level and the $^4I_{11/2}$ level (Svelto and Hanna 1989). Due to splitting of initial and final energy levels, several lasing wavelengths are possible, 1.064 μm being the strongest one. The output of the Nd:YAG laser can be continuous, pulsed, or Q-switched. The light source for pumping depends on the absorption characteristics of the crystal. For continuous operation the laser is excited by continuous krypton-filled or xenon-filled arc lamps or semiconductor diode lasers. Krypton lamps are efficient pumping sources for continuous Nd:YAG laser because the emission lines from Krypton lamps agrees better with absorption lines in Nd:YAG. For pulsed operation, flashlamps are generally used. If the pulses of relatively large-pulse energy are desired, the laser is excited by a flash-lamp, which gives pulses at relatively low pulse repetition rates. Nd:YAG laser is also available in frequency-doubled mode in which the output of the laser is in the green portion of the visible spectrum at 532 nm. In addition to frequency-doubled operation, the laser is also available in frequency-tripled (355 nm) and frequency-quadrupled (266 nm) modes.

1.5.2 Gas Lasers

In gas lasers, as the name suggests, the active laser medium is gas. Gaseous laser materials offer significant advantages over solid material. Some of these advantages are:

1. Gases acts as homogeneous laser medium.
2. Gases can be easily transported for cooling and replenishment.
3. Gases are relatively inexpensive.

However, due to physical nature of the gases (low densities), a large volume of gas is required to achieve the significant population inversion for laser action. Hence, gas lasers are usually relatively larger than the solid-state lasers. Gas lasers can be classified into atomic, ionic, and molecular lasers depending on whether the laser transitions are taking place between the energy levels of atoms, ions, and mole-cules, respectively. There are several laser systems in each class. Only some of the typical gas lasers are explained in this section.

1.5.2.1 CO_2 Laser

CO_2 laser is one of the most important lasers in the laser machining of materials. This is a molecular gas laser consisting of CO_2 gas as its active medium. The CO_2 molecule can undergo three different types of vibrations such as symmetric stretch-ing, bending, and asymmetric stretching (Fig. 1.20). Energy associated with these vibrational modes is quantized (Thyagarajan and Ghatak 1981). In addition to the symmetric and asymmetric vibrations, the molecule can also rotate. However, the energy associated with the rotational modes is much smaller than the vibrational

O - Oxygen   C - Carbon

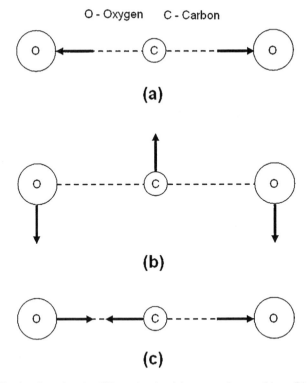

Fig. 1.20 Vibrational modes the CO_2 molecule: **(a)** symmetric stretching, **(b)** bending, and **(c)** asymmetric stretching

modes. This results in the splitting of vibrational energy levels into a number of closely spaced rotational sublevels. The excited state of the CO_2 molecule corresponds to the presence of one or more quanta of energy. The energy level diagram for the operation of CO_2 laser is presented in Fig. 1.21. The operation begins with vibrational excitation of the nitrogen molecules by electrical discharge. As indicated in the figure, the vibrational excitation of the nitrogen molecule closely corresponds to the (001) vibrational levels of CO_2. The excited vibrational levels of nitrogen are metastable. The nitrogen molecule exchanges collisional energy with the CO_2, resulting in vibrational excitation of CO_2 molecules. Laser transition takes place between initial level (001) and final levels (100) and (020), resulting in 10.6 and 9.6 μm laser radiations, respectively. However, the laser radiation at 10.6 μm is the strongest and forms the most usual mode of operation. The practical CO_2 lasers use a mixture of CO_2, nitrogen and helium. The addition of helium increases the output power. The properties of a CO_2 laser are mainly determined by the method of gas flow in which sealed discharge tube, axial flow, and transverse or cross flow are primarily used. CO_2 lasers can be operated in both continuous and pulsed modes (Wilson and Hawkes 1987).

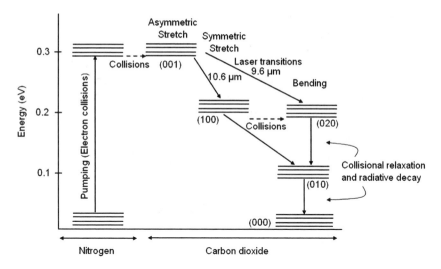

Fig. 1.21 Schematic of the energy levels involved in the operation of CO_2 laser

1.5.2.2 Excimer Laser

The word "excimer" is derived from the term excited "dimer", which means a diatomic molecule that is stable in the excited state and not so stable in the ground state. Since the ground state is unstable, very few dimer molecules remain in ground state, and hence direct excitation from ground state is not possible. However, various indirect excitations by electrical discharge result in laser radiation. When the mixture of argon and fluorine is excited in electrical discharge, the following processes can take place:

1. Electron attachment: $e + F_2 \rightarrow F^- + F$
2. Formation of excited molecule: $Ar^+ + F^- \rightarrow (ArF)^*$
3. Dissociation of excited molecule: $(ArF)^* \rightarrow Ar + F + Photon\ (191\ nm)$

The emission of photon causes the ArF^* molecule to fall to its lowest energy state, in which two atoms repel each other, so that molecule breaks up. Since the ground state is inherently unstable, the population in ground state remains low, giving easier population inversion. Typical characteristics of excimer lasers include average power in the range of 200 W and pulse energy up to 2 J per pulse. Common excimer lasers are ArF, KrF, and XeCl (Ready 1997).

1.5.3 Semiconductor Lasers

Semiconductor lasers, as the name suggests, use semiconductor materials as active medium. Even though it appears at first sight that semiconductor lasers are solid-state lasers, actually they are significantly different. Semiconductor lasers are based on radiative recombination of charge carriers. To understand the operation principles of semiconductor lasers it is necessary to understand the energy band structure of semiconductors.

Typical energy band diagrams of the p–n junction in open circuit, forward bias, and reverse bias are shown in Fig. 1.22. When a p-type semiconductor is brought in contact with the n-type semiconductor at a junction, the electrons from the n-type region diffuse toward the p-type region. Similarly, holes from the p-type region

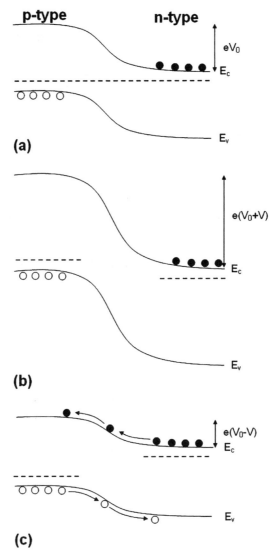

Fig. 1.22 Energy band diagrams for a p–n junction: (**a**) open circuit, (**b**) reverse bias, and (**c**) forward bias

diffuse toward the n-type region. The diffusive flow of charge carriers (electrons and holes) in opposite directions results in the charge separation on the two sides of the junction. The charge separation and the consequent electric field oppose the further diffusion of charge carriers. At equilibrium, an electron in the n-type region must overcome the built-in potential (eV_0) to diffuse into the p-type region (open circuit). When external potential (V) is applied such that the p-type region is made negative and the n-type region is made positive, the external potential adds to the built-in potential resulting in increased potential barrier, $e(V_0 + V)$. Hence, the diffusion of electrons from the n-type region to the p-type region becomes increasingly difficult (reverse bias). However, when the external potential (V) is applied such that the p-type region is made positive and the n-type region is made negative, it reduces the potential barrier from eV_0 to $e(V_0 - V)$. The electrons now face the reduced potential barrier, which they can easily overcome while diffusing from the n-type region to the p-type region. Similarly, holes can diffuse from the p-type region to the n-type region (forward bias). The battery in the circuit can replenish the charges and establish the current flow through the junction. The electrons and holes injected into the p-type and n-type regions, respectively, act as minority charge carriers and undergo recombination (Fig. 1.23). The recombination of minority charge carriers results in the spontaneous emission of photons. This is the basic principle of light emitting diodes (LEDs). Under certain circumstances, lasing action can occur. Beyond certain threshold current, the population inversion becomes high enough such that gain by stimulated emission overcomes the losses, resulting in lasing action. The radiative output increases rapidly with increasing current beyond the threshold current (Fig. 1.24) (Wilson and Hawkes 1987; Chow et al. 1994; Kasap 2002).

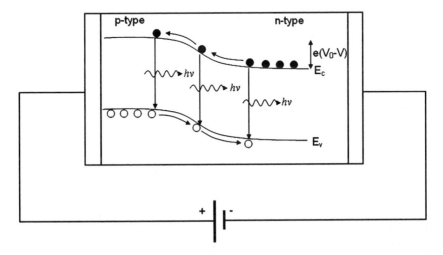

Fig. 1.23 Recombination of minority charge carriers resulting in emission of photons (forward bias)

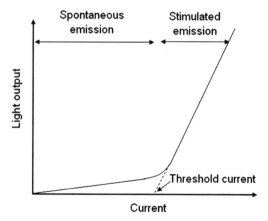

Fig. 1.24 Typical light output-current characteristic of semiconductor laser

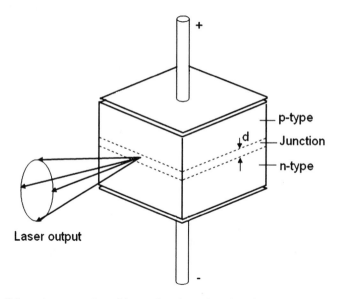

Fig. 1.25 Schematic construction of the semiconductor laser based on a homojunction

In many semiconductor materials such as Si and Ge, the recombination of charge carriers is mostly nonradiative and results in excess heat. A semiconductor such as GaAs is a good emitter of photons by the recombination process. Schematic of the typical structure of a semiconductor diode laser based on a homojunction is presented in Fig. 1.25. Semiconductor lasers based on homojunctions are associated with sideways spreading of the radiation from the gain region resulting in loss. More practical semiconductor lasers are based on double heterostructures having two junctions between two different band gap semiconductors (such as GaAs with band gap of 1.4 eV and AlGaAs with band gap of 2 eV). One of the most important distinguishing characteristic of the semiconductor lasers is that they are associated with very wide beam divergence angles (~40°) (Ready 1997).

1.5.4 Liquid Dye Lasers

Liquid dye lasers consist of liquid solutions (organic dyes dissolved in suitable liquid solvents) as active laser materials. Due to the physical nature (low density, homogeneity, etc.) of the liquid media, the liquid dye lasers are relatively easy to fabricate and are associated with advantages such as ease of cooling and replenishment in the laser cavity. One of the most important characteristics of the dye lasers is the tunability over the wide range of wavelengths (0.2–1.0 µm). This comes from the spectral properties of the organic dye molecules. The dye molecules efficiently absorb radiation over a certain range of wavelength and re-emit over other broad bands at longer wavelengths (Duarte and Hillman 1990). This is illustrated for typical dye material, rhodamine 6G, in Fig. 1.26.

A generalized energy level diagram for the dye molecule is presented in Fig. 1.27. As indicated in the figure, pumping causes the transitions $S_0 \rightarrow S_1$ and $S_0 \rightarrow S_1$. The molecule then rapidly (time $\sim 10^{-11}$ s) relaxes to the lowest levels of the S_1 band. The laser radiations are due to transitions $S_1 \rightarrow S_0$. Since the transitions take place between the bands of energy levels, laser radiations over a wide range of wavelengths are possible. Thus, the tunability of the laser over a wide range of wavelengths is an

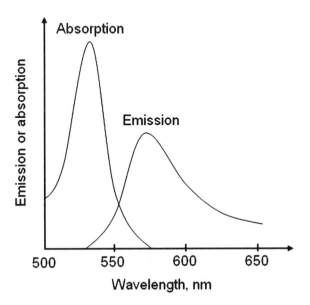

Fig. 1.26 Schematic of the spectral characteristics of rhodamine 6G

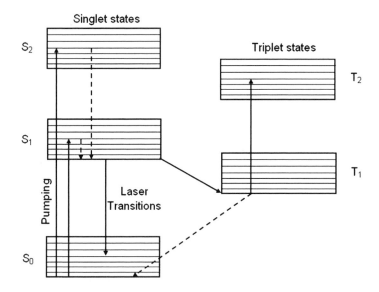

Fig. 1.27 Energy level diagram for dye molecule

important characteristic of dye lasers. The operation of dye lasers is also associated with undesirable transitions such as $S_1 \rightarrow T_1$. This transition represents the loss and lowers the laser efficiency. For better laser efficiency, $S_1 \rightarrow S_0$ transitions should be maximized and $S_1 \rightarrow T_1$ transitions should be minimized. A variety of dye materials is available with a wide tuning range of wavelengths (Duley 1983).

References

Chow WW, Koch SW, Sargent III M (1994) Semiconductor-laser Physics. Springer, Berlin.

Chryssolouris G (1991) Laser Machining-theory and Practice. Springer, New York.

Cullity BD (1978) Elements of X-ray Diffraction. Addison-Wesley, Reading, MA.

Dausinger F, Lichtner F, Lubatschowski H (2004) Femtosecond Technology for Technical and Medical Applications. Springer, Berlin.

Duarte FJ, Hillman LW (1990) Dye Laser Principles with Applications. Academic Press, Boston.

Duley WW (1983) Laser Processing and Analysis of Materials. Plenum, New York.

Einstein A (1917) On the quantum theory of radiation. Physika Zeitschrift 18:121–128.

Haken H (1983) Laser Theory. Springer, Berlin.

Ion J (2005) Laser Processing of Engineering Materials: Principles, Procedures and Industrial Applications. Butterworth-Heinemann, Oxford.

Kasap SO (2002) Principles of Electronic Materials and Devices. McGraw-Hill, New York.

Kogelnik H, Li T (1966) Laser beams and resonators. Applied Optics 5:1550–1567.

Lorrain P, Corson DR (1970) Electromagnetic fields and waves. WH Freeman, San Francosco.

Maiman TH (1960) Stimulated optical radiation in ruby. Nature 187:493–494.

Milloni PW, Eberly JH (1988) Lasers. Wiley, New York.

Ready JF (1997) Industrial Applications of Lasers. Academic Press, San Diego.

Rubahn HG (1999) Laser Applications in Surface Science and Technology. Wiley, Chinchester.

Steen WM (1991) Laser Materials Processing. Springer, London.

Svelto O, Hanna DC (1989) Principles of Lasers. Plenum Press, New York.

Thyagarajan K, Ghatak AK (1981) Lasers: Theory and Applications. Plenum Press, New York.

Wilson J, Hawkes JF (1987) Lasers: Principles and Applications. Prentice-Hall, New York.

Wolbarsht ML (1991) Laser Applications in Medicine and Biology. Plenum Press, New York.

Chapter 2
Laser Materials Interactions

2.1 Introduction

Understanding the capabilities and limitations of laser machining requires the knowledge of physical processes occurring during the laser beam interactions with materials. When the electromagnetic radiation is incident on the surface of a material, various phenomena that occur include reflection, refraction, absorption, scattering, and transmission (Fig. 2.1). One of the most desirable and important phenomena in the laser processing of materials is the absorption of the radiation. Absorption of radiation in the materials results in various effects such as heating, melting, vaporization, plasma formation, etc., which forms the basis of several laser materials-processing techniques (Steen 1991). The extent of these effects primarily depends on the characteristic of electromagnetic radiation and the thermo-physical properties of the material. The laser parameters include intensity, wavelength, spatial and temporal coherence, angle of incidence, polarization, illumination time, etc., whereas the materials parameters include absorptivity, thermal conductivity, specific heat, density, latent heats, etc. The interaction of laser with material is a complex interdisciplinary subject and requires knowledge from several branches of physics. This chapter briefly explains the important laser–material interactions and their effects which are relevant in the laser machining and fabrication of materials.

2.2 Absorption of Laser Radiation

As explained in Chapter 1, laser radiation is essentially electromagnetic waves, which are associated with electric ($\mathbf{E}$) and magnetic field vectors ($\mathbf{H}$). Absorption of light can be explained as the interaction of the electromagnetic radiation (characterized by electric and magnetic vectors) with the electrons (either free or bound) of the material. Electromagnetic radiation can interact only with the electrons of the atoms of the material because the much heavier nuclei are not able to follow the high frequencies of laser radiation. When the electromagnetic radiation passes over

N.B. Dahotre and S.P. Harimkar, *Laser Fabrication and Machining of Materials.*
© Springer 2008

Fig. 2.1 Possible interactions of laser light with material

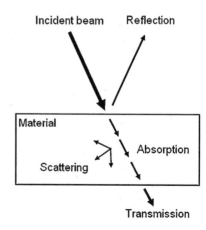

the electrons it exerts a force and sets the electrons into motion by the electric field of the radiation. The force exerted by the electromagnetic radiation of the electron can be expressed as (von Allmen 1987):

$$F = e\mathbf{E} + e\left(\frac{v}{c} \times \mathbf{H}\right), \tag{2.1}$$

where v is the velocity of electron and c is the speed of light. If it is considered that the electric and magnetic field carry the same amount of energy, then, according to above equation, the contribution of magnetic field to the force is smaller than that of the electric field by a factor of the order of v/c. Hence, the most important term in the above equation is $e\mathbf{E}$. The absorbed radiation, thus, results in the excess energy of the charged particles such as kinetic energy of the free electrons, excitation energy of the bound electrons, etc. Eventually, the degradation of the ordered and localized primary excitation energy through various steps leads to the generation of heat. Hence, the absorption process is sometimes referred to as the secondary "source" of energy inside the material and is used to determine the extent of various effects on the material during laser–material interactions.

The absorption of laser radiation in the material is generally expressed in terms of the Beer-Lambert law (Steen 1991):

$$I(z) = I_0 e^{-\mu z}, \tag{2.2}$$

where I_0 is the incident intensity, $I(z)$ is the intensity at depth z, and μ is the absorption coefficient. Thus, the intensity of the laser radiation gets attenuated inside the material. The length over which the significant attenuation of laser radiation takes place is often referred to as the attenuation length and is given by the reciprocal of the absorption coefficient (Welch and Gardner 2002):

$$L = \frac{1}{\mu}. \tag{2.3}$$

For a strongly absorbing material, the absorption coefficients are in the range of 10^5–10^6 cm^{-1} such that the attenuation lengths are in the range 10^{-5}–10^{-6} cm (Duley 1983).

One of the important parameters influencing the effects of laser–material interactions is the absorptivity of the material for laser radiation. It can be defined as the fraction of incident radiation that is absorbed at normal incidence. For opaque materials, the absorptivity (A) can be expressed as (Duley 1983):

$$A = 1 - R, \tag{2.4}$$

where R is the reflectivity of the material. The reflectivity and the absorptivity of the material can be calculated from the measurements of optical constants or the complex refractive index. The complex refractive index (n_c) is defined as:

$$n_c = n - ik \tag{2.5}$$

where n and k are the refractive index and extinction coefficient, respectively. These parameters are strong functions of wavelength and temperature. The reflectivity at normal incidence is then defined as:

$$R = \frac{(n-1)^2 + k^2}{(n+1)^2 + k^2}. \tag{2.6}$$

Since parameters n and k are strong functions of wavelength and temperature, the reflectivity (and hence the absorptivity) of the material is greatly influenced by the wavelength and temperature (Duley 1983).

The variation of reflectivity with the wavelength of some common metallic materials is presented in Fig. 2.2. The wavelengths of two important lasers (Nd:YAG and CO_2) are superimposed on the figure. As indicated in the figure, the reflectivity of the material generally increases with increasing wavelength. Thus, the materials are strong absorbers (less reflective) at shorter wavelengths. For a given material, the radiation from Nd:YAG laser ($\lambda = 1.06\,\mu m$) is absorbed strongly than the CO_2 laser ($\lambda = 10.6\,\mu m$). However, such wavelength dependence of the reflectivity (and hence absorptivity) should be used only as guidelines because there are several other factors which may strongly influence the absorptivity. For example, the reflectivity of a material generally decreases with increasing temperature (Fig. 2.3). Hence, a material which is strongly reflective at low temperature may become strongly absorbing at high temperature. This is of particular importance in the laser processing of materials where laser–materials interaction results in significant increase in the surface temperatures. Other parameters which influence the absorptivity of the material include angle of incidence of the radiation and surface condition of the material. These concepts have been extensively utilized to improve the coupling of the laser radiation to the reflective material by applying antireflective surface coatings and laser irradiating the material at suitable angle of incidence (Steen 1991; Ready 1997).

The laser energy absorbed by the material during laser–material interaction is converted into heat by degradation of the ordered and localized primary excitation

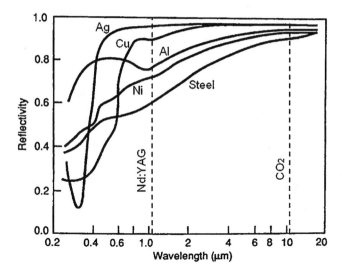

Fig. 2.2 Variation of reflectivity with wavelength for several metallic materials. The wavelengths of two important lasers (Nd:YAG and CO_2) are superimposed on the figure. (Reprinted from Ready 1997. With permission. Copyright Elsevier.)

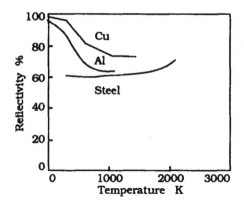

Fig. 2.3 Variation of reflectivity with temperature for 1.06 μm radiation (Reprinted from Steen 1991. With permission. Copyright Springer.)

energy. The typical overall energy relaxation times are of the order of 10^{-13} s for metals (10^{-12}–10^{-6} s for nonmetals). The conversion of light energy into heat and its subsequent conduction into the material establishes the temperature distributions in the material. Depending on the magnitude of the temperature rise, various physical effects in the material include heating, melting, and vaporization of the material. Furthermore, the ionization of vapor during laser irradiation may lead to generation of plasma. In addition to the thermal effects, the laser–material interactions may be associated with photochemical processes such as photoablation of the material. These effects of laser–material interactions are schematically presented in Fig. 2.4.

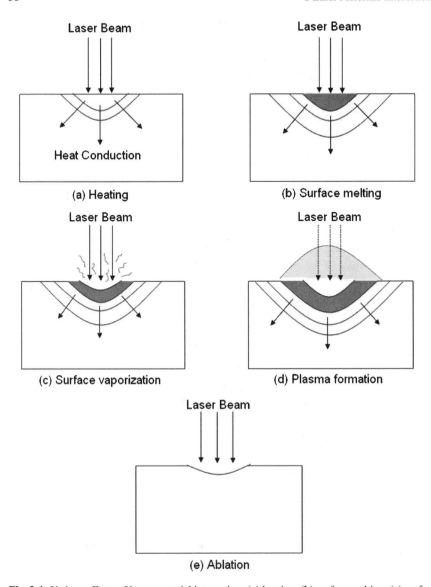

Fig. 2.4 Various effects of laser–material interaction: **(a)** heating, **(b)** surface melting, **(c)** surface vaporization, **(d)** plasma formation, and **(e)** ablation

All of these effects play important roles during laser materials processing. There exist distinct combinations of laser intensities and interaction times where specific effect of laser–material interaction dominates. Figure 2.5 presents one such plot showing the regimes of various laser–material interactions and their applications in materials processing. These effects are briefly explained in the following section in the context of laser materials processing (Bäuerle 2000).

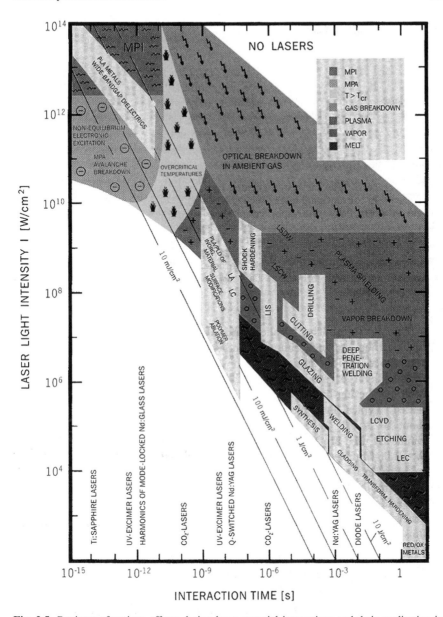

Fig. 2.5 Regimes of various effects during laser–material interactions and their application in laser materials processing (pulsed laser ablation/deposition, PLA/PLD; laser annealing, LA; laser cleaning, LC; laser-induced isotope separation, LIS; multiphoton absorption/ionization, MPA/MPI; laser-supported detonation/combustion waves, LSDW/LSCW; laser-induced CVD, LCVD; laser-induced electrochemical plating/etching, LEC; long pulse or CW CO_2 laser-induced reduction/oxidation, RED/OX). (Reprinted from Bäuerle 2000. With permission. Copyright Springer.)

2.3   Thermal Effects

When a laser beam of intensity I_0 is irradiated on the surface of material, it results in the excitation of free electrons (in metals), vibrations (in insulators), or both (in semiconductors). As mentioned in the previous section, this excitation energy is rapidly converted into heat (time duration in the range 10^{-13} s for metals, 10^{-12} to 10^{-6} s for nonmetals). This is followed by various heat transfer processes such as conduction into the materials, and convection and radiation from the surface. The most significant heat transfer process being the heat conduction into the material. The generation of heat at the surface and its conduction into the material establishes the temperature distributions in the material depending on the thermo-physical properties of the material and laser parameters. If the incident laser intensity is sufficiently high, the absorption of laser energy can result in the phase transformations such as surface melting and evaporation. Generally, these phase transformations are associated with threshold (minimum) laser intensities referred to as melting and evaporation thresholds (I_m and I_v). Melting and evaporation are the efficient material removal mechanisms during many machining processes. In this section, we will deal with the simplified analysis of laser heating, melting, and evaporation of materials. More detailed analyses are presented in the following chapters with reference to specific applications (Bäuerle 2000).

2.3.1   Heating

To understand the effects of laser irradiation on the material, it is necessary to evaluate the temporal and spatial variation of temperature distribution. The most simplified thermal analysis is based on the solution of one-dimensional heat conduction equation with simplified assumptions such as (Carslaw and Jaeger 1959):

1. Material is homogeneous. The thermo-physical properties are independent of temperature.
2. The initial temperature of the material is constant.
3. Heat input is uniform during the irradiation time.
4. The convection and radiation losses from the surface are negligible.

The schematic geometry of laser irradiation and corresponding temperature profiles during laser heating are presented in Fig. 2.6. The governing equation for the one-dimensional heat transfer can be written as:

$$\frac{\partial T(z,t)}{\partial t} = \alpha \frac{\partial^2 T(z,t)}{\partial z^2}, \tag{2.7}$$

where T is the temperature at location z, after time t; and α is the thermal diffusivity.

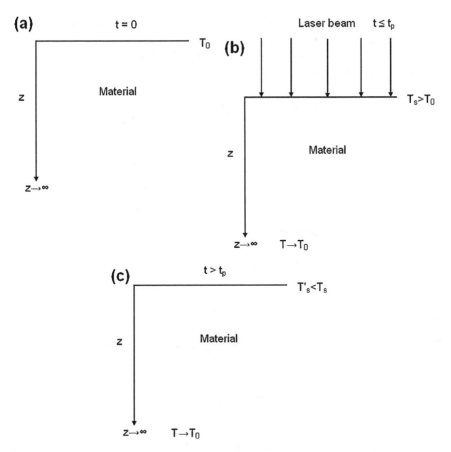

Fig. 2.6 Schematic of the laser irradiation geometry and surface temperatures at various times: (a) initial condition with uniform temperature T_0, throughout the material, (b) laser heating with surface temperature, $T_s > T_0$, and (c) cooling stage (laser off) with surface temperature $T'_s < T_s$ (surface temperatures are less than melting point at all times)

The initial condition can be written as:

$$T(z,0) - T_0, \text{ for } 0 \le z \le \infty, t = 0, \tag{2.8}$$

where T_0 is the initial constant temperature of the material.

The simple boundary condition at the surface ($z = 0$) assuming that laser energy absorbed at the surface equals the energy conducted can be written as:

$$-k \frac{\partial T(0,t)}{\partial z} = \delta H, \tag{2.9}$$

where k is the thermal conductivity and H is the absorbed laser energy. The absorbed laser energy H can be given by the product of absorptivity A and incident laser power density I_0(i.e., $H = AI_0$). If t_p is the irradiation time (pulse on time) then the parameter δ equals unity when the laser is on, i.e., $0 \leq t \leq t_p$. It can be taken as zero when the laser is off, i.e., $t > t_p$.

The solutions of these equations can be obtained as follows:

During heating ($0 < t < t_p$):

$$\Delta T(z,t)_{t<t_p} = \frac{H}{k}(4\alpha t)^{1/2} \,\mathrm{ierfc}\left(\frac{z}{(4\alpha t)^{1/2}}\right). \tag{2.10}$$

During cooling ($t > t_p$):

$$\Delta T(z,t)_{t>t_p} = \frac{2H\alpha^{1/2}}{k}\left[t^{1/2}\mathrm{ierfc}\left(\frac{z}{(4\alpha t)^{1/2}}\right) - (t-t_p)^{1/2}\,\mathrm{ierfc}\left(\frac{z}{\left(4\alpha(t-t_p)\right)^{1/2}}\right)\right] \tag{2.11}$$

The function ierfc(x) is defined as:

$$\mathrm{ierfc}(x) = \frac{1}{\sqrt{\pi}}\left\{\exp(-x^2) - x(1-\mathrm{erf}(x))\right\},$$

$$\text{where } \mathrm{erf}(x) = \frac{2}{\sqrt{\pi}}\int_0^x e^{-\xi^2}d\xi. \tag{2.12}$$

The temperature at the surface during heating and cooling can be obtained by substituting $z = 0$ in Eqs (2.10) and (2.11). Thus

$$\Delta T(0,t)_{t<t_p} = \frac{H}{k}\left(\frac{4\alpha t}{\pi}\right)^{1/2}, \tag{2.13}$$

$$\Delta T(0,t)_{t>t_p} = \frac{H}{k}\left[\left(\frac{4\alpha t}{\pi}\right)^{1/2} - \left(\frac{4\alpha(t-t_p)}{\pi}\right)^{1/2}\right]. \tag{2.14}$$

Typical calculated temperature changes at various depths during laser irradiation of copper using laser power density of 10^{10} W/m^2 and irradiation time of 1 μs are presented in Fig. 2.7. The important characteristics of the temperature changes in a material during laser irradiation can be listed as (Wilson and Hawkes 1987):

1. At the surface ($z = 0$), the temperature increases with increasing irradiation time, reaches maximum corresponding to pulse time (t_p), and then rapidly decreases. Thus, the heating and the cooling parts of the curve are clearly separated at a time corresponding to pulse time.
2. At certain depths below the surface ($z > 0$), the temperature increases with increasing irradiation time, reaches maximum, and then decreases. However, the maximum temperature does not reach exactly at the pulse time (t_p), but at the longer time ($t > t_p$). The time ($t > t_p$) to reach the maximum temperature increases as we go further into the depth below the surface of the material.

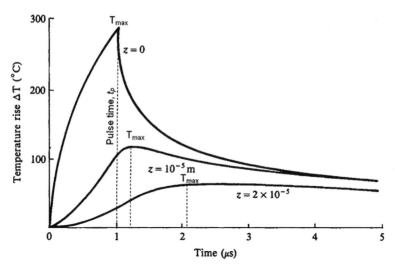

Fig. 2.7 Variation of calculated temperature increases with time at various depths during laser irradiation of copper using laser power density of 10^{10} W/m^2 and irradiation time of 1 μs (Reprinted from Wilson and Hawkes 1987. With Permission. Copyright Pearson.)

2.3.2 Melting

In the preceding discussion, it was considered that the incident laser power density was sufficient to heat the material without any phase transformations. However, the surface temperature may reach the melting point or the boiling point at sufficiently higher laser power densities ($I_0 > 10^5$ W/cm^2). The corresponding laser power densities are often referred to as the melting and boiling thresholds.

Now, let us consider that the surface temperature of the material exceeds the melting point upon irradiation with laser (without surface evaporation). It is important to analyze the temporal evolution of depth of melting during laser irradiation. Figure 2.8 presents the various steps for the determination of the depth of melting during laser irradiation. As indicated in Fig. 2.8a, the temperature of the surface ($z = 0$) increases with increasing irradiation time (t), reaches maximum temperature (T_{max}) at pulse time (t_p), and then decreases. Various heating and cooling steps in this temporal evolution of surface temperature are:

1. Temperature reaches T_1 ($T_1 < T_m$) at time t_1 ($t_1 < t_p$).
2. Temperature reaches melting point (T_m) at time t_2 ($t_2 < t_p$).
3. Temperature reaches maximum, T_{max} ($T_{max} > T_m$) at time t_p.
4. Temperature decreases to melting point T_m at time t_3 ($t_3 > t_p$).
5. Temperature reaches T_1 ($T_1 < T_m$) at time t_1 ($t_1 > t_p$).

The corresponding temperature profiles in the depth of the material are presented in Fig. 2.8b for various times during laser irradiation. By tracing the melting point in the temperature verses depth plots, the positions of the solid–liquid interface can

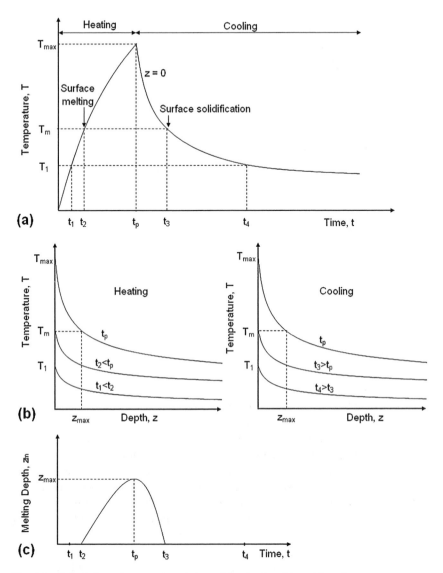

Fig. 2.8 Calculation of temporal evolution of depth of melting: (**a**) surface temperature as a function of time, (**b**) temperature as a function of depth below the surface during heating and cooling, and (**c**) depth of melting as a function of time

be located. For example, at time t_p, the position of solid–liquid interface corresponds to $z = z_{max}$. Similarly, at times t_2 and t_3, these positions can be located at $z = 0$. These positions are schematically plotted in Fig. 2.8c. The figure indicates that during laser irradiation, the melting initiates at time t_2. Below time t_2, the material is simply heated without melting. Beyond t_2, the depth of melting increases with continued irradiation and reaches maximum (z_{max}) at pulse time t_p. This means that the solid–

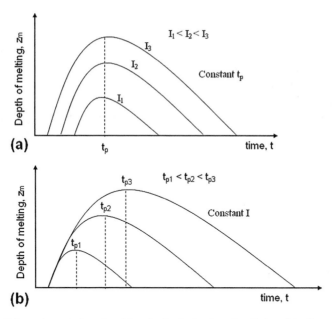

Fig. 2.9 Schematic variation of melting depths during laser irradiation: (**a**) effect of laser power density at constant pulse time, and (**b**) effect of laser pulse time at constant laser power density

liquid interface moves away from the surface during heating phase ($t \leq t_p$). In the cooling phase ($t > t_p$), the surface temperature starts decreasing rapidly and the solid–liquid interface moves towards the surface of the material (start of solidification). The depth of melting decreases beyond t_p and reaches zero at time t_3 which marks the completion of solidification. Beyond t_3, the material simply cools down. Thus, each laser irradiation is characterized by the maximum depth of melting z_{max} corresponding to the cessation of laser power. Figure 2.9 shows the schematic of the influence of important laser processing parameters on the temporal evolution of depths of melting. At constant pulse time, the maximum depth of melting increases with increasing laser power density (Fig. 2.9a). In addition, at constant laser power density, the maximum depth of melting increases with increasing pulse time (Fig. 2.9b). It should be kept in mind that the above generalized trends are valid for the case of laser melting before initiation of surface evaporation.

2.3.3 Vaporization

The depth of melting cannot increase to infinitely large value with increasing laser power density and pulse time because the location of the melting point in the temperature verses depth plot is limited by the maximum achievable surface temperature. Once the surface temperature reaches the boiling point, the depth of melting reaches the maximum value z_{MAX} (Note that z_{max} introduced earlier correspond to the cessation of power where the surface temperature has not yet reached the boiling point). Further increase in the laser power density or the pulse time cause the evaporative material removal from the surface without further increase in the depth of melting. This is illustrated in Fig. 2.10.

The maximum depth of melting (z_{MAX}) at which the surface reaches the boiling point can be calculated as follows (Wilson and Hawkes 1987):

When the temperature reaches melting point (T_m) at some depth z_{MAX}, Eq. (2.10) becomes:

$$T_m = \frac{H}{k}(4\alpha t)^{1/2} \, \text{ierfc}\left(\frac{z_{MAX}}{(4\alpha t)^{1/2}}\right). \tag{2.15}$$

When the surface temperature reaches boiling point (T_b), Eq. (2.13) becomes:

$$T_b = \frac{H}{k}\left(\frac{4\alpha t}{\pi}\right)^{1/2}. \tag{2.16}$$

Taking the ratio of the above equations:

$$\frac{T_m}{T_b} = \sqrt{\pi} \, \text{ierfc}\left(\frac{z_{MAX}}{(4\alpha t)^{1/2}}\right). \tag{2.17}$$

From Eq. (2.16), we get:

$$(\alpha t)^{1/2} = \frac{T_b k \sqrt{\pi}}{2H}. \tag{2.18}$$

Substituting in Eq. (2.17), we get:

$$\text{ierfc}\left(\frac{H z_{MAX}}{k T_b \sqrt{\pi}}\right) = \frac{T_m}{T_b \sqrt{\pi}}. \tag{2.19}$$

Such equations facilitate the calculation of maximum depth of melting (z_{MAX}) at which the surface reaches boiling point during laser irradiation. Such data is particularly useful in the laser welding of materials.

Once the vaporization is initiated at the surface of the material, the continued laser irradiation will cause the liquid–vapor interface to move inside the material. This is accompanied with the evaporative removal of material from the surface above the liquid–vapor interface. If V_s is the velocity of the liquid–vapor interface into the material during the laser irradiation, then the mass of material removed per unit time ($\dot{m}$) and the depth of vaporization (d) will be $V_s \rho$ (where ρ is the density) and $V_s t_p$, respectively. The velocity of the liquid–vapor interface and the depth of vaporization can be calculated by simple energy balance (Steen 1991):

$$V_s = \frac{H}{\rho(cT_b + L_v)}, \tag{2.20}$$

where L_v is the latent heat of vaporization.

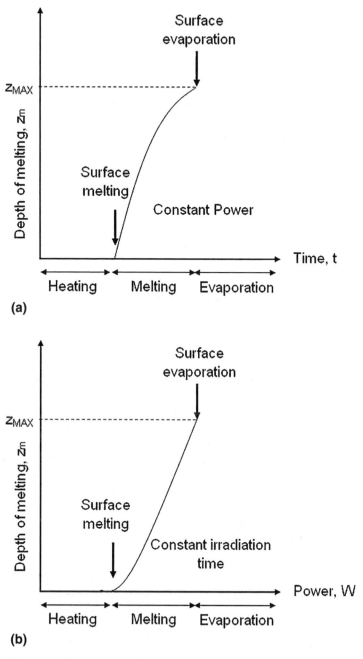

Fig. 2.10 Schematic of the variation of depth of melting with laser irradiation time and powe. The arrows indicate the initiation of surface melting and evaporation during continued laser irradiation

Solving,

$$d = \frac{Ht_p}{\rho\left(cT_b + L_v\right)}. \tag{2.21}$$

Such analysis of vaporization process during laser irradiation is important during material removal processes such laser drilling, cutting, etc.

The two important parameters in the analysis of thermal effects during laser–material interactions are the cooling rate and the temperature gradient. These factors have strong influence on the development of microstructure (such as those formed by dendritic, cellular, or planar growth) during solidification of molten material. From the above thermal analysis, the temperature gradient (G) and the cooling rate (T) can be calculated as:

$$G(z,t) = \frac{\partial T}{\partial z}, \tag{2.22}$$

$$\dot{T}(z,t) = \frac{\partial T}{\partial T}. \tag{2.23}$$

An important relationship in the solidification theory which relates these parameters is (Flemings 1974):

$$\dot{T} = GR, \tag{2.24}$$

where R is the solidification rate.

The calculated variations of temperature gradient (G), the cooling rate ($\dot{T}$), and the solidification rate (R) with fractional melt depth are presented in Figs 2.11–2.13 (Breinan and Kear 1983). As the laser is switched off (cessation of laser power) the depth of melting reaches a certain value z_{max}. As solidification begin the depth of melting decreases and eventually becomes zero corresponding to completion of solidification at the surface. The fractional melt depth used in these plots corresponds to the ratio of instantaneous depth of melting after solidification has started to the depth of melting at the cessation of power. The calculations presented here were performed for pure nickel with fixed initial depth of melting (depth of melting at the cessation of power) equal to 0.025 mm obtained with three absorbed laser power densities. The following important conclusions can be drawn from these plots (Breinan and Kear 1983):

1. The temperature gradient (G) is maximum at the start of solidification (fractional melt depth = 1) and approaches minimum value (zero) at the end of solidification (fractional melt depth = 0). In addition, the temperature gradient strongly depends on the absorbed laser power density. Since the temperature decreases with increasing depth in the material, the actual temperature gradients have a negative sign.
2. At the start of solidification (i.e., at the instance of cessation of laser power), the cooling rate ($\dot{T}$) is zero and then becomes negative as the solidification proceeds towards the surface of the material. Moreover, the cooling rate strongly depends on the absorbed laser power density and has a negative sign.

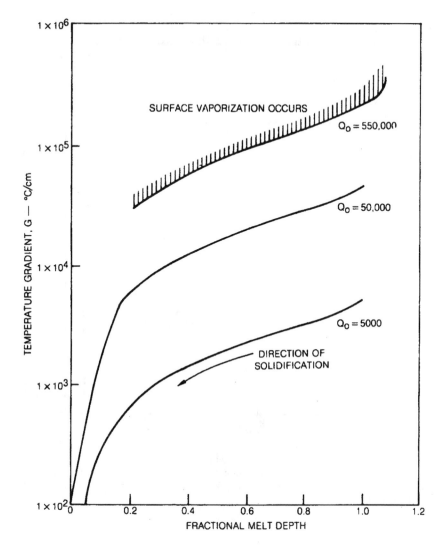

Fig. 2.11 Variation of calculated temperature gradient with fractional melt depth during laser irradiation of nickel. The calculations were performed for initial melt depth of 0.025 mm using three absorbed laser power densities (Q_0, W/cm^2) (Reprinted from Breinan and Kear 1983. With permission. Copyright M. Bass, Editor)

3. At the start of solidification (i.e., at the instance of cessation of laser power), the solidification rate (R) is zero and then approaches infinity as it proceeds towards the surface of the material. The actual solidification rates have a negative sign. The solidification rates exhibit the similar trends for a given melt depth irrespective of the absorbed power density (except for the initial transient).

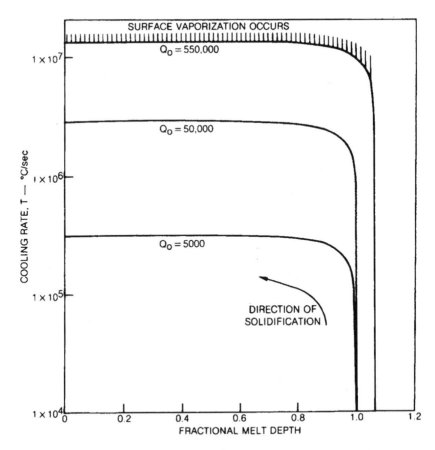

Fig. 2.12 Variation of calculated cooling rate with fractional melt depth during laser irradiation of nickel. The calculations were performed for initial melt depth of 0.025 mm using three absorbed laser power densities (Q_0, W/cm²) (Reprinted from Breinan and Kear 1983. With permission. Copyright M. Bass, Editor)

2.3.4 Important Considerations for Thermal Analysis

Most of the previous discussion assumed that laser is uniformly irradiated on a semi-infinite material and that the heat transfer is one-dimensional. These are very simplified assumptions and facilitate the general understanding of laser interaction with the material. However, in practice, there exist a number of complex parameters which play important roles during laser interaction with material. The refinement of the simple model explained here is likely to result in more realistic results after incorporating these complex parameters. Some of these important considerations for thermal analysis are explained in the following sections.

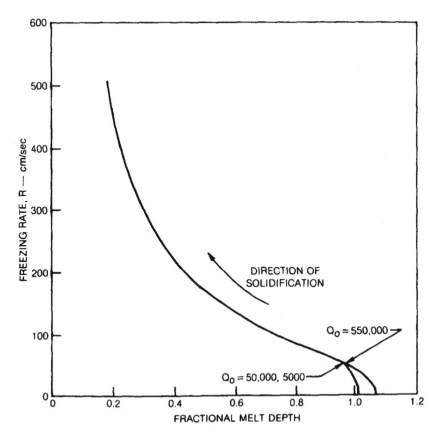

Fig. 2.13 Variation of calculated freezing (solidification) rate with fractional melt depth during laser irradiation of nickel. The calculations were performed for an initial melt depth of 0.025 mm using three absorbed laser power densities (Q_0, W/cm²) (Reprinted from Breinan and Kear 1983. With permission. Copyright M. Bass, Editor)

2.3.4.1 Beam Shapes

In the previous discussion, it was assumed that the large area of the material was irradiated uniformly with the laser beam. This corresponds to the constant laser power density at all the points on the irradiated surface. However, the laser power density (intensity) can be distributed in several distinct shapes. The most common in the laser material processing is the Gaussian distribution of energy. The distribution of energy in the Gaussian beam is given by:

$$I(r) = I_0 \exp\left[\frac{-2r^2}{w^2}\right], \qquad (2.25)$$

where r is the radius of the beam, I_0 is the intensity of the beam at $r = 0$, and w is the radius of the beam at which $I = I_0 e^{-2}$. Various other beam shapes include circular, rectangular, etc. (Bäuerle 2000). Special beam shapes with complex energy distributions can be obtained by beam shaping methods. Due to nonuniform intensity distributions,

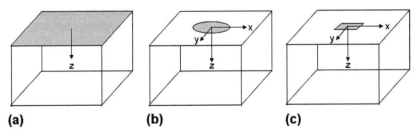

Fig. 2.14 Various beam shapes for laser irradiation of material: (**a**) large area uniform illumination, (**b**) circular beam, and (**c**) rectangular beam

the heat tends to diffuse sideways and the problem can no longer be approximated as one-dimensional. Instead, three-dimensional heat transfer needs to be considered. This is illustrated in Fig. 2.14. Thus, accurate description of intensity distributions is important to obtain the realistic results from thermal modeling. The general equation for three-dimensional heat transfer is given by (Carslaw and Jaeger 1959):

$$\frac{\partial T(x,y,z,t)}{\partial t} = \alpha \left[\frac{\partial^2 T(x,y,z,t)}{\partial x^2} + \frac{\partial^2 T(x,y,z,t)}{\partial y^2} + \frac{\partial^2 T(x,y,z,t)}{\partial z^2} \right]. \quad (2.26)$$

2.3.4.2 Pulse Shapes

The temperature distribution during laser irradiation can be greatly influenced by the temporal variation of laser beam intensity. The continuous wave (CW) laser beams with constant laser intensity with time are easiest to define in the thermal model. The complexity in the thermal analysis arises due to pulsed operation of laser. In such analysis, it is important to define the temporal shape of the pulse. Various single-pulse shapes such as rectangular pulse, triangular pulse, and smooth pulse are shown in Fig. 2.15. Rectangular pulses are generally characterized by the width of the pulse; whereas triangular and smooth pulses are characterized by the width at full width half maxima (FWHM). For the same width (for rectangular pulse) or width at FWHM (for triangular or smooth pulses), the rectangular pulses generally give higher temperature rise compared to triangular and smooth pulses (Bäuerle 2000).

During multipulse operation, temperature of the material increases during each pulse followed by cooling during the time between the adjacent pulses. Since the cooling is not complete during the short duration between the pulses, the initial temperature during heating with the subsequent pulses is always higher than that during heating with preceding pulses. This results in the higher temperature during heating with subsequent pulses. Thus, the pulsed heating of the material is associated with temperature fluctuations (heating and cooling) during each pulse and time interval following the pulse (time between the adjacent pulses). This is schematically shown in Fig. 2.16. The pulsed output of the laser may be approximated by the constant average output such that thermal analysis yields the continuous increase in temperature during heating (dashed curve in Fig. 2.16b).

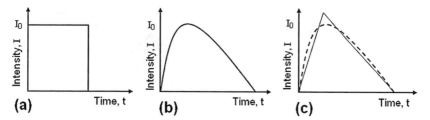

Fig. 2.15 Various single-pulse shapes showing temporal variation of the intensity: **(a)** rectangular pulse, **(b)** smooth pulse, and **(c)** triangular pulse

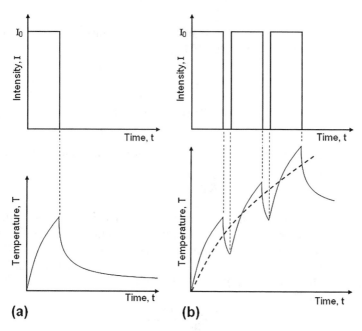

Fig. 2.16 Schematic of temporal evolution of surface temperature during **(a)** single pulse and **(b)** multipulse laser irradiation of material (dotted curve indicate the average temperature)

2.3.4.3 Moving Source of Heat

The thermal model explained in the preceding section considers the irradiation of material where both the laser beam and the material are stationary. Most of the practical laser applications such as welding, cutting, shaping, etc. require the laser beam to move relative to the workpiece. Hence, calculation of temperature distributions around the moving source of heat (laser) becomes important. The theory of moving sources of heat was first advanced by Rosenthal (1946). The schematic of the model geometry for heating with moving point source is presented in Fig. 2.17. For a point heating source moving with a constant velocity (v) in the

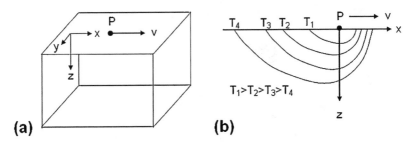

Fig. 2.17 Schematic of the **(a)** heating model geometry with a moving point source of heat, and **(b)** typical temperature distribution (isotherms) in the x–z plane through the point source

x-direction, with point source as origin, the three-dimensional heat transfer Eq. (2.26) can be rewritten as (Rosenthal 1946):

$$-v\frac{\partial T(\xi,y,z,t)}{\partial \xi} + \frac{\partial T(\xi,y,z,t)}{\partial t}$$
$$= \alpha\left[\frac{\partial^2 T(\xi,y,z,t)}{\partial \xi^2} + \frac{\partial^2 T(\xi,y,z,t)}{\partial y^2} + \frac{\partial^2 T(\xi,y,z,t)}{\partial z^2}\right] \qquad (2.27)$$

where ξ is the distance of a considered point from the point source. This distance can be expressed as:

$$\xi = x - vt. \qquad (2.28)$$

For a solid much longer than the extent of heat dissipation, the temperature distribution around the point heat source becomes constant such that an observer located at the moving point source fails to notice the temperature changes around the point source as it moves on. This type of heat flow is generally referred to as quasistationary heat flow. This state of heat flow is defined as:

$$\frac{\partial T(\xi,y,z,t)}{\partial t} = 0 \qquad (2.29)$$

Substitution in Eq. (2.27) yields:

$$-v\frac{\partial T(\xi,y,z,t)}{\partial \xi} = \alpha\left[\frac{\partial^2 T(\xi,y,z,t)}{\partial \xi^2} + \frac{\partial^2 T(\xi,y,z,t)}{\partial y^2} + \frac{\partial^2 T(\xi,y,z,t)}{\partial z^2}\right]. \qquad (2.30)$$

Equation (2.30) can besimplified further by using

$$T = T_0 + e^{-\lambda v \xi}\varphi(\xi,y,z), \qquad (2.31)$$

where T_0 is the initial temperature and φ is the function which needs to be determined to calculate the temperature distributions.

The Eq. (2.30) becomes:

$$-\left(\frac{v}{2\alpha}\right)^2 \varphi(\xi,y,z) = \left[\frac{\partial^2 \varphi(\xi,y,z)}{\partial \xi^2} + \frac{\partial^2 \varphi(\xi,y,z)}{\partial y^2} + \frac{\partial^2 \varphi(\xi,y,z)}{\partial z^2}\right]. \qquad (2.32)$$

For linear flow of heat where $\dfrac{\partial^2\varphi(\xi,y,z)}{\partial y^2}=\dfrac{\partial^2\varphi(\xi,y,z)}{\partial z^2}=0$ Eq. (2.32) reduces to:

$$-\left(\frac{v}{2\alpha}\right)^2\varphi(\xi,y,z)=\frac{d^2\varphi(\xi,y,z)}{\partial\xi^2}. \tag{2.33}$$

These equations can be solved with appropriate boundary conditions to obtain the temperature profiles around the moving point source of heat (Rosenthal 1946).

2.3.4.4 Temperature Dependent Properties

Another important consideration for accurate determination of temperature distribution during laser irradiation is the temperature dependence of the thermo-physical and other properties. Properties such as thermal conductivity, thermal diffusivity, absorptivity, etc. are strongly temperature dependent and expected to influence the temporal and spatial evolution of temperature during laser irradiation.

2.4 Vapor Expansion and Recoil Pressures

As explained in the previous section, surface vaporization is initiated when the laser intensity becomes sufficiently high ($I_0 > 10^5$–10^8 W/cm^2). The vapor plume consists of clusters, molecules, atoms, ions, and electrons. In the steady-state evaporation, the vapor particles escape from the surface (solid or liquid) at temperature T_s. Initially, the vapor particles escaping from the surface have a Maxwellian velocity distribution corresponding to the surface temperature (T_s) with their velocity vectors all pointing away from the surface. Due to collisions among the vapor particles, the velocity distribution in the vicinity of the vaporizing surface (layer of the order of several mean free paths) approaches equilibrium. This region is known as Knudsen layer and is often treated as discontinuity in the hydrodynamic treatment. The detailed analysis of evaporation problem was conducted by Anisimov (1968) to determine the structure of this region and the values of hydrodynamic variables beyond the discontinuity. It was assumed that the laser power density is not excessively large so that there is no significant absorption of laser light by the vapor. Further, Anisimov assumed that the distribution function within discontinuity region can be approximated by the sum of distribution functions before and after the discontinuity with coordinate-dependent coefficients (Anisimov 1968, Anisimov and Khokhlov 1995):

$$f(x,v)=\alpha(x)f_1(v)+\left[1-\alpha(x)\right]f_2(v), \tag{2.34}$$

where

$$f_1 = n_0 \left(\frac{m}{2\pi k_B T_s}\right)^{3/2} \exp\left(-\frac{mv^2}{2k_B T_s}\right), v_x > 0 \tag{2.35}$$

$$f_2 = \beta \, n_1 \left(\frac{m}{2\pi k_B T_1}\right)^{3/2} \exp\left(-\frac{m\left(v_y^2 + v_z^2 + (v_x - u_1)^2\right)}{2k_B T_1}\right), v_x < 0 \tag{2.36}$$

Here, $\alpha(x)$ is a unknown function satisfying $\alpha(0) = 1$ and $\alpha(\infty) = 0$; k_B is the Boltzmann constant; m is the mass of the vapor molecule; T_s and n_0 are the surface temperature and the molecule number density at the evaporating surface, respectively; and T_1 and n_1 are the temperature and molecule number density at the outer boundary of the kinetic layer formed near the evaporating surface, respectively. u_1 is the velocity at the outer boundary of the kinetic layer; β is the coefficient; and v_x, v_y, and v_z are the velocity components on the evaporating surface of the material. Assuming that u_1 is equal to the velocity of sound within the vapor (Jouguet condition), we get:

$$u_1 = \sqrt{\frac{\gamma k_B T_1}{m}}, \tag{2.37}$$

where γ is the adiabatic index of the vapor.

The conservation laws of mass, energy, and momentum hold within the discontinuity region. Thus,

$$\int dv v_x f(x,v) = C_1$$
$$\int dv v_x^2 f(x,v) = C_2$$
$$\int dv v_x v^2 f(x,v) = C_3 \tag{2.38}$$

Solving the above equations for the monoatomic gas with $\gamma = 5/3$, Anisimov obtained:

$$\beta = 6.29,$$
$$T_1 = 0.67 T_s, \tag{2.39}$$
$$n_1 = 0.31 n_0.$$

Thus the vapor is significantly cooler and less dense than the vapor in equilibrium with the surface. Further analysis by Anisimov indicated that approximately 18% of the vapor particles condensed back to the surface. The velocity of the vaporization front is given by:

$$V = \frac{I_a}{\rho\left(L_v + 2.2 k_B T_s / m\right)}, \tag{2.40}$$

where I_a is the absorbed laser power density and L_v is the latent heat of vaporization. The above derivation on the expansion of vapor in the vacuum can also be extended for the case of vapor expansion into ambient gaseous atmosphere by considering the Mach number of the external flow. All the quantities T_1 and n_1 are dependent on the Mach number of the external flow (Anisimov 1968).

The evolving vapor from the surface exerts the recoil pressure on the surface. Based on the above relationships, an equation for calculation of evaporation-induced recoil pressure (p_s) at the evaporating surface is given by (Anisimov 1968):

$$\frac{p_s}{Q_0 / S} = \frac{1.69}{\sqrt{L_v}} \left(\frac{b}{1 + 2.2b^2} \right), \tag{2.41}$$

where Q_0 is the incident laser power; S is the area of laser spot; and $b^2 = K_B T_s / m_v L_v$. For the surface temperatures equal to boiling point, the evaporation-induced recoil pressure according to Anisimov becomes 0.55 p_s, where p_s is the saturated vapor pressure. Under typical materials processing (drilling, cutting, welding, etc.) conditions, this evaporation-induced recoil pressure exceeds the highest possible value of surface tension pressure. Thus evaporation-induced recoil pressure plays an important role in the removal of material in molten state during materials processing. The melt expulsion and surface evaporation processes for material removal are schematically shown in Fig. 2.18. The comparison of the contributions of two material removal processes (melt expulsion and surface evaporation) to the overall material removal rate is presented in Fig. 2.19. The figure indicates that the melt expulsion is a dominant material removal mechanism at low powers; whereas surface evaporation becomes dominant at high powers. In addition, the material removal rate by melt expulsion increases with laser power, reaches maximum, and then decreases; whereas the material removal rate by surface evaporation increases continuously with laser power. Under the nonstationary conditions, the direct effect of recoil pressure is to decrease the thickness of the molten layer by

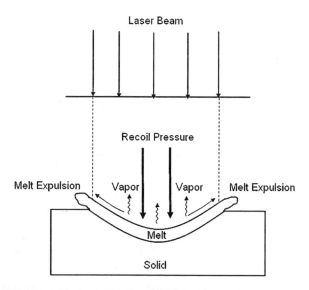

Fig. 2.18 Schematic of surface evaporation and melt expulsion processes of material removal during laser–material interactions

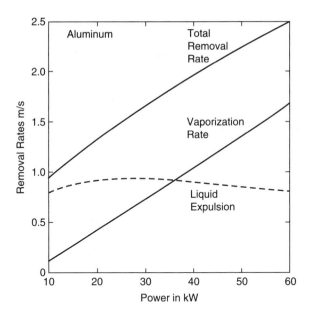

Fig. 2.19 Material removal rates due to melt expulsion and vaporization for aluminum as a function of laser power. The figure also presents the total material removal rate which is the sum of material removal rates due to melt expulsion and vaporization. (Reprinted from Chan and Mazumder 1987. With permission. Copyright American Institute of Physics.)

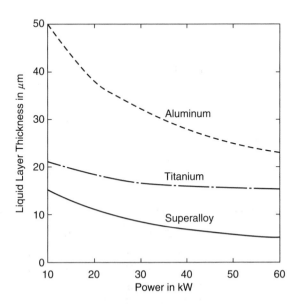

Fig. 2.20 Influence of laser power on the liquid layer thickness during nonstationary conditions with melt expulsion. (Reprinted from Chan and Mazumder 1987. With permission. Copyright American Institute of Physics.)

melt expulsion. At higher absorbed laser power, the recoil pressure becomes high resulting in thinner liquid layer thickness (due to enhanced melt expulsion, Fig. 2.20) (Chan and Mazumder 1987).

2.5 Plasma Formation

When the material is irradiated with sufficiently larger laser intensity (I_v), significant surface evaporation takes place as explained in the previous sections. Once the vaporization is initiated, the interactions between the resulting vapor and the incident laser beam become important in determining the overall effect of the laser irradiation on the material. One of the most important interactions is the ionization of vapor. The highly ionized vapor is termed as plasma. In dynamic equilibrium, the degree of ionization (ξ) in a gas is often expressed by the Saha equation (Bäuerle 2000):

$$\frac{\xi^2}{1-\xi} = \frac{2g_i}{g_a N_g}\left(\frac{2\pi m k_B T}{h^2}\right)^{3/2}\exp\left(-\frac{E_i}{k_B T}\right), \qquad (2.42)$$

with $\xi = N_e/N_g$ and $N_g = N_e + N_a$. Here, N_e and N_a are the number densities of electrons and atoms/molecules, respectively; g_i and g_a are the degeneracy of states for ions and atoms/molecules; and E_i is the ionization energy.

There are two mechanisms of laser-induced breakdown (partial or complete ionization) resulting in dense plasma formation. These are cascade ionization (or avalanche ionization) and multiphoton absorption (Fig. 2.21). The cascade ionization considers the presence of free electrons often termed as "seed" or "priming" electrons in the focal volume at the beginning of the laser pulse. These seed electrons absorb the laser energy by inverse bremsstrahlung absorption process. When the energy acquired by the free electrons exceeds the ionization potential of the molecules, ionization of molecules is initiated by collision (Fig. 2.21a). The ionization of molecules generates the new free electrons. These free electrons continue to absorb the photon energy and ionize the molecules resulting in "avalanche" breakdown. Thus the breakdown process basically consists of electron generation and cascade ionization. Plasma generation by optical breakdown is a threshold phenomenon. It is considered that the breakdown is reached when the free electron density in the plasma corresponds to about $10^{18}/cm^3$. In multiphoton absorption mechanism (Fig. 2.21b), each electron is independently ionized and thus requires no seed electrons, no collision, or particle–particle interactions. Under certain irradiation regimes, both the mechanisms may be significant. In such cases, the multiphoton absorption process not only provides the "seed" electrons for the cascade ionization, but also contributes and accelerates the cascade process (Kennedy 1995, Liu et al. 1997).

The generation of plasma can greatly influence (or interfere with) the interaction of laser radiation with the material. It is convenient to define the laser power density (I_p) at which the significant ionization of the vapor resulting in the formation of plasma takes place. The plasma is generally considered to form near the evaporating

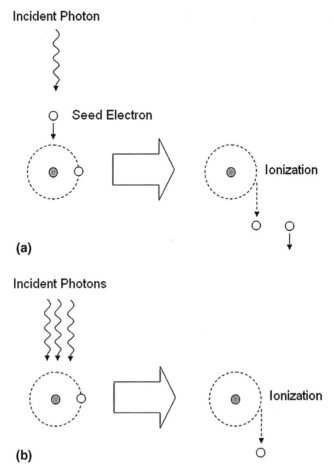

Fig. 2.21 Schematic of the mechanisms of laser induced breakdown: (**a**) avalanche ionization by collision with seed electrons and (**b**) ionization by multiphoton absorption

surface of the target and remain confined to this region during laser irradiation with intensities just above I_p. This confinement of the stationary plasma near the evaporating surface is generally referred to as plasma coupling (Fig. 2.22a). Plasma coupling plays an important role in transferring the energy to the dense phase. The energy transfer may be due to normal electron heat conduction, short-wavelength thermal plasma radiation, or condensation of vapor back to the surface (von Allmen 1987). The plasma coupling is particularly important in conditions where normal laser irradiation is not strongly absorbed by the target material. Such conditions exist during irradiation of highly reflecting materials with infrared (longer wavelength) laser radiation. Plasma coupling results in the significant increase in the absorptivity of laser radiation by the material. When the laser power density is increased significantly beyond I_p, the dynamic interaction of the plasma with the laser radiation causes the rapid expansion and propagation of the plasma away from the evaporating surface

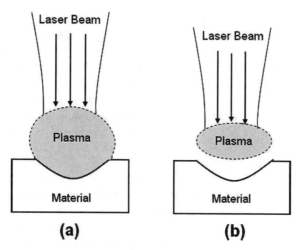

Fig. 2.22 Schematic of (**a**) plasma coupling and (**b**) plasma shielding effects

(i.e., towards the laser beam). Eventually, the plasma gets decoupled from the surface and transfer of energy to the dense phase ceases. The laser radiation is then essentially absorbed in the plasma. This condition is referred to as plasma shielding where the decoupling of the plasma ceases the interaction of the laser radiation with the target material via plasma (Fig. 2.22b). The propagating plasma is often referred to as laser supported absorption wave (LSAW). The LSAWs are generally classified into laser supported combustion waves (LSCWs) and laser supported detonation waves (LSDWs) depending on the speed at which they propagate with respect to gas. The LSAW propagating at the subsonic speed is termed as LSCW, while, it is termed as LSDW when propagating at supersonic speed (Bäuerle 2000).

2.6 Ablation

The term ablation is generally used for material removal processes by photo-thermal or photo-chemical interactions.

In photo-thermal process, the absorbed laser energy gets converted into thermal energy in the material. The subsequent temperature rise at the surface may facilitate the material removal due to generation of thermal stresses. This is more pronounced in the inhomogeneous targets such as coated materials where the thermal stresses cause the explosive ablation of thin films. When the incident laser energy is sufficiently large, the temperature at the surface exceeds the boiling point causing rapid vaporization. These processes of material removal by thermal stresses and surface vaporization are generally referred to as thermal ablation (Bäuerle 2000).

In photoablation, the energy of the incident photon causes the direct bond breaking of the molecular chains in the organic materials resulting in material removal by molecular fragmentation without significant thermal damage. This suggests that for the ablation process, the photon energy must be greater than the bond energy. The ultraviolet radiation with wavelengths in the range 193–355 nm corresponds to the photon energies in the range 6.4–3.5 eV. This range of photon energies exceeds the dissociation energies (3.0–6.4 eV) of many molecular bonds (C–N, C–O, C=C, etc.) resulting in efficient ablation with UV radiation (Vogel and Venugopalan 2003). However, it has been observed that ablation also takes place when the photon energy is less than dissociation of energy of molecular bond. This is the case for far ultraviolet radiation with longer wavelengths (and hence correspondingly smaller photon energies). Such an observation is due to multiphoton mechanism for laser absorption. In multiphoton mechanism, even though the energy associated with each photon is less than the dissociation energy of bond, the bond breaking is achieved by simultaneous absorption of two or photons.

The laser–material interaction during ablation is complex and involves interplay between the photo-thermal (vibrational heating) and photo-chemical (bond breaking) processes. One of the important considerations during the laser–tissue interaction studies is the thermal relaxation time (τ). Thermal relaxation time is related with the dissipation of heat during laser pulse irradiation and is expressed as (Thompson et al. 2002):

$$\tau = \frac{d^2}{4\alpha},\qquad(2.43)$$

where d is absorption depth and α is the thermal diffusivity. For longer pulses (with pulse time longer than thermal relaxation time), the absorbed energy will be dissipated in the surrounding material by thermal processes. To facilitate the photo-ablation of material with minimum thermal damage, the pulse time must be shorter than that of thermal relaxation. For such short pulses (pulse times in the range of microseconds), the laser energy is confined to a very thin depth with minimum thermal dissipation. Thus, efficient ablation of the material during laser–material interactions necessitate the laser operating at shorter wavelengths with microsecond pulses.

The ablation process is generally explained on the basis of "blow-off" model which assumes that ablation process takes place when the laser energy exceeds the characteristic threshold laser energy. Ablation threshold represents the minimum energy required to remove the material by ablation. Figure 2.23 presents the schematic distribution of absorbed energy in the material irradiated with incident laser energy E_0. Above the ablation threshold energy ($\mu_a E_{th}$), the material removal is facilitated by bond breaking; whereas below ablation threshold energy, the thermal effects such as heating takes place. Absorption properties of tissue and incident laser parameters determine the location at which the absorbed energy reaches the ablation threshold thus determining the depth of ablation. The depth of ablation according to this model is given by:

$$\delta = \frac{1}{\mu_a}\ln\left(\frac{E_0}{E_{th}}\right)\qquad(2.44)$$

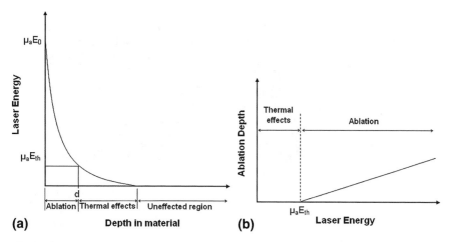

Fig. 2.23 Blow-off model of laser ablation: (a) distribution of absorbed laser intensity in the depth of material and (b) variation of depth of ablation with laser energy

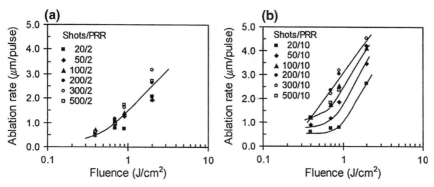

Fig. 2.24 Variation of ablation rate with laser fluence during pulsed laser ablation of PMMA with pulse repletion rate of (a) 2 Hz and (b) 10 Hz. (Reprinted from Liu et al. 2000. With permission. Copyright Elsevier.)

Above ablation threshold, the depth of ablation predicted based on "blow-off" model increases continuously with the laser energy (Fig. 2.23b). However, actual ablation depths depends on a number of other effects such as plasma shielding and radiation-induced changes in the material absorption coefficient which tend to deviate the actual ablation depth from that predicted by simple "blow-off" model (Vogel and Venugopalan 2003).

The ablation rates are primarily determined by the laser fluence, the pulse duration, the number of pulses, and the pulse repetition rates (PRRs). Figure 2.24 presents the influence of some of these parameters on the laser ablation of polymethyl methacrylate (PMMA) using KrF laser (wavelength: 248 nm) and pulse dura-

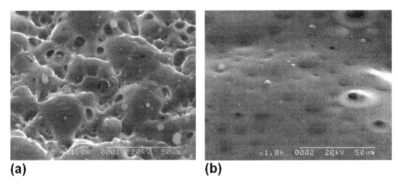

(a)   **(b)**

Fig. 2.25 Morphology of the craters in PMMA ablated by 100 laser pulses with a fluence of 0.7 J/cm² and pulse repetition rate of (**a**) 2 Hz and (**b**) 10 Hz. (Reprinted from Liu et al. 2000. With permission. Copyright Elsevier.)

tion of 30 ns. As indicated in the figure, there exists a threshold fluence below which no ablation is observed. Once the laser energy exceeds the ablation threshold, the ablation rates increase with laser fluence. In addition, for the constant laser fluence, the ablation rate increases with increasing number of pulses before eventual saturation. Higher pulse repetition rates generally results in higher ablation rates. The ablation of material results in the formation of well-defined craters on the surface (Fig. 2.25) (Liu et al. 2000).

Pulsed laser ablation is extensively used in materials processing and medical applications. In materials processing, it can be used for micromachining, marking, grooving, cutting, drilling, and patterning of wide range of materials; while, in medical applications, it can be used for precision ablation of tissues such as human corneal tissues.

References

Anisimov SI (1968) Vaporization of metal absorbing laser radiation. Soviat Physics JETP 27:182–183.

Anisimov SI, Khokhlov VA (1995) Instabilities in Laser–Matter Interaction. CRC Press, Boca Raton, FL.

Bäuerle D (2000) Laser Processing and Chemistry. Springer, Berlin.

Breinan EM, Kear BH (1983) Rapid solidification laser processing at high power density. In: Bass M (ed) Laser Materials Processing, North-Holland, Amsterdam, The Netherlands, pp. 237–295.

Carslaw HS, Jaeger JC (1959) Conduction of Heat in Solids. Oxford University Press, Oxford.

Chan CL, Mazumder J (1987) One-dimensional steady-state model for damage by vaporization and liquid expulsion due to laser-material interaction. Journal of Applied Physics 62:4579–4586.

Duley WW (1983) Laser Processing and Analysis of Materials. Plenum Press, New York.

Flemings MC (1974) Solidification Processing. McGraw-Hill, New York.

Kennedy PK (1995) A first-order model for computation of laser-induced breakdown thresholds in ocular and aqueous media: part I –theory. IEEE Journal of Quantum Electronics 31:2241–2249.

Liu X, Du D, Mourou G (1997) Laser ablation and micromachining with ultrashort laser pulses. IEEE Journal of Quantum Electronics 33:1706–1716.

Liu ZQ, Feng Y, Yi XS (2000) Coupling effects of the number of pulses, pulse repetition rate and fluence during PMMA ablation. Applied Surface Science 165:303–308.

Ready JF (1997) Industrial Applications of Lasers. Academic Press, San Diego, CA.

Rosenthal D (1946) The theory of moving sources of heat and its application to metal treatment. Transactions of the ASME 68:849–866.

Steen WM (1991) Laser Materials Processing. Springer, London.

Thompson KP, Ren QS, Parel JM (2002) Therapeutic and diagnostic application of lasers in ophthalmology. In: Waynant RW (ed) Lasers in Medicine, CRC Press, Boca Raton, FL.

Vogel A, Venugopalan V (2003) Mechanisms of pulsed laser ablation of biological tissues. Chemical Reviews 103:577–644.

von Allmen M (1987) Laser-beam Interactions with Materials. Springer, Berlin.

Welch AJ, Gardner C (2002) Optical and thermal response of tissue to laser radiation. In: Waynant RW (ed) Lasers in Medicine, CRC Press, Boca Raton, FL, pp. 27–45.

Wilson J, Hawkes JF (1987) Lasers: Principles and Applications. Prentice-Hall, New York.

Part II
Laser Machining

Chapter 3
Manufacturing Processes: An Overview

3.1 Introduction

Manufacturing refers to the processes of converting the raw materials into useful products. This is normally accomplished by carrying out a set of activities such as product design, selection of raw material, and materials processing (Kalpakjian and Schmid 2001). There exist a large number of conventional manufacturing processes which are used for the manufacturing of common products. However, the manufacturing engineering is a dynamic field marked with continuous advancement in the traditional approaches and the incorporation of novel approaches for manufacturing advanced products. Not all manufacturing processes can produce a product with equal ease, quality, and economy. Each manufacturing process is generally characterized by some advantages and limitation over the other processes. On the same lines, manufacturing using lasers may offer extraordinary benefits in some cases or may be a total failure in others. In order to keep the manufacturing of materials using lasers in the correct context, this chapter intends to give a brief overview of the various manufacturing processes.

3.2 Manufacturing Processes

There are a large number of manufacturing processes currently used in industries. It is convenient to discuss the manufacturing processes by grouping them into certain classes based on some characteristic common features. The manufacturing processes can be classified in various ways based on factors such as geometry of workpiece, temperature of the workpiece, and type of deformations (Kalpakjian 1967):

1. *Primary and secondary manufacturing processes*: Primary manufacturing processes involve the initial conversion of the raw materials into the semifinal product stage. The output of primary manufacturing processes is then subjected to secondary manufacturing processes to obtain the final or finished product geometry. Various primary manufacturing processes include casting, forging,

rolling, extrusion, etc., whereas secondary manufacturing processes involve various machining and forming processes.

2. *Hot working and cold working processes*: When the manufacturing process is carried out at temperatures above the recrystallization temperature of the material, it is referred to as the hot working process, whereas below the recrystallization temperature it is referred to as the cold working process. Hot working processes are generally carried out at elevated temperatures. For example, rolling of steel may be a hot working (hot rolling) or cold working (cold rolling) process.

3. *Metal forming and metal removal processes*: Metal forming processes involve the manufacturing of a product by deforming the raw material, whereas metal removal processes, as the name suggests, involves the removal of material from the workpiece to obtain the desired shapes. Various rolling, forging, and bending operations can be regarded as metal forming processes. Material removal processes include various machining operations.

In the following sections important manufacturing processes are briefly discussed. Each manufacturing process generally involves complexity of equipment, processing mechanisms, processing parameters, quality considerations, etc. This chapter is not designed to provide the detailed discussion of all these aspects of each manufacturing process. Instead, only important aspects of the main manufacturing processes are discussed so that the capabilities and the limitations of the laser-based manufacturing processes can be appreciated. Readers interested in a detailed discussion of each manufacturing process are suggested to study the standard textbook by Kalpakjian and Schmid (2001).

3.2.1 Casting Processes

Casting process is one of the earliest manufacturing processes and involves the production of simpler shapes from the molten material. The material is generally melted separately and subsequently poured into the mold cavity (Fig. 3.1). The mold cavity consists of the shape of the object to be manufactured. After solidification of molten material in the mold, the object is removed for subsequent processing. Casting processes are widely used as primary processes to produce the prefinished products of a wide range of metallic materials (Heine et al. 1967). Important characteristics of the casting processes are:

1. Complex products with internal cavities, hanging structures, and hollow sections can be produced by careful design of gates, risers, mold, cores, etc.

2. Large parts of sound quality can be produced. The largest size of the part that can be produced is related with the solidification characteristics of the metal.

3. The process is economical when the large number of parts needs to be manufactured.

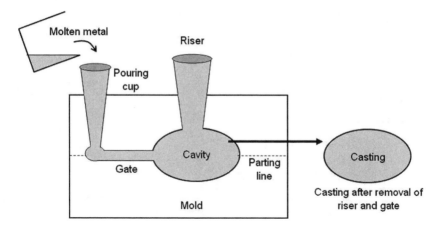

Fig. 3.1 Schematic of the metal casting process

The casting processes are often associated with casting defects such as blowholes, porosity, shrinkages, folds, flash, and inclusions due to excess solidified materials (flash), incomplete filling of mold cavity, solidification shrinkages, and chemical reactions. To enhance the soundness of the castings and minimize the scrap, importance must be given to the following considerations:

1. Mold material and mold cavity design. The mold material influences the heat transfer characteristics, surface finish of the part, and the ease of part removal after casting. The mold cavity design considerations include location of the parting line and the compensation for various allowances (solidification shrinkages, machining allowances, etc.).
2. Design of gates, risers, and sprues to fill the molten material in the various sections of the mold cavity and the shrinkages.
3. Geometric aspects of the part such as cross-sectional thicknesses, overhanging sections, hollow sections, sharp cornets, etc.
4. Solidification and fluidity characteristics of the material.
5. Processing conditions such as pouring temperature, pouring rate, and liquid head.

3.2.2 Forming Processes

In forming processes, the starting material may be in the form of continuous solid workpiece, solid particles, or molten material. The part is manufactured by deforming the initial workpiece, compacting and sintering the powder, or molding the material into desired shapes. Various deformation processes include rolling, forging, extrusion, drawing, and sheet metal forming processes for metallic materials.

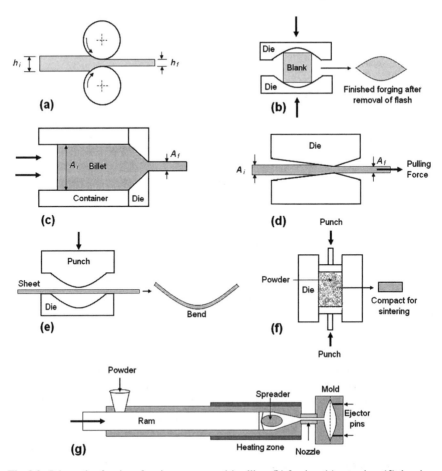

Fig. 3.2 Schematic of various forming processes: (**a**) rolling, (**b**) forging, (**c**) extrusion, (**d**) drawing, (**e**) sheet metal forming, (**f**) powder processing (compaction and sintering), and (**g**) molding

Powder-based manufacturing processes are generally used for the materials with difficulty in forming by deformation (e.g., ceramics), or to achieve specific properties in the product (controlled densification, metallurgical properties, etc.). The molding processes are generally used for polymeric (nonreinforced or reinforced) materials. A schematic of these forming processes is presented in Fig. 3.2 (Tschaetsch 2006; Kalpakjian and Schmid 2001; Dieter 1986).

Rolling process involves the passing of a workpiece through a set of rollers where the compressive deformation of the workpiece causes the change in shape of the workpiece. The rolling process is generally carried out in a series of rolling operations with stepwise reduction in the thicknesses of the workpiece. Hot rolling operations are used for the initial thickness reductions followed by cold rolling operations for final thickness reductions. The flat or profiled rollers are used depending on the desired cross-sectional shape of the product. The products of rolling processes include plates, sheets, bars, and rods which are extensively used for structural applications (Roberts 1983).

Forging process is another compressive deformation process in which the workpiece is pressed in a die consisting of a shape of the product. The forging process may be classified as the open and closed die forging depending on the design of the die. The critical parameters in the design of forging are the temperature, amount of initial material, shape and size of the object, forging forces, etc. The excess material escapes from the die as flash and may influence the forging forces and the quality of the forged product. Formation of flash generally increases the forging load required for further deformation and inhibits the flow of material in the sharp and intricate areas of the die cavity. Forging is widely used for manufacturing semifinished products which are subjected to further processing to obtain the final useful shapes (Semiatin 1988).

Extrusion process involves the reduction in the cross section of the workpiece by forcing it through a die in a container. The die consists of an orifice (opening) of desired cross-sectional profile. The extrusion processes are broadly classified into two types depending on the direction in which the product is extruded with respect to the force application: direct (forward) and indirect (backward) extrusion. Indirect extrusion processes require less energy compared with direct extrusion processes due to the absence of frictional forces between the billet and the container. Extrusion processes are used for manufacturing of rods, tubes, and other products with complex cross sections (Raghupathi 1988).

Drawing is a material forming operation in which a workpiece in the form of rod, wire, or tube is pulled through a die opening to reduce its cross section. The important die design parameter which influences the forces during extrusion and drawing operations is the die angle. The length of the drawn product depends on whether the product can be coiled or not. Long lengths of wires can be produced by coiling the product, whereas for thicker sections, the length is limited by the length of the drawbench. The drawing processes are also generally carried out in a series of operations with stepwise reduction in cross section (Dieter 1986).

Sheet metal forming includes a large number of processes in which flat sheets are converted into various shapes by bending, stretching, shearing, etc. Often, a number of sheet metal forming operations are carried out in a sequence to obtain the desired shape of the product. The common equipments for mechanical sheet metal forming processes include various dies and punches. The important considerations during most of the sheet metal forming operations are the spring-back effect, surface defects, sheet buckling, and wrinkling. Other material-specific effects such as non-uniform deformations (stretcher strains) in case of low-carbon steels are important (Hu et al. 2005).

Powder-based manufacturing (powder metallurgical) processes involve the compaction of the fine powder into a desired shape followed by high temperature sintering. The compaction may be carried out by pressing powder into dies or by continuously feeding the powder into rolls. The compacts are subsequently heated to sintering temperature, which is generally less than the melting point of the material. Powder metallurgy can be used to produce parts from a wide range of metal or alloy powders with close tolerances. Important factors in the powder metallurgy processes include the ease of part removal from the compaction dies, density and green strength of the compacts, sintering temperatures, sintering mechanisms, shrinkage allowances, strength of the sintered products, etc. (German 1998).

Molding processes generally refer to the manufacturing processes for polymeric products. These include processes like injection molding, blow molding, and rotational molding. The initial polymeric material is in the form of solid pellets which are melted and subsequently transferred into the mold. The molten polymeric material may be blown into the mold to produce hollow products. The polymeric materials can also be formed by casting, thermal forming, etc. The selection of manufacturing process for polymeric materials depends on the type of polymeric material (thermoplastics, thermosets, elastomers) and desired shape of the product. Special processes are required for manufacturing reinforced polymer matrix composites and laminates (Crawford 2001).

3.2.3 Joining Processes

In many cases, the manufacturing of a product as a one-piece with acceptable quality is not feasible or not economical and requires joining of many simple/complex shapes which are manufactured separately. Various joining processes include welding, mechanical joining, adhesive bonding, etc. Important consideration during most of the joining process is the achievement of properties in the joined product equivalent to or better than the product that would have been manufactured as one-piece shape. In many cases joining allows the fabrication of composite structures from a wide range of materials and shapes (Nicholas 1998).

Welding is a group of joining processes such as fusion welding, solid-state welding, brazing, and soldering. In *fusion welding*, two workpieces are sufficiently melted at the joint such that subsequent solidification of the melt pool causes the joining. Filler materials may or may not be used in fusion welding. *Brazing* and *soldering* are generally considered as low temperature processes where two parts are joined together by using filler material which is subsequently melted by an external heating source to form a joint after solidification. In *solid-state welding*, the two parts are joined without fusion. This includes processes like friction welding, diffusion bonding, resistance welding, etc. The properties of the workpieces need to be considered during selection of welding process. Even though welding processes are commonly used, joining of workpieces with widely different properties by welding still presents significant challenges. Other consideration include strength of the joints and welding defects (cracks, heat affected zones, spatter, incomplete fusion, shrinkage, etc.) (Geary 1999).

Mechanical joining methods involve obtaining generally detachable joints using fasteners, bolts, nuts, rivets, etc. Additional preprocessing steps such as drilling of workpieces are generally required to position the nuts, bolts, etc. Mechanical joining methods are well-suited for shapes which undergo design changes quite frequently such that detachment and assembly can be done easily.

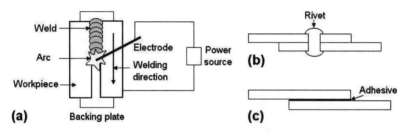

Fig. 3.3 Schematic of various joining processes: **(a)** arc welding, **(b)** mechanical joining, and **(c)** adhesive bonding

Adhesive bonding involves the joining of the parts by using adhesives which holds the parts together. Adhesives can be formulated with a wide range of properties such as strength, corrosion resistance, thermal stability, etc. Wide ranges of natural and synthetic adhesives are commercially available for various joining applications. Figure 3.3 presents the schematic of various joining processes.

3.2.4 Machining Processes

The manufacturing processes such as casting, forming, joining, etc. are extensively used in the industry. However, most of the shapes manufactured by these processes must be subjected to secondary processes for converting them into useful fully finished products. One of the most important secondary processes is the machining process. Machining involves removal of material from a workpiece for converting it into the desired shape with size and tolerances per design specification. It is difficult to imagine a manufactured product which does not require machining for at least one of its components. Since the machining adds value to the finished product, material removal processes have gained considerable importance to control the overall economics of the process. Large number of machining processes have been developed and used with varying degree of success for machining different materials. Still, one machining method might be exceptionally good for one material and it may well be totally unacceptable for another in terms of machining cost and quality.

The machining processes have been broadly categorized into traditional and nontraditional machining processes based on the mechanism of material removal (Walker 1999).

3.2.4.1 Traditional Machining Processes

Conventional machining involves material removal due to mechanical stresses exceeding the strength of the material, induced by the tool. The basic machining mechanism involves the localized shear deformation on the workpiece material

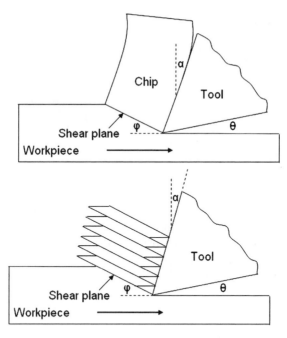

Fig. 3.4 Schematic of the orthogonal geometry and chip formation during machining

ahead of the cutting edge of the tool. The material is removed in the form of chip formed by fracture along the successive shear planes which are inclined to the cutting direction. The process is analogous to the sliding of cards ahead of the cutting edge of the tool (Fig. 3.4). The two deformation zones in metal cutting are identified as primary and secondary deformation zones. The primary deformation zone is due to shear deformation ahead of the cutting tool due to compression of the workpiece material during cutting action, whereas the secondary deformation zone is due to shear and sliding of the chip as it passes over the rake face of the cutting tool (Black 1989).

A large variety of material removal processes are used in industrial practices primarily based on processes like turning, drilling, milling, shaping, abrasive machining, sawing, etc (Fig. 3.5). A brief summary of these processes is given below (DeVries 2006; King 1985; Pandey and Shan 1980).

Turning is the most commonly used industrial process for producing cylindrical components by removing the material from the external surfaces of the rotating workpieces. The process can be well adapted to perform various operations such as straight turning, taper turning, facing, threading, form turning, grooving, and cut off by varying the relative feed motions between the workpiece and the cutting tool. *Boring* is similar to turning operation except that the material is removed from the internal surfaces of the workpiece.

In *drilling*, holes are produced by removing the material from stationary workpiece by longitudinal feeding of rotary end cutting tool (drill). The drills are usually

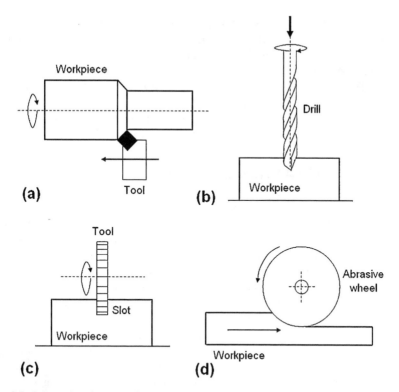

Fig. 3.5 Schematic of some of the common traditional machining processes: **(a)** turning, **(b)** drilling, **(c)** milling, and **(d)** grinding

provided with one or more flutes for the removal of chips and supply of cutting fluids. The most commonly drilled holes ranges from 3.2 to 38 mm in diameter, however, special techniques are available to drill the holes as small as 0.025 mm and as large as 152 mm in diameter. The various drilling operations include reaming, tapping, spot facing, counterboring, and countersinking.

In *milling*, the material is generally removed by a rotating multiple-tooth cutter by feeding the workpiece perpendicular to the axis of cutter. Commonly milled materials include steels with hardness of 35 HRC, however, the harder materials have also been successfully milled. Various milling operations include hobbing, generating, and thread milling.

Shaping and *planning* processes are used for removal of material from the surfaces of the workpiece to produce flat surfaces by relative linear reciprocating motion of the cutting tool. The process can be readily configured to machine contours and irregular surfaces with grooves and slots.

Various *abrasive machining* operations include grinding, honing, and lapping. In grinding, material is removed from the workpiece by the cutting action of abrasive grains of grinding wheel surface which is rotated relative to the workpiece surface. Grinding efficiency is greatly influenced by the sharpness of the abrasive grains on the surface of wheels. Single-point diamond dresser is frequently used to resharpen the grinding wheel surface by exposing the new grits by selectively removing the material from dull grinding wheel surface. Honing is abrasive machining operation specially used for precision finishing of inside cylindrical surfaces. The material is removed by shearing action of the abrasive grains of the honing stones or sticks which are mounted on the periphery of cylindrical rotating mandrel oscillating inside the bore to be finished. The bores with diameters as small as 1.6 mm and as large as 1,270 mm can be honed successfully.

3.2.4.2 Nontraditional Machining Processes

Nontraditional machining processes are evolved to meet the special machining needs for which the conventional machining proves unsatisfactory both in terms of economics and achievable machining quality. Most of these needs come from the rapid technological advancement in the areas of materials development for advanced applications in aerospace, automotive, and nuclear power industries. Increasing development and utilization of super-hard, high-strength, high-temperature, and high-performance materials in these applications is increasingly demanding complex machining requirements for such difficult-to-machine materials. Some of these difficult-to-machine materials are titanium, nimonics, metal matrix composites, advanced ceramics, aluminides, etc. Conventional machining is limited mainly due to unavailability of ultrahard tool material for economical machining of these difficult-to-machine materials. In most of the cases, nontraditional machining is the most economical and effective way of machining these materials (Springborn 1967).

Nontraditional machining processes are generally considered to be manufacturing processes that use common energy forms in new ways or that applies new forms of energy. Nontraditional machining processes are categorized based on the form of energy employed such as mechanical, electrical, thermal, and chemical. Table 3.1 lists the details of nontraditional machining processes (Benedict 1989).

Table 3.1 Various nontraditional machining processes

Class	Mechanism	Examples
Mechanical	Materials removal due to abrasive action of powder particles, water, slurry, etc.	Abrasive jet machining
		Waterjet machining
		Abrasive waterjet machining
		Ultrasonic machining
Electrical	Material removal in the electrolyte due electrochemical reactions	Electrochemical machining
		Electrochemical grinding
		Electrochemical discharge grinding
Thermal	Material removal using electrical sparks, electrons, photons, etc.	Electrical discharge machining
		Electron beam machining
		Laser beam machining
Chemical	Material removal due to chemical action	Chemical milling
		Photochemical machining

In most of the *nontraditional mechanical machining* methods, the material removal is due to progressive erosion of the workpiece carried out by waterjet or abrasive grains suspended in high-velocity gas or water streams. The method is particularly suitable for ceramics, composites, and organic materials which are difficult to be machined using electrical and thermal methods due to low electrical conductivity and susceptibility to thermal damage such as burning, charring, or cracking.

In *electrical methods*, the material removal from the workpiece is carried out through the electrolytic action by constructing an electrolytic cell with the workpiece as an anode. The electrolytic action not only removes the material from the workpiece but also change the shape of it. The electrical processes are limited to the machining of electrically conducting materials, however, the processes finds utilization in specialized applications where conventional techniques find difficulty in machining.

In *thermal methods* of machining, high energy source concentrates energy on a small area of the workpiece so that the material removal takes place by localized melting and/or evaporation. The various energy sources may be electrons, photons, electrical sparks, etc. The material removal rate is independent of the hardness and strength of the material and hence the thermal methods are often applied for machining extremely hard and difficult-to-machine materials.

Chemical method involves the material removal by chemical action without any force acting on the surface. The most common chemical methods are controlled etching techniques which can be used to uniformly remove the workpiece material from all the surfaces or thin the specific areas of surfaces. The parts to be machined are immersed in a tank containing etching solutions without electrical assistance. The selective machining applications, the remaining areas are protected by special coatings and masks.

3.3   Lasers in Manufacturing

Lasers are finding continuously increasing utilization in the manufacturing processes. The applications of lasers have been demonstrated in many casting, forming, joining, and machining processes. Some of these processes are still in the stage of development. Currently, there are several laser-based manufacturing processes which are commonly used for specific applications. This section outlines these laser-based manufacturing processes. These processes are explained in detail in the subsequent chapters.

3.3.1   Laser Casting

The conventional casting methods are still the most commonly used methods for producing initial shapes of the products. Lasers have been demonstrated to be used in casting of materials. These laser casting methods are primarily derived from laser cladding processes. In laser cladding, generally a surface of highly corrosion and/or wear-resistant material is created on some substrate material. Laser is used to completely melt the preplaced powder and partially melt the substrate such that a strong bond is formed at the interface. In laser casting, metal powder of predefined thickness is placed on the mold followed by laser melting. The most important consideration for the success of laser casting is the selection of laser parameters such that fusion at the mold–powder interface is avoided (Fig. 3.6). This facilitates the removal of casting from the mold (Gedda et al. 2005). The process is still under development and there are significant challenges with respect to size, shape, and quality of the castings produced by such laser casting processes.

3.3.2   Laser Forming/Shaping

As mentioned in the previous sections, the common forming processes includes a variety of solid-state bulk plastic deformation processes (such as extrusion, rolling, and forging), sheet metal forming processes, and powder-based forming processes. The two areas where lasers are increasingly used are laser forming and rapid prototyping.

Laser forming primarily refers to the sheet metal forming processes such as bending (Fig. 3.7). The laser bending of sheet is primarily based on establishing steep temperature gradients in the sheet by laser heating (during scanning) such that differential thermal expansion results in thermal stresses. When the thermal stress exceeds the temperature-dependent yield stress, plastic deformation of the material causes the bending of sheet. In general, the laser forming methods offer significant advantages over the conventional mechanical forming processes such as rapid processing and bending without spring-back effects. However, there are specific applications where laser forming have been proved particularly useful and efficient. Laser forming (bending) processes are primarily suitable for rapid fabrication of low-quantity parts such as prototypes and special shapes where conventional mechanical forming processes are often uneconomical due to longer design time, high cost of dies, and longer fabrication times. Due to noncontact nature of the laser processing, lasers may be used for bending complex shapes using complex laser scanning strategies. Recently, laser forming have attracted significant interests in the microfabrication for precision adjustment of components. The major application area of laser forming is the correction of bend angles obtained by conventional forming methods.

Lasers are at the forefront of the rapid prototyping technologies for the fabrication of three-dimensional objects. Laser-based rapid prototyping processes such as stereolithography (SL), selective laser sintering (SLS), laminated object manufacturing (LOM), laser engineered net shaping (LENS), etc. are now extensively used for the fabrication of a variety of complex shapes for a wide range of materials. These processes are based on various mechanisms such as photopolymerization of liquid polymeric resins, sintering of solid particles (metals, ceramic, polymers, etc), stacking of laser-cut adhesive bonded sheets, melting and solidification of powder fed in laser beam, etc. Schematics of the various laser-based rapid prototyping processes are presented in Fig. 3.8.

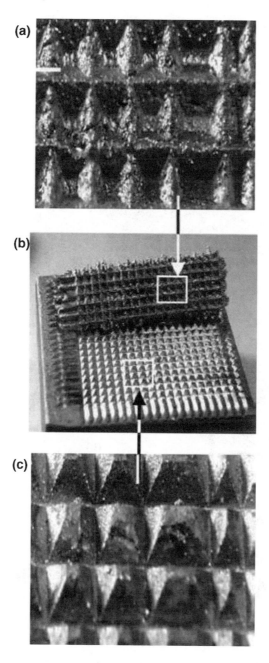

Fig 3.6 Laser casting process: **(a)** casting, **(b)** casting and mold, and **(c)** mold. (Reprinted from Gedda et al. 2005. With permission. Copyright Springer.)

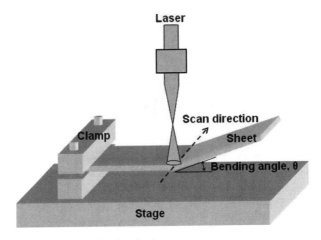

Fig. 3.7 Schematic of laser forming (bending) process

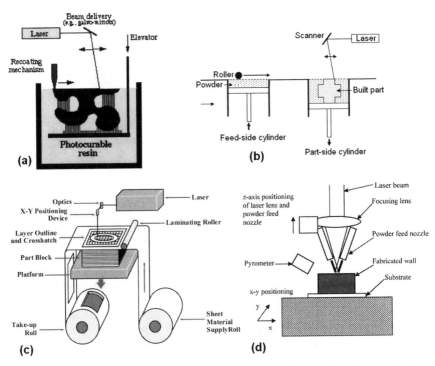

Fig. 3.8 Laser-based rapid prototyping processes: (**a**) stereolothography, (**b**) selective laser sintering, (**c**) laminated object manufacturing, (**d**) laser engineered net shaping. (Fig. 3.8a reprinted from Dutta et al. 2001. With permission. Copyright American Society of Mechanical Engineers.; Fig. 3.8d reprinted from Ye et al. 2006. With permission. Copyright Elsevier; Fig 3.8c courtesy of Cubic Technologies, Inc.)

3.3.3 Laser Joining

Lasers have been extensively used for joining (welding and soldering) of variety of materials. The various laser welding processes involve spot welding, seam welding, and deep penetration welding. Laser welding generally involve the formation of keyhole by the surface vaporization of material (Fig. 3.9). This keyhole facilitates the absorption of light energy and distribution of heat such that melting of the workpieces at the joint results in the formation of weld. The laser welding offer significant advantages over the conventional welding such as high welding speed, possibility of dissimilar welding, welding of difficult-to-weld materials, microwelding, precision welding of components, narrow heat affected zone (HAZ), etc. There exist some specialized applications (such as inside welding, welding of devices, etc.) where conventional welding may not produce quality welds. Such applications are well handled with laser welding processes. Laser welding is particularly economical for high production volumes compared to other competent welding processes. There may be some limitations on the thickness of the workpieces and the speed of laser welding. The selection of laser welding for a given application must be derived from the welding capabilities of the various laser

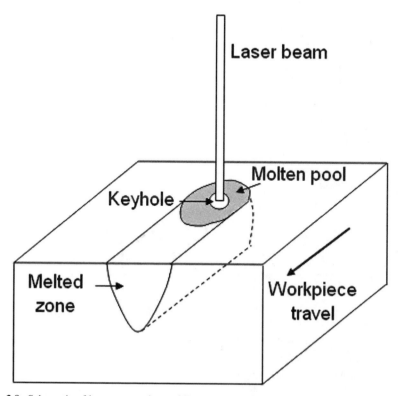

Fig. 3.9 Schematic of laser penetration welding

sources, overall economics of the manufacturing, and the quality of the welds produced. The flexibility of lasers in the manufacturing processes is due to ability of the lasers to sequentially weld and machine (cutting, drilling, etc.) by optimizing the laser processing parameters.

3.3.4 Laser Machining

Another major area of laser applications in the manufacturing processes is the laser machining. By controlling the motion and geometry of the workpiece, and laser beam delivery system, lasers can be used for a variety of one-, two-, and three-dimensional machining applications (Chryssolouris 1991). Schematics of the basic laser machining processes are presented in Fig. 3.10.

Lasers are now extensively used in the industry for drilling, cutting, and shaping of materials. The laser machining operations are efficient and economical in a number of industrial applications where large production rates are desired. Laser drilling is particularly useful for high-aspect ratio microdrilling applications where conventional mechanical drilling is not applicable or less efficient. Laser drilling can be used with a variety of material systems ranging from, metal, polymers,

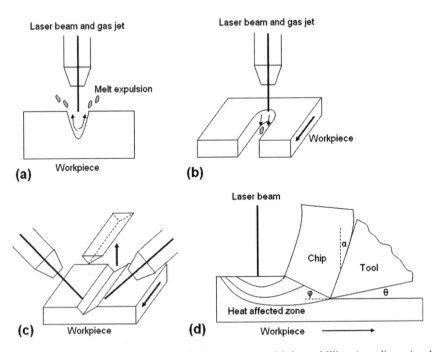

Fig. 3.10 Schematic of basic laser machining processes: **(a)** laser drilling (one-dimensional machining), **(b)** laser cutting (two-dimensional machining), **(c)** laser milling (three-dimensional machining), and **(d)** laser-assisted machining

ceramics, to composites. There may be certain limitations on the diameter and the depth of the laser-drilled holes. However, with the careful design of the laser drilling systems (such as trepanning and percussion drilling), big and wide holes may be produced. In addition, the laser-drilled holes may be associated with certain undesirable geometric and metallurgical defects (such as taper, recast layer, etc.). Again, careful design of laser parameters (pulse shapes, beam profiles, laser power, etc.) and other process parameters (application of coating, assist gases, nozzle designs, etc.), may minimize or eliminate these laser drilling defects. Thus, lasers can be used for drilling high-aspect ratio holes in variety of materials. The overall drilling economics and the desired quality levels must be considered before selecting laser drilling for a given application.

Laser cutting is an established technology for machining difficult-to-machine materials such as advanced ceramics. Laser cutting can be used in a number of ways to remove the materials (vaporization, melt expulsion, chemical reactions, etc.), thus, extending the limits of machining capabilities. Both pulsed and continuous lasers have been used for laser cutting, which offers significant advantages such as cutting of complex geometries, faster processing speeds, cutting of wide range of materials, clean cuts, etc. Laser cutting may also be associated with certain defects such as striations (periodic pattern on the cut edge), dross, heat-affected zones, etc. Hence, selection of the laser cutting process for a given application primarily involves careful consideration of laser cutting capabilities, quality considerations, and overall economics of manufacturing.

3.4 Selection of Manufacturing Processes

As explained in the previous sections, lasers can be used in a number of manufacturing processes. In many cases, laser manufacturing seems to offer significant advantages over the conventional manufacturing processes. However, the final selection of the best manufacturing process for a given application is a complex process and involves the careful consideration of several factors such as desired shape and size of the product, properties of materials, quality of the manufactured product, and the cost of manufacturing. Often, there exists a trade-off between these factors. For example, in critical applications, highest quality of the product is desired irrespective of the cost of manufacturing; whereas in less critical applications, low-cost manufacturing is desired at acceptable product quality levels.

3.4.1 Properties of Materials

The selection of a manufacturing process for a given material is influenced by a number of factors, the most important being the physical properties of the desired product material. Materials are broadly classified into four categories: metals,

Table 3.2 Relative capabilities of various conventional and laser manufacturing processes

Material	Conventional manufacturing				Laser manufacturing		
	Casting	Forming	Joining	Machining	Forming	Joining	Machining
Metals	E	E	E	E	G	G	E
Ceramics	P	G	P	P	G	P	E
Polymers	G	E	P	P	G	P	G
Composites	P	G	P	P	P	P	G

E: Excellent, G: Good, P: Poor

polymers, ceramics, and composite materials. All materials cannot be manufactured by a single process with equal ease, economy, and quality (Kalpakjian and Schmid 2001). Table 3.2 provides the manufacturing processes suitable for various broad classes of materials. This suitability of various manufacturing processes is determined by the materials properties such as melting point, deformation resistance, strength, ductility, etc. The suitability of laser-based manufacturing processes is primarily determined by the light absorption characteristics of the material.

Metallic materials are probably the most extensively used materials for the manufactured products. Manufacturing processes such as casting, forming, joining, and machining can be effectively used for producing metallic products. The conventional casting methods (such as sand casting) can be used to produce the parts of low-melting point metallic materials (such as aluminum, magnesium, cast irons, copper, etc. with melting point less than ~1,800 K) over a wide range of sizes. Similarly, forming methods such as rolling, extrusions, and forging are used for low-melting point metallic materials. For high-melting point metallic materials (such as tungsten, molybdenum, tantalum, etc. with melting point greater than ~2,000 K), the conventional casting methods cannot be used and the advanced casting methods such as electron beam casting may be required. Powder-based forming methods are most popular for the high melting point metallic materials; however, low-melting point materials can also be formed (Fig. 3.11) (Ashby 1999). Large numbers of joining processes are available for joining similar and dissimilar metallic materials. Conventional machining of metallic materials using variety of cutting tool materials such as high speed steels, carbide, diamond, ceramics, and cubic boron nitride are widely used material removal method. Lasers are effectively used for the manufacturing of metallic materials. Metallic materials are generally characterized by very high reflectivities to the laser radiations. Hence, the application of lasers in the manufacturing of metallic materials requires the careful consideration of laser wavelengths such that efficient laser–material interactions can be obtained. Various useful laser interactions with the metallic materials during laser manufacturing involve heating, melting, vaporization, ablation, plasma generation, etc.

Plastics are normally used in as-formed condition produced by molding processes. However, for the small production volume, complex and accurate shapes machining is necessary. In most cases, conventional machining techniques are used for machining plastics despite of difficulties due to the complexity of plastic grades and

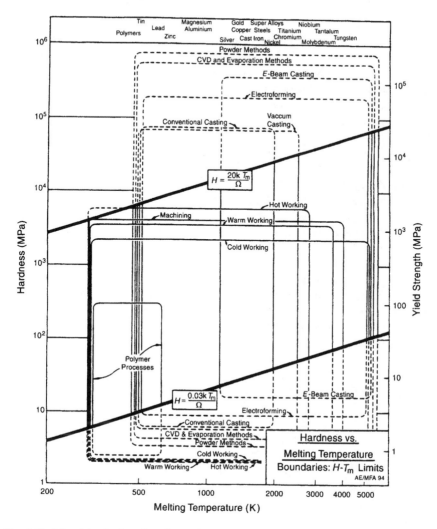

Fig. 3.11 Material selection chart for casting. (Reprinted from Ashby 1999. With permission. Copyright Elsevier.)

abrasive filler/fiber contents in these materials. In addition to the machining conditions such as rake angle, tip radius, depth of cut and cutting speed, and machining of plastics is greatly influenced primarily by their thermal, mechanical, and rheological properties. Lasers can be used for the machining and forming of the polymeric materials. In many applications, the laser interactions for the machining of the polymeric materials involve the direct ablation of material. The forming of three-dimensional polymeric shapes is facilitated by the photo-polymerization of resins as used in the laser-based rapid prototyping techniques.

Ceramics are widely used in a wide range of applications such as structural, automotive, and electronic applications due to their unique combination of properties such as high hardness, high thermal strength, chemical inertness, attractive high temperature abrasion resistance, and low density. Despite these remarkable properties, the actual utilization of advanced ceramics has been quite limited mainly because of high cost of the machining and inability to achieve the dimensional control and workpiece quality required for these applications. The production methods for ceramics like compaction and sintering are still not developed enough to produce dimensional accuracy of the finished part, in addition to the processing defects like material shrinkage and distortion. Conventional machining includes most widely used grinding or diamond machining which are very expensive and time consuming. Machining cost for ceramics by conventional machining accounts for 60–90% of the total cost of the final product (Chryssolouris et al. 1997). For such applications laser machining offers economical solutions. Lasers are well-suited for processing of most of the ceramic materials due to very high absorptivities of ceramics to laser radiations. Lasers are now extensively used for machining and forming of ceramic materials.

The composite materials represent a class of advanced materials with unique properties obtained by combining two or more constituent materials. Due to widely differing properties of the constituent phases in the composite materials, the manufacturing composites present significant challenges. Depending on the type of constituent materials (matrix and reinforcing phases), the composite materials have widely differing manufacturing routes. The application of lasers in the manufacturing of composite materials is primarily in the areas of machining and forming.

3.4.2   Geometrical Complexity of Product

In the simplest of terms, this consideration refers to the capability of the process to produce the part of desired shape, size, and thickness. Each manufacturing process is often characterized by a range of sizes and thicknesses, and complexity of shapes that can be produced. For example, the largest size of the part produced by casting is often limited by capabilities of the casting process to efficiently fill the various sections of the cavity such that sound casting can be produced. Similarly, the complexity of the shape in casting depends on the ability to completely fill the sharp corners and intricate cavities of the mold. Figure 3.12 presents the process selection chart showing the regions of part size and the shape complexities for various manufacturing processes (Ashby 1999). With the recent interests in microfabrication, the capabilities of the manufacturing processes are also assessed in terms of the size of the smallest part that can be produced. Thus, there are limits on the maximum and minimum sizes of the part that can be effectively produced by a manufacturing process. This is illustrated in Fig. 3.13 for the fabrication of a web-shaped component by various fabrication

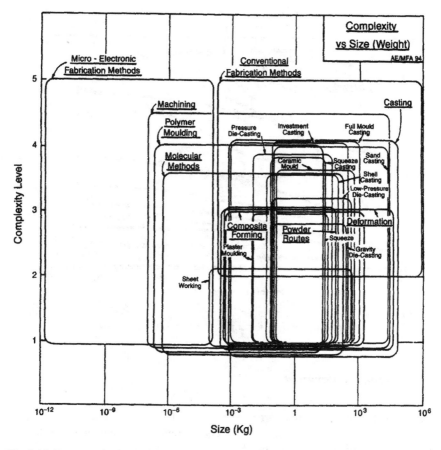

Fig. 3.12 Process selection chart showing the relationship between the part size and shape complexity for various manufacturing processes. (Reprinted from Ashby 1999. With permission. Copyright Elsevier.)

processes (Schey 1997). As indicated in the figure, the size of the part that can be produced by a given process is also influenced by the material properties. In addition, the design of the manufacturing equipment may limit the size of the part that can be produced. In case of laser processing, due to the noncontact nature of the process, the largest size of the part that can be produced seems to be limited primarily by the equipment design. For example, lasers can be efficiently used for the forming of the large sheet metal structures provided the beam delivery systems must be adequate to scan the laser along the longer distances. Lasers are currently used in a number of microfabrication applications. Thus, laser processing is significantly more flexible than the conventional processes for the bulk- and microfabrication.

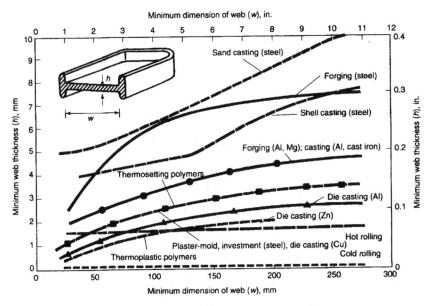

Fig. 3.13 Capabilities of various manufacturing processes for obtaining minimum part dimensions. (Reprinted from Schey 1987. With permission. Copyright McGraw-Hill.)

3.4.3 Quality Parameters

Quality of the manufactured part is related with the geometric and metallurgical soundness of the part. This includes the considerations of manufacturing defects, dimensional tolerances, surface finish, etc.

Every process is inadvertently associated with certain manufacturing defects. For example, casting process is often associated with defects like blowholes, microporosity, etc.; whereas rolling process is often associated with wavy edges and edge cracking. Table 3.3 presents the list of possible defects in various manufacturing processes. Some of these defects may not be critical enough to deteriorate the functional performance of the part. However, in some cases, these defects may seriously influence the performance. For example, defects like surface flaws are significantly related with the strength degradation of the parts. One of such strength degradations is the fatigue failure which can be strongly correlated with the surface flaws introduced during machining since the crack initiation lifetime is a negligible fraction of total life. Secondary processing may be used to minimize/remove the defects in the primary manufacturing processes. The key in the selection of the manufacturing process for a given product is to assess the severity of the common manufacturing defects on the functional performance of the part.

The other important quality considerations are related with the dimensional tolerances and surface finish. Dimensional tolerances are particularly important in the operations involving assembly of many parts or the operations where the

Table 3.3 Possible defects in various manufacturing processes

Manufacturing process	Defects
Conventional manufacturing	
Casting	Blowholes, porosity, segregation, hot tears, shrinkage, incomplete filling, and inclusions
Rolling	Wavy edges, edge cracks, alligatoring, and center cracks
Forging	Internal and external cracks, tears, cold shuts, seams, laps, bursts, and poor grain structure
Extrusion	Surface cracking, center burst, internal oxide stringers, internal pipes, and axial hole (funnel)
Sheet metal forming	Spring-back, stretcher strains, orange peels, wrinkling, cracking, and residual stresses
Powder metallurgy	Lamination cracking, blowouts, excessive porosity, shrinkage, and incomplete filling
Molding	Distortion, shrinkage, porosity, incomplete filling, etc.
Machining	Surface and subsurface cracking, roughness, stresses, poor surface finish, etc.
Welding	Spatter, incomplete fusion, incomplete penetration, longitudinal cracking, undercutting, porosity, etc.
Laser manufacturing	
Laser forming	Nonuniform bending angle, heat-affected zone, excessive section thickening, and strain hardening
Laser machining	Heat affected zone, taper, recast layer, cracking, striations, geometric irregularities, etc.
Laser joining	Geometric irregularities, incomplete penetration, etc.

functional performance is closely linked with the dimensions of the part. Dimensional tolerances refer to the allowable deviations from the designed specification. To achieve the designed performance of the part, closer tolerances should be met. Figure 3.14 presents the tolerances obtained in various common machining processes (Kalpakjian and Schmid 2001). As indicated in the figure, some machining processes give tight tolerances of the products primarily due to characteristic material removal mechanisms associated with the process. In addition to the requirements of the tighter dimensional tolerances, the surface finish of the manufactured product is also important. The surface finish is related with the geometric irregularities of the surfaces of manufactured product. Surface finish is influenced by the number of parameters such as surface defects, phase transformations, residual stresses, reactions, etc. The undesirable surface finish may necessitate the additional finishing operations. In case of high-temperature manufacturing processes, the surface oxidation may be removed by additional machining operations. However, maintaining the required designed dimensions of the part after finish machining is important. Similarly, machining of materials may introduce surface cracks resulting in strength degradation. One of the important parameters used to characterize the surface finish is the surface roughness. Figure 3.15 presents the ranges of roughness values obtained with a number of

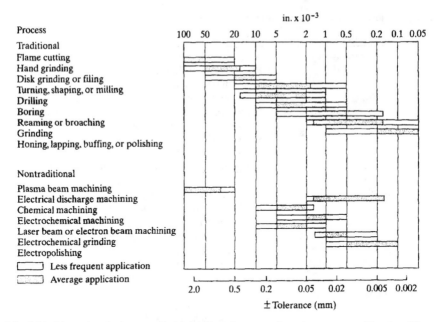

Fig. 3.14 Dimensional tolerance obtained with various manufacturing processes (Reprinted from Kalpakjian and Schmid 2001. With permission. Copyright Pearson Education, Inc.)

manufacturing processes (ASME Report 2003; Field et al. 1989). As indicated in the figure, the nontraditional processes generally give better surface finish compared to the traditional manufacturing processes. In many cases, surface finish of the product can be improved by controlling the process parameters. For example, heat generation during machining may influence the surface finish of the machined product. During machining of polymeric materials, if the heat generation increases the temperature of the workpieces above glass transition temperature, good surface finish can be achieved due to material removal in ductile manner. Else, rough surface may appear as a result of brittle fracture.

3.4.4 Manufacturing Economics

In many cases, final selection of the manufacturing process for a given application is dominated by the economic considerations. This is targeted towards minimization of production costs at acceptable quality of the product. There are many factors such as production volumes and production rates which influence the overall economics of the process. The procedures for calculating the manufacturing costs differ significantly for various manufacturing processes due to differing mechanisms associated with them. In general, the manufacturing costs

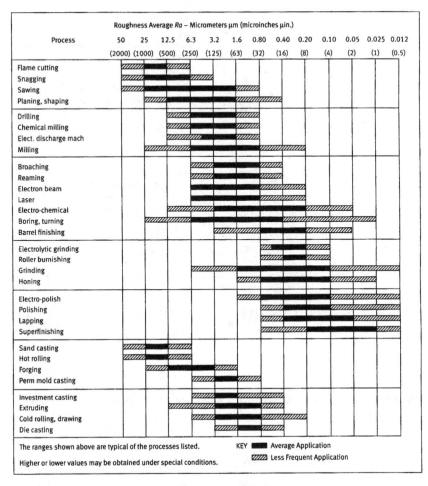

Process	Roughness Average Ra – Micrometers μm (microinches μin.)												
	50 (2000)	25 (1000)	12.5 (500)	6.3 (250)	3.2 (125)	1.6 (63)	0.80 (32)	0.40 (16)	0.20 (8)	0.10 (4)	0.05 (2)	0.025 (1)	0.012 (0.5)

Fig. 3.15 Surface roughness obtained with various manufacturing processes. (Reprinted from ASME Report 2003. With permission. Copyright American Society of Mechanical Engineers.)

include material cost, fixed cost, tooling cost, variable cost, labor cost, etc. This is illustrated for the case of machining of materials. The important factor which influences the machining economics is the cutting speed. As the cutting speed increases, the machining time decreases resulting in reduced labor cost. However, increase in cutting speed is also accompanied with rapid tool wear resulting in higher costs of tools and tool changing. Due to these opposing factors, there exists some optimum range of cutting speed which results in minimum cost of machining per piece. Thus total cost of machining per piece is a sum of various costs such as machining cost (depends on machining time, labor cost, overhead cost, etc.), tool cost, tool changing cost, and other nonmachining cost (set up cost, fixtures cost, idle cost, etc.). Schematic of the variations of these costs with

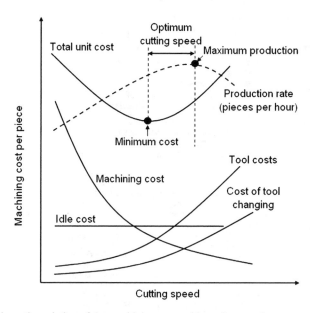

Fig. 3.16 Schematic variation of the machining costs with cutting speed

Table 3.4 Relative economic comparison of various machining processes (Springborn 1967; Pandey and Shan 1980)

Machining process	Parameter influencing economy				
	Capital investment	Toolings/ fixtures	Power requirements	Removal efficiency	Tool wear
Conventional machining (CM)	Low	Low	Low	Very low	Low
Ultrasonic machining (USM)	Low	Low	Low	High	Medium
Abrasive jet machining (AJM)	Very low	Low	Low	High	Low
Electrochemical machining (ECM)	Very high	Medium	Medium	Low	Very low
Chemical machining (CHM)	Medium	Low	High	Medium	Very low
Electric discharge machining (EDM)	Medium	High	Low	High	High
Plasma arc machining (PAM)	Very low	Low	Very low	Very low	Very low
Laser beam machining (LBM)	Medium	Low	Very low	Very high	Very low

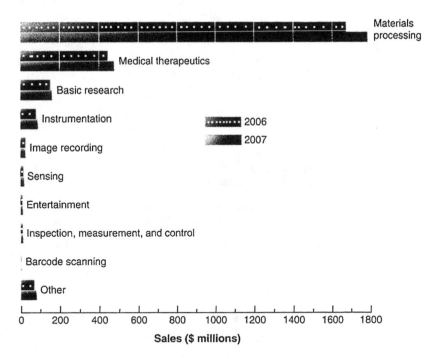

Fig. 3.17 Worldwide sales of nondiode lasers for various applications (Reprinted from Kincade and Anderson 2007. With permission. Copyright Pennwell.)

cutting speed is presented in Fig. 3.16. The figure indicates that the optimum range of cutting speed lies between the cutting speeds which give minimum unit cost and maximum production (Dieter 1986). Relative economic comparison of various machining processes is presented in Table. 3.4.

In summary, laser processing have demonstrated the capabilities to be used in a number of manufacturing processes such as casting, forming, joining, and machining for a wide range of materials. However, laser processing is currently used for highly specialized applications. In many of these applications laser processing offer significant benefits not only from the viewpoint of technical capabilities but also from the economic considerations. Figure 3.17 presents the data of the worldwide sale of nondiode lasers for various applications (Kincade and Anderson 2007). As indicated in the figure, the application of lasers is dominated in the field of materials processing (welding, cutting, drilling, semiconductor and microelectronics manufacturing, marking, rapid prototyping, etc.). Furthermore, in line with the continuous advancement in the laser technology, the applications of lasers are rapidly increasing with more and more laser systems being employed in the production lines. The selection of laser manufacturing in a given application must be derived from careful consideration of various aspects of manufacturing such as process capabilities, economy, product quality, etc.

References

Ashby MF (1999) Materials Selection in Mechanical Design. Butterworth-Heinemann, Oxford.

ASME Report (2003) Surface texture (surface roughness, waviness and lay). ASME Report No. B46.1-2002, American Society of Mechanical Engineers.

Benedict GF (1989) Non-traditional machining processes: introduction. In: Davis JR (ed) ASM Handbook: Machining, ASM International, Ohio, vol. 16, pp. 509–510.

Black JT (1989) Mechanics of chip formation. In: Davis JR (ed) ASM Handbook: Machining, ASM International, Ohio, vol. 16, pp. 7–12.

Chryssolouris G (1991) Laser Machining: Theory and Practice. Springer, New York.

Chryssolouris G, Anifantis N, Karagiannis S (1997) Laser assisted machining: an overview. Journal of Manufacturing Science and Engineering, Transactions of ASME 119:766–769.

Crawford RJ (2001) Plastic Engineering. Butterworth-Heinemann, Oxford.

DeVries WR (2006) Analysis of Materials Removal Processes. Springer, New York.

Dieter G (1986) Mechanical Metallurgy. McGraw-Hill, New York.

Dutta D, Prinz FB, Rosen D, Weiss L (2001) Layered manufacturing: current status and future trends. Journal of Computing and Information Science in Engineering, Transactions of the ASME 1:60–71.

Field M, Kahles JF, Koster WP (1989) Surface finish and surface integrity. In: Davis JR (ed) ASM Handbook: Machining, ASM International, Ohio, vol. 16, pp. 19–36.

Geary D (1999) Welding. McGraw-Hill, New York.

Gedda H, Kaplan A, Powell J (2005) Melt-solid interactions in laser cladding and laser casting. Metallurgical and Materials Transactions B 36:683–689.

German RM (1998) Powder Metallurgy of Iron and Steel. Wiley, New York.

Heine RW, Loper CR, Rosenthal PC (1967) Principles of Metal Casting. McGraw-Hill, New York.

Hu J, Marciniak Z, Duncan J (2005) Mechanics of Sheet Metal Forming. Butterworth-Heinemann, Oxford.

Kalpakjian S (1967) Mechanical Processing of Materials. Van Nostrand, New Jersey.

Kalpakjian S, Schmid S (2001) Manufacturing Engineering and Technology. Prentice-Hall, New Jersey.

Kincade K, Anderson SG (2007) Laser industry navigates its way back to profitability. Laser Focus World 43:82–100.

King RI (1985) Handbook of high-speed machining technology. Chapman & Hall, New York.

Nicholas MG (1998) Joining Processes. Kluwer Academic, The Netherlands.

Pandey PC, Shan HS (1980) Modern Machining Processes. McGraw-Hill, New Delhi.

Raghupathi PS (1988) Cold extrusion. In: Davis JR (ed) ASM Handbook: Forming and Forging, ASM International, Ohio, vol. 14, pp. 299–312.

Roberts WL (1983) Hot Rolling of Steel. CRC Press, Boca Raton, FL.

Semiatin SL (1988) Introduction to forming and forging processes. In: Davis JR (ed) ASM Handbook: Forming and Forging, ASM International, Ohio, vol. 14, pp. 15–21.

Schey JA (1997) Introduction to Manufacturing Processes. McGraw Hill, New York.

Springborn RK (1967) Non-traditional machining processes. American Society of Tool and Manufacturing Engineers, Michigan.

Tschaetsch H (2006) Metal Forming Practise: Processes–Machines–Tools. Springer, Berlin.

Walker J (1999) Machining Fundamentals: From Basic to Advanced Techniques. Goodheart-Willcox, Illinois.

Ye R, Smugeresky JE, Zheng B, Zhou Y, Lavernia EJ (2006) Numerical modeling of the thermal behavior during the LENS process. Materials Science and Engineering A 428:47–53.

Chapter 4
Laser Drilling

4.1 Introduction

Laser drilling is one of the earliest applications of lasers in materials processing. Output energies of the first ruby laser were often described in terms of the number of razor blades which could be penetrated by the focused laser beam. Laser drilling is most extensively used in the aerospace, aircraft, and automotive industries (Tam et al. 1994; Benes 1996; Giering et al. 1999). The most important application of laser drilling in the aerospace industry is the drilling of a large number of closely spaced effusion holes with small diameter and high quality to improve the cooling capacity of turbine engine components. In addition, laser drilling of diamond drawing dies and gemstones have been extensively used (Nagano 1978; Meijer 2004). The common industrial applications of laser drilling include cooling holes in aircraft turbine blades, optical apertures, flow orifices, and apertures for electron beam instruments (Knowles 2000, Wu et al. 2006).

Laser drilling is a noncontact, precise, and reproducible technique that can be used to form small diameter (~100 μm) and high-aspect ratio holes in a wide variety of materials. The advantages of laser drilling include the ability to drill holes in difficult-to-machine materials such as superalloys, ceramics, and composites without high tool wear rate normally associated with conventional machining of these materials (Voisey and Clyne 2004). Conventional mechanical drilling is often a slow process (drilling time ~60 s/hole) and associated with difficulties of drilling at high angles. Drilling rates as high as 100 holes/s can be achieved in production environment by coordinating the workpiece motions with pulse period of pulsed laser source. Laser drilling does not pose substantial problems at high angles of incidences. Laser drilling is also well suited for the nonconducting substrates or metallic substrates coated with nonconducting materials where the electric discharge machining is limited. For example, the drilling of thermal barrier-coated superalloys in aerospace applications can be well achieved by laser drilling instead of electrical discharge machining (Kamalu et al. 2002). In addition, recently the laser drilling of composite materials such as multilayer carbon fiber composites for aircraft applications is attracting increasing interest due to potential advantages of

N.B. Dahotre and S.P. Harimkar, *Laser Fabrication and Machining of Materials.*   97
© Springer 2008

rapid processing, absence of tool wear, and ability to drill high-aspect ratio holes at shallow angles to the surface (Rodden et al. 2002).

4.2   Laser Drilling Approaches

In laser drilling, the high intensity, stationary laser beam is focused onto the surface at power densities sufficient to heat, melt, and subsequently eject the material in both liquid and vapor phases. The erosion front at the bottom of the drilled hole propagates in the direction of the line source in order to remove the material (Chryssolouris 1991). In general, there are three approaches to laser drilling, namely, single pulse, trepanning, and percussion drilling. These are shown in Fig. 4.1 (Verhoeven 2004). A coaxial assist gas is almost always used while drilling in order to shield the laser optics from contamination from ejected debris and also to facilitate the material removal.

Single pulse drilling is used for drilling narrow (less than 1 mm) holes through thin (less than 1 mm) plates. High pulse energies are supplied in drilling with single pulse because the irradiated energy levels must be sufficient to vaporize the material in single pulse.

In trepanning, wider holes (less than 3 mm) in thicker plates (less than 10 mm) are produced by drilling a series of overlapping holes around a circumference of a circle so as to cut a contour out of the plate. Trepanning can be performed by translating either the workpiece or the focusing optic. The process is much similar to contour cutting and can be performed by the laser operating in the continuous wave (CW) or pulsed mode. CO_2 and Nd:YAG lasers are most commonly used in trepanning.

In percussion drilling, a series of short pulses (10^{-12} to 10^{-3} s) separated by longer time periods (10^{-2} s) are directed on the same spot to form a through hole. Each laser pulse contributes to the formation of hole by removing a certain volume of material (Verhoeven et al. 2003). Pulsed Nd:YAG lasers are most commonly used for percussion drilling because of their higher energy per pulse. Percussion drilling is used to produce narrow holes (less than 1.3 mm) through relatively thicker (up to 25 mm) metal plates (Elza and White 1989). High speed of the percussion drilling makes it the most cost-effective method in applications such as drilling of combustion chambers having 40,000–50,000 holes and other applications such as

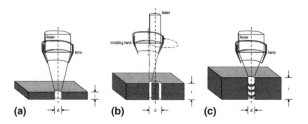

Fig. 4.1 Approaches for laser drilling **(a)** single pulse drilling, **(b)** trepanning drilling, and **(c)** percussion drilling. (Courtesy of Bob Mattheij, Eindhoven University.)

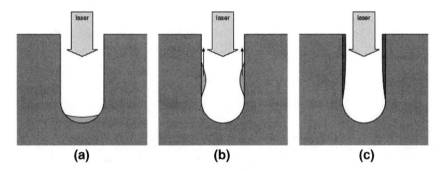

Fig. 4.2 Physical processes during percussion laser drilling (**a**) Melt pool formation (**b**) splashing out of molten material, and (**c**) formation of recast layer. (Reprinted from Verhoeven et al. 2003. With permission. Copyright Elsevier.)

drilling of turbine and guide vanes having 50–200 holes, given the large number of total components involved (French et al. 1998, McNally et al. 2004).

The physical processes occurring during laser percussion drilling are shown in Fig. 4.2 (Verhoeven et al. 2003). The process can be split up into three stages. Initially a thin region of molten material is formed by absorption of laser energy at the target surface. After some time the surface of this melt pool reaches vaporization temperature. The sudden expansion of the vapor evaporating from the surface eventually leads to the splashing stage when the melt pool is pushed radially out by the recoil pressure. Melt expulsion occurs when the pressure gradients on the surface of the molten layer are sufficiently large to overcome the surface tension forces and expel molten material from the hole (Basu and DebRoy 1992). On its way of escaping out some part of this molten material may resolidify at the walls. The timescales for these three stages are between 10^{-5} and 10^{-4} s for melting and 10^{-5} s for splashing and solidification (Verhoeven 2004).

Since laser drilling requires the temperature of the target to be raised above boiling point, lasers with pulse lengths in the region of several hundred microseconds, or even less, are most commonly used in laser drilling applications. In addition, the holes with smaller diameter and larger aspect ratio require the laser with shorter wavelengths. This makes the CO_2 and Nd:YAG lasers primary candidates for laser drilling. The range of hole diameters possible for typical thicknesses of metal, with Nd:YAG laser is 0.001–0.060 in., whereas with CO_2 laser the range is 0.005–0.050 in. (Elza and White 1989).

4.3   Melt Expulsion During Laser Drilling

The significant material removal mechanisms during laser drilling are vaporization and the physical expulsion of the melt. The dominant mechanism depends on the irradiation conditions and material properties. Holes drilled with purely evaporative material removal mechanisms are generally marked with clean surface and sharp boundaries without recast layer, spatter, and associated dross. On the other hand, material removal by melt expulsion is an energetically efficient mechanism. Generally, the energy required to remove the material via melt expulsion is about one quarter of that required to vaporize the same volume (Schoonderbeek et al. 2004, Voisey et al. 2003). For example, energy required to remove $1 \, m^3$ of iron by vaporization is 65.8 GJ; whereas the same amount of material can be removed in molten state with 12.3 GJ energy (Voisey et al. 2000). However, the material removal by melt expulsion is generally irregular and may result in asymmetric and irregular hole shapes. In the context of these two mechanisms, ideal material removal mechanism should be purely evaporative, i.e., ablation without melting. However, such a thermally induced laser ablation is possible only for much selected materials such as graphite, diamond, polymethyl methacrylate (PMMA), etc., that can be vaporized or thermally decomposed without melting (Tokarev and Kaplan 1999).

Melt expulsion during the drilling may be due to pressure generated from inside the melt (explosive melt expulsion) or from outside the melt (hydrodynamic melt expulsion). Explosive melt expulsion can take place in some of the following cases (Tokarev and Kaplan 1999):

1. Increase in the volume of the chemical decomposition reaction products due to subsurface overheating (Kitai et al. 1990).
2. Vaporization of the lowest melting point components in a multicomponent system while the other components are still in molten state.
3. Increase in volume of the dissolved contaminant gases at high temperature at subsurface layer of metallic systems (Weaver and Lewis 1996).

Hydrodynamic melt expulsion takes place along the surface of the laser-drilled hole towards the periphery of laser spot and is due to differential vapor (plasma) plume pressure on the molten surface. Figure 4.3 shows the radiation pressure distribution and the fluid motion caused by a laser beam having Guassian energy distribution assuming a V-shaped molten region. The radiation pressure which is highest at the center of the cavity accelerates the material at the apex of the V-shaped cavity. The condition of continuity causes the material near the edges of the cavity where the radiation pressure is weakest to be accelerated in opposite direction. Analysis of melt expulsion process shows that the molten material flowing along the wall of the hole breaks up into discrete droplets while leaving the hole due to surface tension effects. At the end of the drilling process, the molten material which could not be ejected in form of droplets form a thin layer of recast material at the periphery of hole (Wagner 1974).

Hydrodynamic melt expulsion is the primary mechanism for the accumulation of the recast material around the periphery of the laser-drilled holes and thus deteriorating the quality of holes in terms of dimensions and shape (Zhang and Faghri 1999). Hence the analysis of hydrodynamic melt expulsion is important so that the control of the quality of the laser-drilled holes can be achieved with this energetically favorable process (Low et al. 2002, Yilbas and Sami 1997).

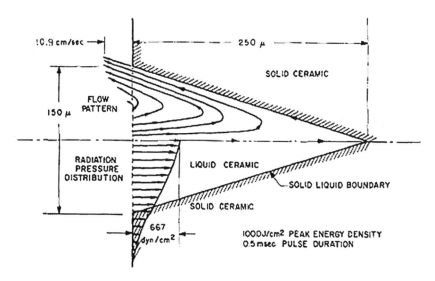

Fig. 4.3 Effects of radiation pressure on the cavity of molten material during laser drilling. (Reprinted from Wagner 1974. With permission. Copyright American Institute of Physics.)

Melt expulsion phenomenon have extensively been analyzed by quantifying the melt ejection fraction (MEF), melt ejection velocity, and particle size and angular distribution of ejected droplets. MEF is measured by positioning a thin glass slide in the beam path, above the surface to be drilled. During drilling, the ejected droplets collide with and adhere to the surface of the glass slide. MEF is then given by the ratio of mass gained by the slide due to melt ejection to the total mass lost by the substrate due to all the materials removed (Voisey et al. 2003):

$$\text{MEF} = C_{\text{XRD}} \frac{\text{Mass gain of collection slide}}{\text{Mass lost by drilled sample}}$$

where C_{XRD} is the correction factor to account for mass changes due to in-flight oxidation of ejected droplets and is determined by Rietveld refinement of x-ray diffraction spectra. Figure 4.4 shows the variation of MEF with laser pulse energy (pulse period of 0.5 ms) for steel and aluminum substrates. The initial increase in the MEF with pulse energy is due to increase in the pressure gradients causing the more complete ejection of melt. However, increase in pulse energy also increases the temperature gradients such that the rate of vaporization from the surface increases resulting in reducing the thickness of molten layer. This results in lowering the MEF at high pulse energies because very less material is available for ejection (Voisey et al. 2000). In addition, the variation of velocity of ejected droplets as a function of laser power density is shown in Fig. 4.5 for aluminum, nickel, and titanium. The velocity of ejected droplets is determined by tracking the trajectory

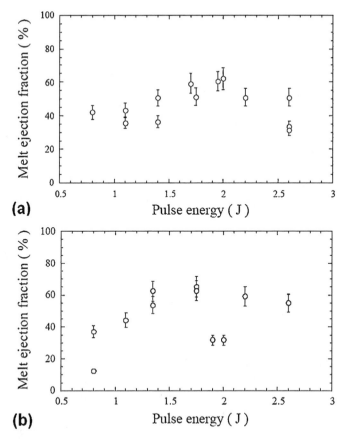

Fig. 4.4 Variation of melt ejection fraction with pulse energy for (**a**) steel and (**b**) aluminum substrates. (Reprinted from Voisey et al. 2000. With permission. Copyright Materials Research Society.)

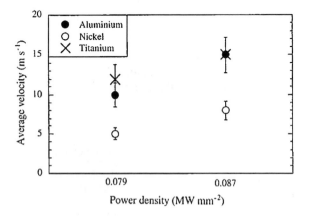

Fig. 4.5 Variation of velocity of ejected droplets measured by high speed photography with laser power density. (Reprinted from Voisey et al. 2003. With permission. Copyright Elsevier.)

of the droplets through successive frames of high-speed photography. The figure shows the increase in velocity of ejected droplets with increase in power density. Such quantitative studies are important in facilitating its incorporation into existing numerical models for more accurate prediction of temperature distribution and hole profile during laser drilling (Voisey et al. 2003).

4.4 Analysis of Laser Drilling Process

Irradiation of the material with high intensity laser source is a complex phenomenon and results in various thermally induced effects such as heating, melting, vaporization from the surface, dissociation and ionization of the vaporized material, and shock waves in both the vaporized material and the solid. Some of these events in the context of laser drilling are schematically shown in Fig. 4.6 (Chryssolouris 1991). The most important effects which contribute to the material removal during drilling process are the vaporization and melt expulsion.

Considerable research has been carried out to develop the model of laser drilling process. Earliest analytical model of laser drilling consider the material removal by vaporization process. The model deals with the determination of drilling velocity, i.e., the velocity of material removal by vaporization from solid surface and the

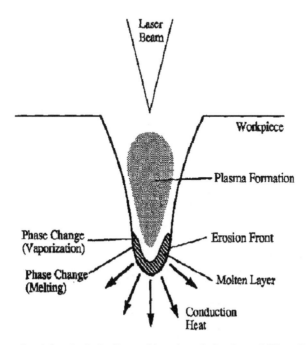

Fig. 4.6 Schematic of the physical effects taking place during laser drilling. (Reprinted from Chryssolouris 1991. With permission. Copyright Springer.)

temperature distribution in the solid. The following assumptions are made (Dabby and Paek 1972):

1. The laser beam intensity I (W/cm²) is sufficient to cause vaporization of the front surface of the material.
2. The gas created by the vaporization of material is transparent to the incident laser energy.
3. Heat losses due to reradiation are negligible.
4. The thermal constants and optical absorption coefficient b (cm⁻¹) are independent of laser beam intensity and the temperature of the solid.
5. The effects of radial heat conduction and the liquid phase can be ignored.

The energy balance at the front surface requires the energy given to the vaporized material equals the energy conducted from the solid (Dabby and Paek 1972):

$$\rho L_v \, \dot{Z} = k \left(\frac{\partial T}{\partial z} \right)_{z=Z} , \qquad (4.1)$$

where k is thermal conductivity, ρ is density of solid, L_v is the heat of vaporization, and Z is the depth to which the material has been removed. $\dot{Z}$ is the velocity of the front surface which is the time differential of Z.

Referring to Fig. 4.7, the temperature distribution within the material ($z > Z$) is determined using the heat conduction equation with a distributed laser heat source moving with the front surface (Dabby and Paek 1972). Thus,

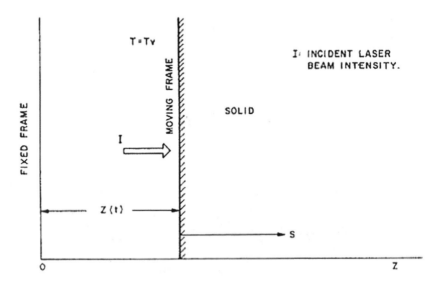

Fig. 4.7 Coordinate system used for the laser vaporization problem. (Reprinted from Dabby and Paek 1972. With permission. Copyright Institute of Electrical and Electronics Engineers.)

$$\rho C_p \frac{\partial T}{\partial t} = bIe^{-b(z-Z)} + k\frac{\partial^2 T}{\partial z^2}, z > Z \tag{4.2}$$

where, C_p is the specific heat (J/g °K) and t is time (s).

The solution of Eqs. (4.1) and (4.2) with appropriate boundary conditions will yield the temperature and velocity. The appropriate boundary conditions are

$$T = T_v \text{ at } z = Z \tag{4.3}$$

and

$$T = 0 \text{ at } z = \infty \tag{4.4}$$

The two initial conditions are required for Eqs. (4.1) and (4.2). The first assumes that the drilling process begins at $t = 0$. Thus,

$$Z = 0 \text{ at } t = 0 \tag{4.5}$$

The second initial condition depends on how the material is heated before the drilling process has begun. The preheating is generally done by the earlier portions of the laser pulse. The solution of the heat equation with distributed source could be written in terms of exponentials so that it would be convenient to write the second initial condition describing the temperature profile at $t = 0$ in terms of exponential function. Instead of expanding formally, the initial condition is approximated with an exponential term $(1 + qz)e^{-qz}$ and allow the constant q to be a fitted parameter that would most closely match the determined temperature initial temperature profile. Thus, at $t = 0$

$$T = T_v(1 + qz)e^{-qz} \tag{4.6}$$

Transforming Eqs. (4.1)–(4.6) into a moving dimensionless reference frame gives,

$$\frac{\partial \theta}{\partial \tau} - u\frac{\partial \theta}{\partial s} - \frac{\partial^2 \theta}{\partial s^2} = \frac{b}{\lambda}e^{-Bs}, \tag{4.7a}$$

and

$$u = \lambda\left(\frac{\partial \theta}{\partial s}\right)_{s=0} \tag{4.7b}$$

subject to the following conditions:

$$\text{at } \tau = 0, \theta = (1 + Qs)e^{-Qs} \tag{4.7c}$$

$$\text{at } s = 0, \theta = 1 \tag{4.7d}$$

$$\text{at } s = \infty, \theta = 0 \tag{4.7e}$$

where

$$\theta = T/T_v \tag{4.8a}$$

$$s = (IC_p/kL_v)(z - Z) \tag{4.8b}$$

$$u = (\rho L_v/I)\dot{Z} \tag{4.8c}$$

and

$$\tau = (I^2 Cp/\rho kL_v^2)t \tag{4.8d}$$

A heating parameter λ, absorption parameter B, and a normalized initial parameter Q are defined as follows:

$$\lambda = (C_p T_v)/(L_v) \tag{4.9a}$$

$$B = (kbL_v)/(IC_p) \tag{4.9b}$$

and

$$Q = (kqL_v)/(IC_p) \tag{4.9c}$$

Initially $(\partial\theta/\partial S)_{S=0}$ is zero. Thus the normalized inner velocity u_i is zero for small values of τ. Substituting this value of u_i into Eq. (4.7a) gives

$$\frac{\partial\theta_i}{\partial\tau} - \frac{\partial^2\theta_i}{\partial s^2} = \frac{B}{\lambda}e^{-Bz} \tag{4.10}$$

where θ_i is the temperature based upon the initial velocity and is subjected to the conditions given in Eqs. (4.7c)–(4.7e).

The solution of Eq. (4.10) can be obtained by taking the Laplace transform of Eq. (4.10) with respect to τ, solving the resulting second-order ordinary differential equation, and taking the inverse transform. Using this technique, θ_i is found to be (Dabby and Paek 1972):

$$\theta_i(\tau, s) = \left[1 + \frac{1}{\lambda B}\right] \text{erfc}\left[\frac{s}{2\sqrt{\tau}}\right]$$

$$-\frac{e^{B^2\tau}}{2\lambda B}\left\{e^{-Bs}\,\text{erfc}\left[\frac{s}{2\sqrt{\tau}} - B\sqrt{\tau}\right] + e^{Bs}\,\text{erfc}\left[\frac{s}{2\sqrt{\tau}} + B\sqrt{\tau}\right]\right\}$$

$$+\frac{e^{Q^2\tau}}{2}\left\{\begin{array}{l}\left[2Q^2\tau - Qs - 1\right]e^{-Qs}\,\text{erfc}\left[\frac{s}{2\sqrt{\tau}} - Q\sqrt{\tau}\right] + \\[2mm] \left[2Q^2\tau + Qs - 1\right]e^{Qs}\,\text{erfc}\left[\frac{s}{2\sqrt{\tau}} + Q\sqrt{\tau}\right]\end{array}\right\}$$

$$-\frac{e^{-Bs}}{\lambda B}\left[1 - e^{B^2\tau}\right] - e^{Q^2\tau - Qs}\left[2Q^2\tau - Qs - 1\right] \tag{4.11}$$

The dimensionless inner velocity u_i can then be recalculated as

$$u_i = \lambda\left(\frac{\partial\theta_i}{\partial s}\right)_{s=0} \tag{4.12a}$$

Therefore,

$$u_i = 1 - e^{B^2\tau}\,\text{erfc}\left[B\sqrt{\tau}\right] + 2\lambda Q^3\tau e^{Q^2\tau}\,\text{erfc}\left[Q\sqrt{\tau}\right] - \frac{2\lambda}{\sqrt{\pi}}Q^2\sqrt{\tau} \tag{4.12b}$$

The steady-state temperature distribution can be obtained by setting $\partial\theta / \partial\tau = 0$ in Eq. (4.7a) and solving with u constant. The steady-state normalized temperature profile θ_{ss} in the moving normalized reference frame is found to be

$$\theta_{ss} = e^{-vs} - \frac{1}{\lambda(B-v)}\left(e^{-Bs} - e^{-vs}\right) \tag{4.13}$$

where v is steady-state value of u.

Thus, the steady-state temperature is a traveling temperature wave independent of the initial temperature distribution. The value of v can be determined by replacing u with v and θ with θ_{ss}, in Eq. (4.7b) solving for v. The normalized steady-state velocity is

$$v = 1/(1 + \lambda) \tag{4.14}$$

The normalized temperature distribution (θ_0) correct for large values of τ, can be determined by replacing u with v in Eq. (4.7a). The resulting equation can be solved for θ_0 in a manner similar to that used in solving Eq. (4.10). Knowing θ_0, the dimensionless first outer velocity u_0 can be recalculated as:

$$u_0 = \lambda \left(\frac{\partial \theta_0}{\partial s} \right)_{s=0}. \tag{4.15}$$

Figure 4.8 shows the temperature profiles within the material at different normalized times. The figure shows that the temperature inside the material is higher than the surface temperature T_v. The distributed heat generation by the laser raises the temperature throughout the solid; however, the vaporization of the front surface removes the energy from the region near the front surface. This cooling mechanism causes the maximum temperature to lie below the surface. If the temperature reached is high enough to cause the vaporization at depths below the surface, tremendous pressures arising from the vaporized material would explosively remove

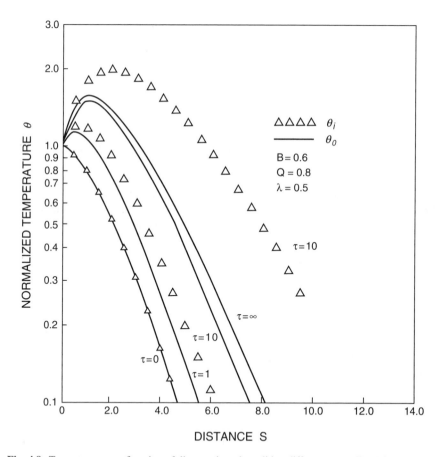

Fig. 4.8 Temperature as a function of distance into the solid at different normalized times. (Note that the maximum temperature is reached below the surface of the material during transient conditions). (Reprinted from Dabby and Paek 1972. With permission. Copyright Institute of Electrical and Electronics Engineers.)

the intervening material. The material removal based on such an explosive technique would be considerably more efficient than the material removal based on vaporization of all material. These predictions are in good qualitative agreement with experimental results (Dabby and Paek 1972).

In addition to the dimensional control, most of the practical applications of drilling require stringent control over the shape of the laser-drilled holes. Quantitative models have been reported which predict the depth and shape of laser-drilled holes. The shapes of the laser-drilled holes in alumina are found to be similar to the shapes of energy density distributions both at the high energy and near threshold energy for scribing ceramics (Fig. 4.9) (Wagner 1974).

Laser drilling produces large temperature gradients in the material which can induce the thermal stresses. This is of particular importance in case where large numbers of closely spaced holes have to be drilled. The thermal stresses reduce the strength of the material and may lead to failure if the stresses exceed the fracture limit of the material. A three-dimensional model have been developed to calculate the temperature profile and the thermal stress distribution for laser-drilled holes in alumina ceramic substrate using a continuous, distributed, and a moving heat source. The analysis involves the solution of unsteady heat equation (Paek and Gagliano 1972):

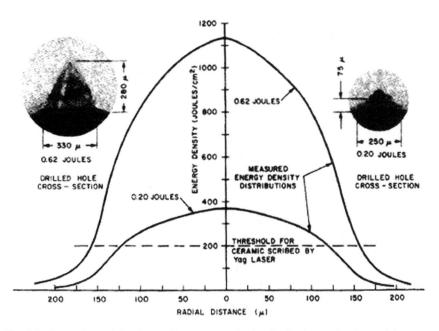

Fig. 4.9 Comparison of the shape of laser energy density distribution to the shape of the corresponding laser-drilled hole. (Reprinted from Wagner 1974. With permission. Copyright American Institute of Physics.)

$$\frac{\partial T}{\partial t} = K\left(\frac{\partial^2 T}{\partial x^2} + \frac{\partial^2 T}{\partial y^2} + \frac{\partial^2 T}{\partial z^2}\right), \tag{4.16}$$

where K is thermal diffusivity. For infinite medium, the instantaneous source solution of Eq. (4.16) located at the point (x', y', z') can be written as

$$T(x, y, z, t) = \frac{Q(x', y', z')}{8(\pi K t)^{3/2}} \exp\left\{-\frac{(x-x')^2 + (y-y')^2 + (z-z')^2}{4Kt}\right\} \tag{4.17}$$

where Q is defined as instantaneous point source strength at (x', y', z') and is expressed as a "temperature" to which a unit volume of material would be raised by the amount of heat liberated. The method of images can be applied to obtain the desired solutions in polar coordinates for two distinct cases.

For semi-infinite body, the solution is

$$T(r, z, t) = \int_0^a \int_0^t \frac{P(r', t')}{4\rho C_p \sqrt{\pi\left[K(t-t')\right]^{3/2}}} \exp\left\{-\frac{r^2 - r'^2}{4K(t-t')}\right\} I_0\left(\frac{rr'}{2K(t-t')}\right) r'$$

$$\left[\exp\left\{\frac{[z-f(t')]^2}{4K(t-t')}\right\} + \exp\left\{-\frac{[z+f(t')]^2}{4K(t-t')}\right\}\right] dt' dr' \tag{4.18}$$

For finite body with thickness d, the solution is

$$T(r, z, t) = \int_0^a \int_0^t \frac{P(r', t')}{4\rho C_p \sqrt{\pi\left[K(t-t')\right]^{3/2}}} \exp\left\{-\frac{r^2 + r'^2}{4K(t-t')}\right\} I_0\left(\frac{rr'}{2K(t-t')}\right)$$

$$\left[1 + \sum_{n=1}^\infty \cos\frac{n\pi[z+f(t')]}{d} \exp\left\{-\frac{Kn^2\pi^2(t-t')}{d^2}\right\}\right.$$

$$\left. + \sum_{n=1}^\infty \cos\frac{n\pi[z-f(t')]}{d} \exp\left\{-\frac{Kn^2\pi^2(t-t')}{d^2}\right\}\right] r' dt' dr' \tag{4.19}$$

where r' is the radius of the ring in which the instantaneous point source is distributed; $P(r', t')$ is the laser intensity (J/cm²s); $f(t')$ is the location of the moving source at $t = t'$; and a is radius of the instantaneous disk source (beam radius).

The components of thermal stress can be calculated from the stress–strain relationships coupled with the equations of temperature obtained from the solution of heat conduction equation. The tangential component of stress (σ_θ), which is particularly important for determining the fracture behavior of materials, is found to be

$$\sigma_\theta = \left(\frac{\alpha E}{1-\upsilon}\right)\frac{1}{r^2}\left[\int_g^r T_r d_r - T_r^2\right] \tag{4.20}$$

where α is coefficient of linear thermal expansion; E is Young's modulus; v is Poisson ratio; and g is radius of drilled hole as a function of z. The calculated temperature profile and tangential stress distribution on the surface of Al_2O_3 as a function of radial distance are shown in Figs. 4.10 and 4.11, respectively (Paek and Gagliano 1972).

As mentioned in the preceding sections, melt expulsion during laser drilling is a significant material removal mechanism. Hence, many attempts have been made to incorporate the melt expulsion phenomenon in the vaporization models to accurately predict the laser drilling parameters such as drilling velocity, drilling efficiency, and shape of the drilled holes. One such model includes expulsion of liquid and allows the calculation of drilling velocity and drilling efficiency as functions of the absorbed intensity (Von Allmen 1976). This one-dimensional model is based on energy balance between the absorbed intensity Φ and the energy flux carried away with the expulsed material in vapor and liquid form (Von Allmen 1976):

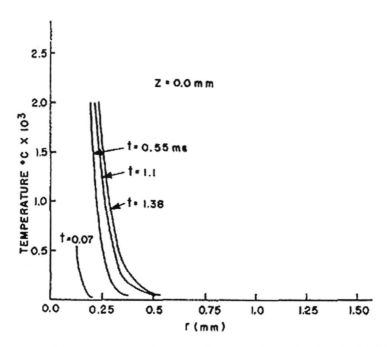

Fig. 4.10 Calculated temperature profile on the surface of alumina as a function of radial distance and given pulse lengths for ruby radiation (laser intensity = 3.2×10^7 W/cm^2 and spot size = 0.006 cm). (Reprinted from Paek and Gagliano 1972. With permission. Copyright Institute of Electrical and Electronics Engineers.)

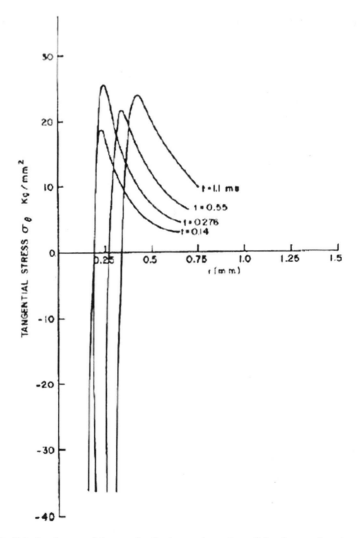

Fig. 4.11 Calculated tangential stress distribution on the surface of alumina as a function of radial distance for ruby radiation at different times within pulse length of 1.38 ms. (Reprinted from Paek and Gagliano 1972. With permission. Copyright Institute of Electrical and Electronics Engineers.)

$$\Phi = j_v L_v + j_l L_l, \tag{4.21}$$

where j_v and j_l are expulsion rates (g/cm^2 s) for vapor and liquid, respectively; and L_v and L_l are specific absorbed energy in expulsed vapor and liquid, respectively. The drilling velocity is then given by:

$$u = (1/\rho)(j_v + j_l), \tag{4.22}$$

where ρ is the density of metal. The rate of evaporation j_v at a hot surface at temperature T_0 is given by:

$$j_v(T_0) = (1 - \alpha)p_s(T_0)(m / 2\pi k_B T_0)^{1/2}, \qquad (4.23)$$

where $\alpha \sim 0.2$ is a mean particle reflection coefficient of the metal surface, m is the particle mass, and k_B is the Botzmann constant. The saturation pressure (p_s) is approximately given by:

$$p_s(T_0) = p_0 \exp\left[(\lambda_e / k_B T_e)(1 - T_e / T_0)\right], \qquad (4.24)$$

where T_e, λ_e, and p_0 are the evaporation temperature, the heat of evaporation per particle, and the ambient pressure, respectively. The liquid expulsion rate with respect to the irradiated surface is given by:

$$j_l = \left[(2K / r)\ln(T_0 / T_m)\right]^{1/2} p_s^{1/4} \rho^{3/4}, \qquad (4.25)$$

where K and T_m are thermal diffusivity and melting temperature, respectively. Figure 4.12 shows the drilling velocity as a function of laser intensity. The measured drilling velocity (solid line) is calculated from hole depth versus pulse duration curves

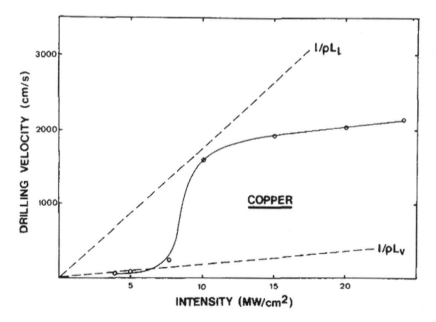

Fig. 4.12 Measured and calculated drilling velocity (given by pure evaporation or liquefaction) for copper. (Reprinted from Von Allmen 1976. With permission. Copyright American Institute of Physics.)

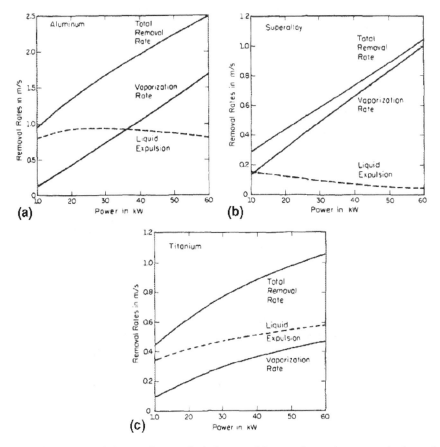

Fig. 4.13 Total material removal rate and relative material removal rates due to vaporization and liquid expulsion as a function of laser power for (**a**) aluminum, (**b**) superalloy, and (**c**) titanium. (Reprinted from Chan and Mazumder 1987. With permission. Copyright American Institute of Physics.)

at different laser intensities. The fictitious drilling velocities corresponding to pure evaporation $(1/\rho)L_v$ or pure liquefaction $(1/\rho)L_l$ are indicated by dashed lines. At lower laser intensities, the measured velocities are close to that given by pure evaporation and then reach the liquefaction limit suggesting the drastic reduction in normal metallic reflectivity in the starting phase of interaction process (Von Allmen 1976).

Chan and Mazumder proposed a one-dimensional steady-state model taking into account both the heat transfer and gas dynamics to predict the thermal damage done by vaporization and liquid expulsion during laser–material interaction. The problem was treated as a conventional Stefan problem with two moving boundaries (solid–liquid and liquid–vapor interfaces). A Knudsen layer of few molecular mean free paths is assumed to be present close to the wall of the hole near vapor and is modeled by a Mott-Smith-type solution. Figure 4.13 presents the predictions of the

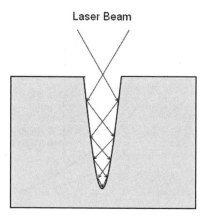

Fig. 4.14 Schematic of multiple reflection from the surface of hole walls during laser drilling

model for total material removal rates and the relative material removal rates in liquid and vapor phases for aluminum, superalloy, and titanium. The material removal rates increase with increasing laser intensity (Chan and Mazumder 1987).

The dimensions and metallurgical quality of the laser-drilled holes can be significantly influenced by the phenomena such as multiple reflections of the laser beam and the liquid metal flow caused by shear stress at the interface of the assist gas and the liquid metal. Laser light undergoes a series of reflections from the hole walls at shallow angles of incidence (Fig. 4.14). A part of the laser energy is absorbed by the reflecting medium at the point of reflection, and the rest is reflected. Such multiple reflections channel the optical energy to the bottom of the hole resulting in deeper holes than the depth of focus of laser beam (Yeo et al. 1994). Moreover, during laser drilling, in addition to the effects of recoil pressure due to vapor generated at the liquid surface, the flow of liquid metal may be influenced by the shear stresses induced by the impinging assist gas on the hole cavity (Chen et al. 2000).

Extensive research efforts were directed towards developing the two-dimensional axisymmetric mathematical model for material removal by taking into account the multiple reflections of laser beam and the assist gas-induced liquid metal flow during laser irradiation (Chan and Mazumder 1987, Mazumder 1991, Kar et al. 1992). Figure 4.15 shows the effects of multiple reflection and shear stress at the interface of the assist gas and liquid metal on the drilled cavity depth and the thickness of recast layer during laser irradiation of IN 718 with a spatially Gaussian and temporally triangular pulse laser. The following observations can be made from the figure (Kar et al. 1992):

1. The cavity depth of hole decreases with increasing beam radius which corresponds to decreasing laser intensity (Fig. 4.15a).
2. For a given laser power and beam quality, the cavity depth increases for the case in which multiple reflection takes place due to higher deposition of energy at the bottom of the cavity (Fig. 4.15a).

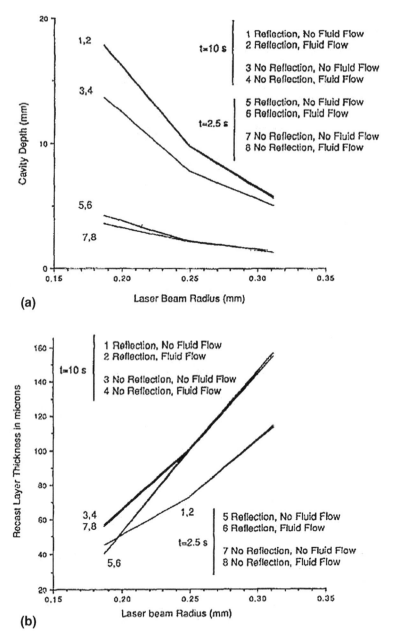

Fig. 4.15 Effects of beam radius and multiple reflections on (**a**) cavity depth and (**b**) thickness of recast layer in laser-drilled holes with laser power of 250 W and laser pulse-on time of 1.7 ms (t = irradiation time). (Reprinted from Kar et al. 1992. With permission. Copyright American Institute of Physics.)

3. The thickness of recast layer decreases as the beam radius decreases (laser intensity increases). This is in agreement with the fact that at high laser intensity the rate of evaporation from the surface increases to such an extent that the thickness of liquid metal decreases resulting in thinner recast layer (Fig. 4.15b).
4. For a given laser power and beam quality, recast layer thickness decreases for the case in which multiple reflection takes place (Fig. 4.15b).
5. The effects of fluid flow due to shear stress induced by impinging assist gas on the cavity depth and thickness of recast layer is insignificant (curves 1 and 2, 3 and 4, 5 and 6, and 7 and 8 are almost coinciding) (Fig. 4.15 a, b).

Of all the models mentioned above, almost all of them does not take into account all the processes, namely the free surface flow of the melt, multiphase heat transfer, vaporization gas dynamics, and the melting and solidification of the substrate, simultaneously in more than one dimension. Recently, a numerical model with significant improvements in the melting and solidification submodels has been developed (Ganesh and Faghri 1997). This improved model which emphasizes on melting and solidification submodels relaxes many assumptions that were made in earlier models. This included the consideration of properties like thermal conductivity (k) and specific heat (c_p) as different in solid and liquid phases, melting over a range of temperature spanning the entire mushy zone exhibited by alloys and mixtures going through the phase change and also the consideration of latent heat of fusion which is neglected in earlier models.

Various models of laser drilling have been reviewed in light of distinct categories, namely the dimensionality of the problem, whether or not the energy is taken into account in addition to fluid flow, whether or not the free surface compatibility exists, whether or not recast formation are considered in phase change model, the type of methodology adopted, and finally whether or not experimental comparison exists. Some of these models are listed in Table 4.1 (Ganesh and Faghri 1997).

4.5 Quality Aspects

Among various industries utilizing laser drilling applications, it is particularly in the aerospace industry where the geometric and metallurgical quality of holes becomes important. Laser drilling, in addition to the ability to produce high-aspect ratio holes in noncontact manner, is often associated with the geometric defects related with hole size, taper, circularity, and the metallurgical defects related with spatter, heat-affected zone, recast layer, and microcracking (Bandyopadhyay et al. 2002). A typical geometry of the laser-drilled hole is shown in Fig. 4.16 (Ghoreishi et al. 2002).

Hole size is best defined in terms of the aspect ratio, a ratio of hole depth to diameter at the midspan of the hole. For a particular material, the limiting aspect ratio depends on both the optical characteristics of the laser beam and the

Table 4.1 Summary of thermal models of laser drilling process [Reproduced with permission from Ref. Ganesh and Faghri 1997. Copyright, Elsevier]

Reference	Dimension	Fluid Flow	Energy	Free Surface	Phase Change	Comments
Paek and Gagliano (1972)	3-D transient	No	Yes	No	No	Temperature profile, tangential stress distribution
Von Allmen (1976)	1-D transient	Yes	Yes	No	Vaporization	Drilling velocity, drilling efficiency
Chan and Mazumder (1984)	2-D transient	Yes	Yes	No	No	Surface tension driver flow
Kou and Wang (1986)	3-D quasi-Steady state	Yes	Yes	No	No	Surface tension and buoyancy driven flow
Mazumder et al. (1991)	3-D quasi-Steady State	Yes	Yes	Yes	No	Point by point partially vectorized scheme
Chan and Mazumder (1987)	1-D steady state	Yes	Yes	No	Vaporization	Immobilization transformation, Newton-Raphson method
Basu and Srinivasan (1988)	2-D steady state	Yes	Yes	No	Vaporization	Vorticity stream function formulation, finite difference method
Zacharia et al. (1989)	3-D transient	Yes	Yes	Yes	No	Arc-Welding, moving arc, DEA method
Kar and Mazumder (1990)	2-D transient	No	Yes	No	Vaporization	Energy, Stefan bc, Runge-Kutta method
Mazumder et al. (1991)	2-D axisymmetric Q-S state	Yes	Yes	No	No	Effect of convection and Pr on pool geometry
Mazumder et al. (1991)	3-D transient	Yes	Yes	No	No	Scanning, FDM, vectorization
Patel and Brewster (1991)	2-D steady Axisymmetric	Yes	Yes	No	No	Gas-assist, 1-D heat conduction, top-heat Runge-Kutta method
Bellantone et al. (1991)	2-D transient Axisymmetric	Yes	Yes	Yes	Vaporization	Spatial and temporal variation of free surface temperature and pressure
Kar and Mazumder (1992)	2-D transient	No	Yes	No	Vaporization	Gas-assist and reflections, Runge-Kutta method
Gorden et al. (1994)	2-D transient axisymmetric	No	Yes	No	Vaporization	CVFEM, Landau transformation
Ho et al. (1994)	1-D transient	No	Yes	No	Vaporization	2-D axisymmetric gas flow, Crank-Nicholson method

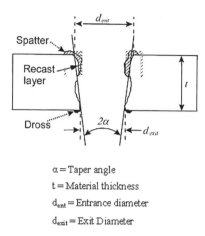

α = Taper angle
t = Material thickness
d_{ent} = Entrance diameter
d_{exit} = Exit Diameter

Fig. 4.16 Typical geometrical features of laser-drilled hole. (Reprinted from Ghoreishi et al. 2002. With permission. Copyright Elsevier.)

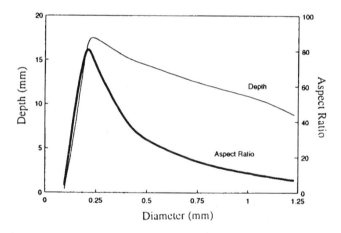

Fig. 4.17 Dependence of hole depth and aspect ratio on the hole diameter showing the limitation of achieving maximum hole depth and aspect ratio with increasing hole diameter for laser drilling of high nickel alloys. (Reprinted from Yeo et al. 1994. With permission. Copyright Elsevier.)

thermo-physical properties of the material. Figure 4.17 shows the relationship between the hole diameter, hole depth, and aspect ratio in the drilling of high nickel alloys with a pulsed YAG laser (Belforte and Levitt 1987). The figure shows that there is a limitation on achieving the maximum depth and aspect ratio with increasing hole diameter. Laser beam focused with shallow convergence angle and low thermal diffusivity tends to produce holes of high-aspect ratio. However, high-aspect ratio holes are generally associated with larger entrance diameter than the mean hole diameter (Yeo et al. 1994).

500 μm

Fig. 4.18 Vertical section of laser-drilled hole in 4 mm thick IN718 alloy showing typical taper in laser-drilled hole. (Reprinted from Bandyopadhyay et al. 2005. With permission. Copyright Elsevier.)

In contrast to the mechanically drilled holes, laser-drilled holes are rarely parallel walled and are often associated with taper. Taper is defined as (Ghoreishi et al 2002):

$$\text{Taper}(^0) = \frac{d_{\text{entrance}} - d_{\text{exit}}}{2t} \times \frac{180}{2\pi}. \tag{4.26}$$

The typical taper of the laser-drilled hole is shown in Fig. 4.18 (Bandyopadhyay et al. 2005). The taper in laser-drilled holes is caused by the expulsion of molten and vaporized material from the hole. Laser percussion drilling is a complex process with number of variables involved. In general, shorter pulses give higher taper. In addition, the degree of taper reduces with increasing material thickness (Fig. 4.19) (Yeo et al. 1994).

Spatter is one of the inherent defects associated with laser-drilled holes in which the molten and vaporized material formed during drilling is not completely ejected but resolidifies and adheres around the periphery of the hole. Extensive systematic studies were conducted by Low and coworkers to achieve the spatter-free holes in laser percussion drilling of Nimonic 263 and other aerospace materials. It was observed that a large proportion of the spatter (approximately >70%) was deposited due to initial laser pulses required to drill a through hole (Fig. 4.20). Moreover, short pulse widths, low peak powers, and high pulse frequencies generated smaller spatter deposition areas (Low et al. 2000a, 2001a). The effects of different assist gases, O_2, Ar, N_2, and compressed air on the physical features (viz. thickness and surface geometry) and bonding strength of the spatter were investigated (Fig. 4.21).

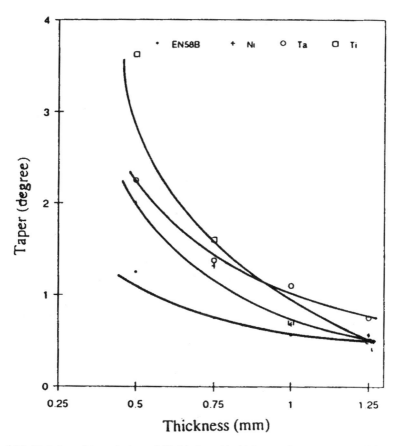

Fig. 4.19 Variation of taper in laser-drilled holes with thickness of aerospace materials drilled with pulsed YAG laser with a pulse energy of 15 J and a pulse duration of 1.4 ms. (Reprinted from Yeo et al. 1994. With permission. Copyright Elsevier.)

The figure shows that spatter thickness deposited using oxygen assist gas was only 10–20% of those drilled using the other three assist gases. The spatter deposited using oxygen assist gas was also fragmented with many peaks of similar heights as compared to that using argon assist gas which showed singular crater-like shape sloping towards the hole centre and ending normally to the material surface (Fig. 4.22). It was also found that the bonding strength of the material is associated with the "inertness" of the assist gas and increases progressively in the order of O_2, compressed air, and N_2 and Ar (Low et al. 2000b). In addition, the spatter deposition area was found to be reduced significantly with the temporal pulse train modulation technique known as sequential pulse delivery pattern control (SPDPC) as compared to the normal delivery pattern (NDP) due to reduced upward material removal fractions obtained with SPDPC (Fig. 4.23) (Low et al. 2001b). A method of spatter prevention had been developed based on the application of antispatter composite

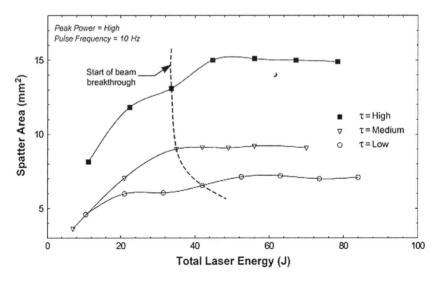

Fig. 4.20 Variation of spatter area with laser energy and pulse width at pulse frequency of 10 Hz for laser drilling of Nimonic alloys. The figure shows that the a large proportion (approximately > 70%) of the spatter was deposited due to initial pulses, prior to beam breakthrough. (Reprinted from Low et al. 2001a. With permission. Copyright Elsevier.)

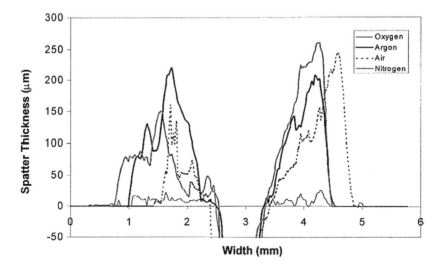

Fig. 4.21 Effect of various assist gas on the spatter thickness/geometry of the laser-drilled holes in nimonic 263 alloy with Nd:YAG laser with pulse energy of 16.8 J/pulse. (Reprinted from Low et al. 2000b. With permission. Copyright Elsevier.)

coating (ASCC) on the workpiece surface prior to laser percussion drilling (Fig. 4.24) (Low et al. 2001c, 2003). When the ejected material resolidifies on the wall surface of the laser-drilled hole, it is often referred to as recast layer.

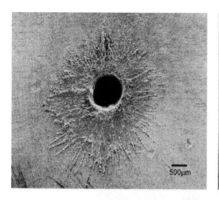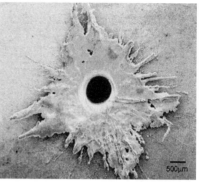

Fig. 4.22 Scanning electron micrographs showing the surface geometries of spatter deposited during laser drilling of nimonic 263 alloy using oxygen *(left)* and argon assist gases *(right)*. (Reprinted from Low et al. 2000b. With permission. Copyright Elsevier.)

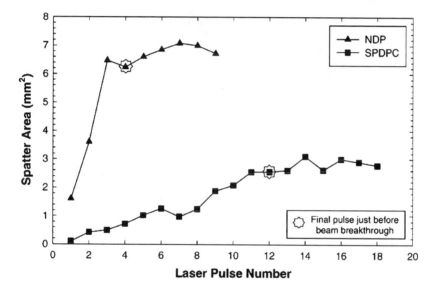

Fig. 4.23 Effects of pulse train modulation on spatter deposition area during laser drilling of nimonic alloy sheets with Nd:YAG laser. The figure shows significant reduction in spatter deposition area using linearly increasing sequential pulse delivery pattern control (SPDPC) compared to that using normal delivery pattern (NDP). (Reprinted from Low et al. 2001b. With permission. Copyright Elsevier.)

Due to complexity of the laser drilling process, a number of statistical studies had been directed towards selection and optimization of laser processing and material parameters and its effect on the quality of the laser-drilled hole (Ghoreishi et al. 2002; Bandyopadhyay et al. 2005; Yilbas 1986, 1987, 1997). In one of these

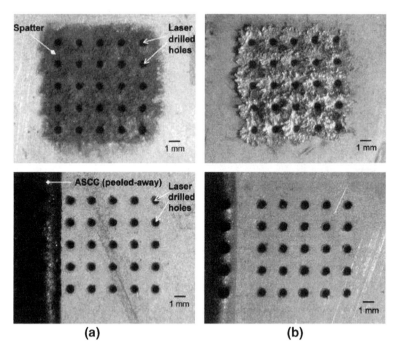

Fig. 4.24 Optical micrographs of laser-drilled closely spaced hole arrays in uncoated (upper micrographs) and coated (bottom micrographs) Nimonic 263 alloy using (**a**) oxygen and (**b**) air assist gases showing the effect of antispatter composite coating (ASCC) on the spatter deposition. (Reprinted from Low et al. 2001c. With permission. Copyright Elsevier.)

studies, Yilbas had designed a factorial experiment including four factors at four levels to investigate all possible combinations of the factors over the operating regions. The responses of these factorial designs were studied by evaluating the geometry of the resulting hole.

4.6   Practical Considerations

Laser drilling is a complex process involving parameters such as laser energy, pulse shape, focusing conditions, and assist gas. To optimize the processing parameters for a particular drilling application, it is necessary to understand the effect of these parameters on the geometric and metallurgical features of the laser-drilled holes. Hence, this section discusses the effects of some of the most important laser processing parameters on the quality of drilled holes.

4.6.1 Effect of Laser Parameters

In general, increasing laser energy increases both depth and diameter of the hole. Figure 4.25 shows the cross-sectional profiles of the laser-drilled holes in aluminum by ruby laser (Ready 1997). The figure also illustrates the typical geometric characteristics of the laser-drilled holes such as the entrance diameter, exit diameter, and taper. At high laser pulse energy the exit diameter increases due to pressure exerted by the vaporization of the material in the hole.

In case of percussion laser drilling, the numbers of laser pulses are delivered successively at the same spot and the material is removed after each pulse. The depth of the hole increases with increasing number of pulses until a certain number of pulses; thereafter, the depth remains essentially constant. Figure 4.26 presents the variation of hole depth with number of pulses (for smaller number of pulses) for various laser powers during drilling of mild steel. The figure indicates that the drilling rate increases substantially after certain initial pulses which may be due to efficient impulsive removal of material caused by confinement of high pressure vapor within the walls of the crater. At very high number of pulses, the defocusing of the laser beam at the bottom of the drilling front leads to inefficient material removal and hence significant increase in the depth may not be observed with further increase in the number of pulses (Hamilton and Pashby 1979).

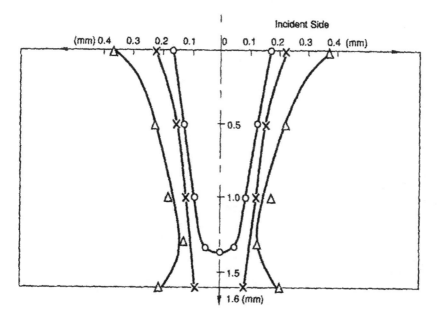

Fig. 4.25 Cross-sectional profiles of laser-drilled holes in 1.6 mm aluminum plate with ruby laser pulses at three different pulse energies. (Pulse energies: O –0.36 J; × –1.31 J; Δ –4.25 J). (Reprinted with permission from Ref. Ready 1997. Copyright, Elsevier)

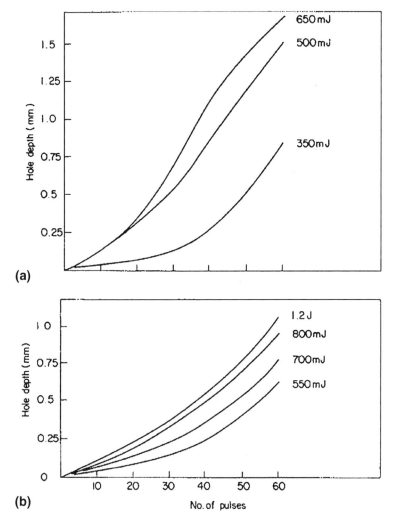

Fig. 4.26 Variation of hole depth with number of pulses during laser drilling of mild steel with two different pulse shapes: **(a)** Pulse time 40 μs, peak power 330 kWJ^{-1}; **(b)** pulse time 15 μs, peak power 1.3 MWJ^{-1}. (Reprinted from Hamilton and Pashby 1979. With permission. Copyright Elsevier.)

The characteristics of laser-drilled holes are significantly affected by the laser pulse characteristics such as pulse shape and pulse duration. It is observed that in case of drilling with normal pulse, the hole made by the relaxation oscillations in the beginning of the pulse gets immediately clogged by the molten material generated in the later part of the pulse. Hence, for efficient laser drilling, it is preferable to use a pulse having "spikes" along the whole width of the pulse. Figures 4.27 and 4.28 show the pulse shape and the corresponding hole drilled in 0.1 mm aluminum

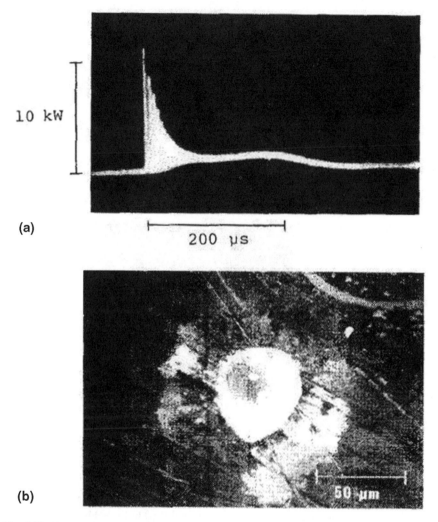

Fig. 4.27 Normal pulse shape and the corresponding hole drilled in 0.1 mm aluminum foil with pulse energy of 30 J. (Reprinted from Roos 1980. With permission. Copyright American Institute of Physics.)

foil with pulse energy of 30 mJ. The normal pulse gives a small crater, whereas the pulse train with each spike width of 0.5 μs gives an excellent quality hole (Roos 1980).

Apart from temporal shaping of the individual laser pulse, there is growing interest in the shaping of the entire train of pulses (interpulse shaping) by a technique called SPDPC to obtain through holes with excellent geometric and metallurgical quality (Low et al. 2001b).

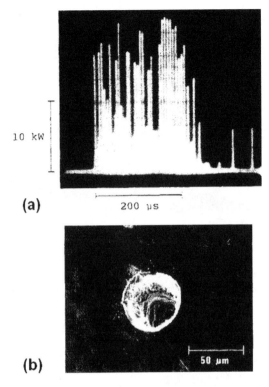

Fig. 4.28 Pulse train and the corresponding hole drilled in 0.1 mm aluminum foil with pulse energy of 30 J. (Reprinted from Roos 1980. With permission. Copyright American Institute of Physics.)

4.6.2 Effect of Focusing Conditions

The focusing optics determines the focal length, depth of focus, and focal spot size. The size of the single pulse and percussion laser-drilled holes is highly determined by the spot size of the laser beam on the surface of the workpiece. The minimum spot diameter (d (mm)) depends on the focal length of the lens (f (mm)) and is given by

$$d = f\theta, \tag{4.27}$$

where θ is the total beam divergence angle (radians). In addition, the beam diameter along the length of the beam is given by the beam divergence. The depth of focus is given by

$$\Delta f = \frac{2df}{D} = \frac{2\theta f^2}{D}, \tag{4.28}$$

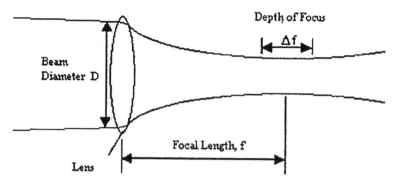

Fig. 4.29 Schematic of focal pattern of converging lens. (Reprinted from McNally et al. 2004. With permission. Copyright Maney Publishing.)

where D is the beam diameter (mm) and Δf is the depth of focus (mm). Various optical parameters are shown in Fig. 4.29 (McNally et al. 2004).

From Eq. (4.28), smaller focal spot diameter can be obtained by reducing the focal length, thus increasing the laser energy density (intensity). Higher laser energy density increases the material removal rates, thus increasing the drilling speed. However, reduction in focal length is also associated with decrease in depth of focus, which may undesirably affect the flatness and perpendicularity of the laser-drilled holes. The smaller depth of focus may also lead to disruption of the process if the workpiece to be drilled is not flat. Hence, longer depth of focus is desired to allow the workpiece to be positioned more easily. The effects of the beam focus on the depth and shape of the laser-drilled holes is shown in Fig. 4.30. As shown in the figure good quality holes in terms of geometry (straightness and depth) are obtained when the beam waist is positioned just below the surface of the workpiece (Yeo et al. 1994).

4.6.3 Effect of Assist Gas Type, Gas Pressure, and Nozzle Design

Almost all the laser drilling operations use an assist gas jet coaxial to the laser beam. The coaxial gas jet is used to facilitate the material removal and prevent the contamination of the focusing lens from the impingement of ejected debris during drilling operation. At low laser powers, where the melting process dominates, the assist gas jets increase the material removal and reduce the recast layer by increasing the shear forces on the molten material which is subsequently ejected during drilling. At high laser powers, where the vaporization dominates the material removal, the assist gas increases the drilling efficiency by removing the absorptive vapors and debris that can prevent the incident laser energy from reaching the target. Various assist gases such as compressed air, oxygen, argon, and nitrogen have been

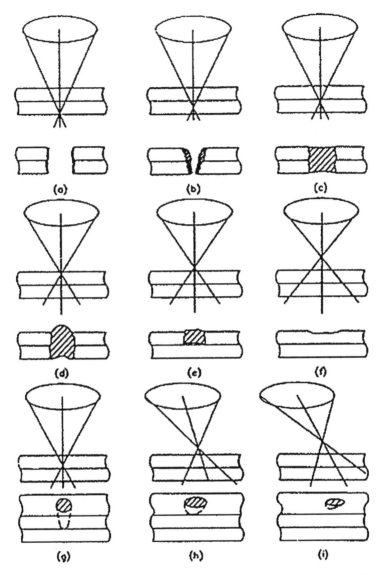

Fig. 4.30 Effect of beam focus on the depth and shape of laser-drilled holes. (Reprinted from Yeo et al. 1994. With permission. Copyright Elsevier.)

used for drilling different materials. Inert gases are used as assist gas when the laser-drilled holes free from oxide layer are desired. Air can also produce acceptable quality holes in some metals such as aluminum. Oxygen gas is used as assist gas to enhance the material removal by exothermic reaction. Laser drilling using oxygen as an assist gas contributes energy to the process by the oxidation of metal vapors and liquid. Oxygen-assisted laser drilling is even more effective for the materials whose oxides act as a flux and reduces the viscosity of the liquid metal thereby enhancing the material ejection.

The two most important effects during laser drilling with oxygen as an assist gas are the change in the absorptivity of the surface due to oxidation and the change in the temperature required to expel the molten material because of difference in the melting points of the oxide and metal. Table 4.2 gives the values of room temperature and the maximum oxygen-enhanced absorptivities for metallic materials. Increase in the absorptivity favorably affect the drilling time. However, if the products of oxidation have higher melting point, the surface temperature required to melt the oxide and expel the melt will be increased, thus increasing the drilling time. The balance among these competing factors will finally determine whether the oxygen assist is helpful to the laser drilling. This is substantiated by the experimental observation (Fig. 4.31) of longer drilling time for AL 6061 (change in

Table 4.2 Absorptivity values for metallic materials (Patel and Brewster 1991)

Material	Room temperature value	Oxidation-enhanced value
Al6061	0.28	0.4
Cu	0.02	0.2
304 stainless steel	0.32	0.32
Low C steel	0.45	0.6

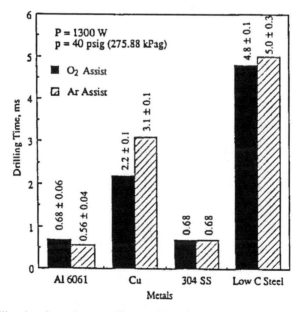

Fig. 4.31 Drilling time for various metallic materials using oxygen and argon assist gases during laser drilling with incident laser power of 1300 W and assist gas pressure of 40 psig. (Reprinted from Patel and Brewster 1991. With permission. Copyright American Institute of Aeronautics and Astronautics.)

absorptivity from 0.28 to 0.4 and change in melting point from 933 K for Al to 2,345–3,125 K for oxides) and shorter drilling time for Cu (change in absorptivity from 0.02 to 0.2 and change in melting point from 1,326 K for Cu to 1,508–1,599 K for oxides). The effect of change in melting point is more dominant in Al6061, whereas the effect of change in absorptivity is more dominant in Cu (Patel and Brewster 1991a, b).

Studies have shown that the best quality holes are produced in iron, nickel, and their alloys by using oxygen as an assist gas because it helps burn the metal away. Special care must be taken during laser drilling metals such as titanium with oxygen and nitrogen as assist gas because it burns easily. Figure 4.32 shows the effects of gas composition on the hole dimensions (exit and entrance diameters of the holes) during single-pulse (pulse width of 0.1 ms) laser drilling of 0.8 mm thick titanium with Nd:YAG laser. The figure indicates that the entrance diameter of the hole increases with the oxygen content, while the exit diameter of the hole remains unaffected. Increase in the entrance diameter with the oxygen content is associated with the improved material removal due to energy of exothermic reaction when titanium reacts with oxygen (Rodden et al. 2001).

The analysis of effect of assist gas pressure on the drilling parameters is interesting because it contradicts with the intuition that drilling time would decrease with increasing assist gas pressure due to the efficient removal of molten material during drilling. However, it observed that high gas pressures leads to the formation of density gradient fields which changes the refractive index of the medium between the workpiece and the gas nozzle, thus defocusing the laser beam. The defocusing

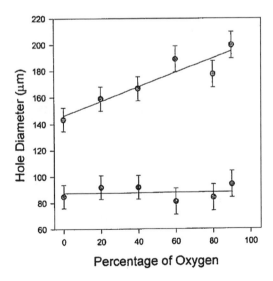

Fig. 4.32 Variation of entrance diameter (upper curve) and exit diameter (lower curve) of laser-drilled hole with oxygen content during oxygen-assisted laser drilling of 0.8 mm thick titanium (Reprinted from Rodden et al. 2001. With permission. Copyright Laser Institute of America.)

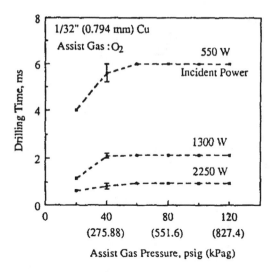

Fig. 4.33 Variation of drilling time with assist gas pressure for copper for different laser powers. (Reprinted from Patel and Brewster 1991. With permission. Copyright American Institute of Aeronautics and Astronautics.)

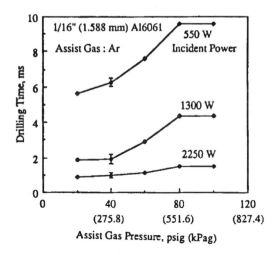

Fig. 4.34 Variation of drilling time with assist gas pressure for Al6061 alloy for different laser powers. (Reprinted from Patel and Brewster 1991. With permission. Copyright American Institute of Aeronautics and Astronautics.)

of the laser beam associated with high pressures reduces the flux density of the surface resulting in increase in minimum drilling time. Figures 4.33 and 4.34 show the dependence of drilling time on assist gas pressures for copper and Al6061. It is observed that at all incident powers and for both argon and oxygen, the drilling time increases with assist gas pressures up to a certain pressure beyond which the drilling time remains essentially constant (Patel and Brewster 1991a, b). Typical oxygen gas pressures range from 100 to 350 kPa. Compressed air and inert gas pressures range from 200 to 620 kPa (Elza and White 1989).

In addition to the type of assist gas, the geometry and quality of laser-drilled holes is also affected by nozzle design, nozzle–material standoff, and nozzle gas supply pressure. Nozzle designs differ significantly for various laser drilling setups, but typical orifice openings range from 2.5 to 6.0 mm and 0.75 to 2.5 mm for percussion drilling and trepanning, respectively. Nozzle–material standoff is the distance between the nozzle tip and the surface of the workpiece and often equals the nozzle–focal point standoff when the beam is focused on the surface of the workpiece. Nozzle–materials standoff depends on the laser type, nozzle type, and gas flow rates and often range from 4 to 40 mm and 0 to 15 mm for percussion drilling and trepanning, respectively. Nozzle standoff with Nd:YAG laser is about 5 mm. In gas-assisted laser drilling, the hole depth is found to be decreased with decreasing the nozzle–material standoff. For best laser drilling results the nozzles must be aligned coaxially with the focused laser beam (Elza and White 1989).

4.7 Laser Drilling Applications

Laser drilling is currently used in many industrial applications for large-scale production. In addition, laser drilling is increasingly replacing conventional drilling techniques for various applications. This section gives an overview of the established applications of laser drilling along with a brief mention of the applications where laser drilling is likely to be a dominant technology. By no means is this a complete account of all the applications of laser drilling.

4.7.1 Drilling of Cooling Holes

Laser drilling of cooling holes in aerospace gas turbine parts such as turbine blades, nozzle guide vanes, combustion chambers, and afterburners is an established technology. These parts are typically made of nickel-based superalloys which are difficult to machine using conventional techniques. Figure 4.35 shows a laser-drilled component of gas turbine. Such closely spaced cooling holes increase the efficiency of the engines and allow the higher operating temperatures. Table 4.3 presents the typical hole dimensions of different engine components. The most preferred drilling technique in such applications uses Nd:YAG laser in a trepanning drilling mode (French et al. 2003). However, other laser sources can also be used in various different laser drilling modes based on the size, depth, and quality of the desired holes. Increased drive towards improving the efficiency of the engines has led to the development of plasma-sprayed thermal barrier coatings (TBCs) for the engine

Fig. 4.35 Laser-drilled holes in an aerospace component. (Reprinted from French et al. 2003. With permission. Copyright Laser Institute of America.)

Table 4.3 Typical dimensions of engine hole components. (Reproduced from French et al. 2003. With permission. Copyright Laser Institute of America.)

Component	Diameter (mm)	Wall thickness (mm)	Angle to surface (degrees)	No. of holes
Blade	0.3–0.5	1.0–3.0	15	25–200
Vane	0.3–1.0	1.0–4.0	15	25–200
Afterburner	0.4	2.0–2.5	90	40,000
Baseplate	0.5–0.7	1.0	30–90	10,000
Seal ring	0.95–1.05	1.5	50	180
Cooling ring	0.78–0.84	4.0	79	4,200
Cooling ring	5.0	4.0	90	280

components. Even though laser drilling is well-suited for drilling TBC superalloys (Fig. 4.36), it is faced with multiple challenges to meet the growing needs of the aerospace industry. These areas are the prevention of delamination in TBCs in the drilled areas, drilling of low angle cooling holes, and drilling through complicated and smaller cavities (Voisey and Clyne 2004).

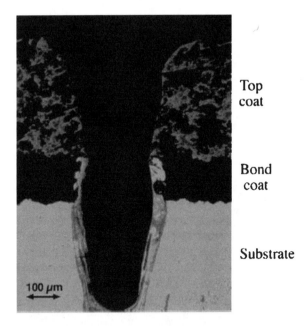

Fig. 4.36 Laser-drilled cooling hole in a plasma-sprayed thermal barrier-coated (TBC) nickel superalloy, which consisted of partially stabilized zirconia (PSZ) top coat and NiCrAlY bond coat deposited on Ni-superalloy substrate. (Reprinted from Voisey and Clyne 2004. With permission. Copyright Elsevier.)

4.7.2 Drilling of Diamonds

Diamonds, being highly transparent and the hardest known material, presents unique difficulties for machining. Drilling of natural and synthetic diamonds is of particular interest for many applications. Pulsed solid-state lasers are used for drilling diamond wire drawing dies to produce high finish, precisely profiled holes with minimum heat affected zone. In addition, laser drilling is used in jewelry to enhance the clarity grade of diamonds by helping in removing the trapped inclusions. Lasers are used to bore a small hole in the diamond from the surface to the targeted inclusion, which is subsequently bleached out or burned away by forcing a strong acid through the drilled hole.

4.7.3 Microdrilling

Lasers are now extensively used in the micromachining applications where feature resolution in the range of micron and submicron is required. Many of these microdrilling applications are discussed in the following chapter dealing with micromaching.

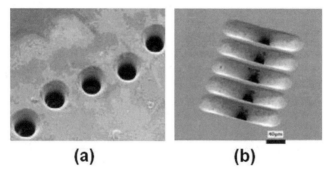

Fig. 4.37 Some geometries of laser-drilled holes in inkjet printer head. (Reprinted from Gower 2002. With permission. Copyright Optical Society of America.)

One such application is the drilling of holes in the head plate of ink jet printers. Ink jet printer heads consists of an array of tapered holes through which the ink droplets are squirted on the paper. The resolution of the prints can be increased by increasing the number of holes and decreasing the hole diameter and pitch of the holes in the head. Modern printers use excimer lasers to drill 300 holes of $28\,\mu m$ diameter compared to earlier printers which used electroforming to produce 100 holes of $50\,\mu m$ diameter. In addition to the increased resolution (600 dots per inch), laser drilling gives higher production yields ($> 99\%$) compared to conventional resolution of 300 dots per inch and production yield of 70–85%. Figure 4.37 presents some geometries of laser-drilled holes in inkjet printer heads (Gower 2000).

Drilling of fuel injector nozzles is traditionally done by electrical discharge machining (EDM) and the hole diameters in the range $150–200\,\mu m$ are achieved. Increasingly, stringent environmental issues to reduce emissions necessitate the employment of smaller injection holes and/or specially shaped holes which are beyond the capabilities of the EDM. Laser drilling is capable of drilling smaller holes (as small as $20\,\mu m$ in diameter) in specially engineered configurations such as shape and taper (Herbst et al. 2004).

Table 4.4 presents a summary of industrial applications of laser drilling.

4.8 Advances in Laser Drilling

The continuously increasing precision requirements in micromachining applications led to the development of laser drilling processes using ultrashort laser pulses. At such ultrashort pulse durations ($\sim 100\,fs$), the material removal mechanism causes direct ablation of the solid to vapor. This leads to negligible heating and correspondingly negligible heat-affected zone (damage zone) resulting in clean holes without spatter, resolidified layer, and microcracking. Hence no postdrilling finishing operations are generally required. For example, typical femtosecond ($100\,fs$) laser pulses in low fluence approximation produces thermal diffusion

Table 4.4 Summary of applications of laser drilling in various industrial fields

Industrial field	Applications
Aerospace	Turbine component cooling
	Engine silencing
	Aerofoil laminar flow
Microelectronics	Inkjet printer nozzle
	PCB via interconnects
	IC test vertical probe card
	Optical switching
Automotive	Fuel injection nozzle
	Fuel filter
	Car brake sensors
	Connecting rod lubrications
Biomedical/MEMS	Catheter sensors
	Aerosol spray atomizers
	DNA sampling
	Vaccine Production
	Lab-on-a-chip
Environmental/Renewable energy	Toxic gas sensors
	Solar cell technology
	Fuel cell
	Particulate filters
Others	Food packaging
	Gemstone drilling
	Digital finger printing

length of around 1 nm in steel. Due to inherent advantages of femtosecond pulse processing, this technology have a great potential to produce micron and submicron features in medical devices, photonic devices, micromechanical devices, and photomasks.

Laser drilling is generally carried out using Gaussian beams. Due to nonuniform energy distribution, drilling with Gaussian beams is often associated with poor edge quality, wider heat-affected zones, and waste of energy carried in the wings of Gaussian distribution which does not contribute significantly to the drilling process. Hence, considerable research efforts are directed towards shaping the laser beam to redistribute the energy in more uniform profile. In general, the percussion laser drilling is also carried out using normal pulses having relaxation oscillations (relaxation spikes) in the beginning of the pulse which decline and merge into continuous emission. Temporal modulation of the individual pulse to obtain the spikes continuously along the pulse has beneficial effects on the drilling performance. In addition to the temporal modulation of the individual pulse, the efforts are done to modulate the entire pulse train by SPDPC. Figure 4.38 shows a normal pulse delivery (NPD) pattern and SPDPC, which gives rise to better quality of holes in percussion drilling in terms of the deposited spatter area (Low et al. 2001b).

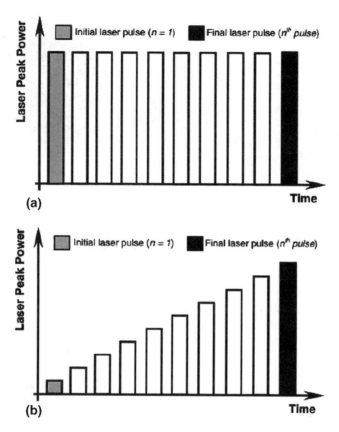

Fig. 4.38 Schematic of (**a**) normal delivery pattern (NPD) and (**b**) linearly increasing pulse delivery pattern (SPDPC). (Reprinted from Low et al. 2001b. With permission. Copyright Elsevier.)

Recently, laser drilling under various surrounding media have attracted increased interests. One such effort is underwater laser drilling which uses water as a surrounding medium. The increase in efficiency and quality of laser drilling with water compared to that in air is due to a combination of the following effects (Lu et al. 2004):

1. During drilling, the laser–material interaction generates high temperature and high pressure plasma which causes the explosive removal of the material. However, if water is present in the surrounding, the expansion of the generated plasma is confined to such an extent that the recoil pressure against solid increases tremendously resulting in enhanced laser drilling efficiency. One of the studies have reported that the impact induced by plasma confined in water is about four times higher and the shock-wave duration is about 2–3 times longer than that in direct regime (in air) at the same laser intensity.

2. The size of the plasma is much smaller in water environment; hence, the detrimental effects of plasma on the drilling efficiency such as absorption of incident

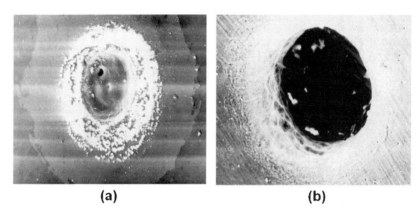

(a)   **(b)**

Fig. 4.39 Hole drilled in 0.3 mm thick aluminum plate by pulse laser in (**a**) surrounding air and (**b**) water. (Reprinted from Lu et al. 2004. With permission. Copyright American Institute of Physics.)

energy and reduction in laser coupling to the surface are minimized. In addition, the plasma lifetime in water is one half of that in air.

3. Surrounding water in underwater drilling improves the surface quality of the laser-drilled holes by preventing the molten metal to solidify along the edges of the hole and also by conducting away the excessive heat from the surface.

4. Cavitation bubbles which form during laser matter interaction in liquids collapse against the wall and thus produce impulse force.

Fig. 4.39 shows the SEM micrograph of laser-drilled holes in surrounding air and water using Nd:YAG laser and pulse duration of 20 ns. Underwater laser drilling gives more regular, cleaner, and smoother surface morphology (Lu et al. 2004).

References

Bandyopadhyay S, Sundar JK, Sundararajan G, Joshi SV (2002) Geometrical features and metallurgical characteristics of Nd:YAG laser drilled holes in thick IN718 and Ti-6Al-4V sheets. Journal of Materials Processing Technology 127:83–95.

Bandyopadhyay S, Gokhale H, Sundar JK, Sundararajan G, Joshi SV (2005) A statistical approach to determine process parameter impact in Nd:YAG laser drilling of IN718 and Ti-6Al-4V sheets. Optics and Lasers in Engineering 43:163–182.

Basu S, DebRoy T (1992) Liquid metal expulsion during laser irradiation. Journal of Applied Physics 72:3317–3322.

Basu B, Srinivasan J (1988) Numerical study of steady-state laser melting problem. International Journal of Heat and Mass Transfer 31:2331–2338

Belforte DA, Levitt MR (1987) Laser materials processing data and guidelines. SPIE Proceedings 919:20–29.

Bellantone R, Ganesh RK, Hahn Y, Bowley WW (1991) A model of laser hole drilling: calculation with experimental comparison. Proceedings of the 10th International Invitational Symposium on the Unification of Numerical, Analytical and Experimental methods, WPI, Worcester, MA 01609, pp. 317–339.

Benes JJ (1996) Technology adds a new twist to difficult drilling. American Machinist 140:78–79.

Chan CL, Mazumder J (1987) One-dimensional steady-state model for damage by vaporization and liquid expulsion due to laser-material interaction. Journal of Applied Physics 62:4579–4586.

Chan C, Mazumder J, Chen MM (1984) A two-dimensional transient model for convection in laser melted pool. Metallurgical Transactions A 15:2175–2184.

Chen K, Yao YL, Modi V (2000) Gas jet-workpiece interactions in laser machining. Journal of Manufacturing Science and Engineering 122:429–438

Chryssolouris G (1991) Laser Machining, Theory and Practice. Springer, New York.

Dabby FW, Paek UC (1972) High-intensity laser-induced vaporization and explosion of solid material. IEEE Journal of Quantum Electronics 8:106–111.

Elza D, White G (1989) Laser beam machining. In: Davis JR (ed) ASM Handbook: Machining, ASM International, Ohio, vol. 16, pp 572–578.

French PW, Hand DP, Peters C, Shannon, Byrd P, Steen WM (1998) Investigations of the Nd: YAG laser percussion drilling process using high speed filming, Science-Lasers for Science Facility-Physics, CLF Annual Report, Laser Institute of America, Orlando, FL.

French PW, Naeem M, Watkins KG (2003) Laser percussion drilling of aerospace material using 10 kW peak power laser using 400 µm optical fibre delivery system. Proceedings of the International Congress on Laser Applications and Electro-optics, (ICALEO 2003) 95:1–9.

Ganesh RK, Faghri A (1997) A generalized thermal modeling for laser drilling process-?. Mathematical modeling and numerical methodology. International Journal of Heat and Mass Transfer 40:3351–3360.

Ghoreishi M, Low DK, Li L (2002) Comparative statistical analysis of hole taper and circularity in laser percussion drilling. International Journal of Machine Tools and Manufacture 42:985–995.

Giering A, Beck M, Bahnmuller J (1999) Laser drilling of aerospace and automotive components. Proceedings of the International Congress on Laser Applications and Electro-optics, ICALEO'99, Laser Institute of America, Orlando, FL, vol. 87, pp. 80–87.

Gordon MH, Touzelbaev M, Xiao M, Goforth RC (1994) Numerical solution of diamond film ablation under irradiation by a laser beam. ASME Proceedings, Chicago, IL, pp. 73–77.

Gower MC (2000) Industrial applications of laser micromachining. Optics Express 7:56–67.

Hamilton DC, Pashby IR (1979) Hole drilling studies with a variable pulse length CO_2 laser. Optics and Laser Technology 8:183–188.

Herbst L, Lindner H, Heglin M, Hoult T (2004) Targetting diesel engine efficiency. Industrial Laser Solutions 10:1–4.

Ho JR, Grigoropolous CP, Humphrey JA (1994) Computational study of heat transfer and gas dynamics in the pulsed laser evaporation of metals. Journal of Applied Physics 78:4696–4709.

Kamalu J, Byrd P, Pitman A (2002) Variable angle laser drilling of thermal barrier coated nimonic. Journal of Materials Processing Technology 122:355–362.

Kar A, Rockstroh T, Mazumder J (1992) Two-dimensional model for laser-induced material damage: effects of assist gas and multiple reflections inside the cavity. Journal of Applied Physics 71:2560–2569.

Kar A, Mazumder J (1990) Two-dimensional model for laser-induced material damage due to melting and vaporization during laser irradiation. Journal of Applied Physics 68:3884–3891.

Kitai MS, Popkov VL, Semchishen VA (1990) Dynamics of UV excimer laser ablation of PMMA, caused by mechanical stresses: theory and experiment. Makromolekulare Chemie. Macromolecular Symposia 37:257–267.

Knowles M (2000) Micro-ablation with high power pulsed copper vapor lasers. Optics Express 7:50–55.

Kou S, Wang YH (1986) Three-dimensional convection in laser melted pools. Metallurgical Transactions A 17:2265–2270.

Low DL, Li L, Byrd PJ (2000a) The effects of process parameters on spatter deposition in laser percussion drilling. Optics and Laser Technology 32:347–354.

Low DK, Li L, Corfe AG (2000b) Effects of assist gas on the physical characteristics of spatter during laser percussion drilling of NIMONIC 263 alloy. Applied Surface Science 154–155:689–695.

Low DK, Li L, Corfe AG (2001a) Characteristics of spatter formation under the effects of different laser parameters during laser drilling. Journal of Materials Processing Technology 118:179–186.

Low DK, Li L, Byrd PJ (2001b) The influence of temporal pulse train modulation during laser percussion drilling. Optics and Lasers in Engineering 35:149–164.

Low DK, Li L, Corfe AG, Byrd PJ (2001c) Spatter-free laser percussion drilling of closely spaced array holes. International Journal of Machine Tools and Manufacture 41:361–377.

Low DK, Li L, Byrd PJ (2002) Hydrodynamic physical modeling of laser drilling. Journal of Manufacturing Science and Engineering 124:852–862.

Low DK, Li L, Byrd PJ (2003) Spatter prevention during laser drilling of selected aerospace materials. Journal of Materials Processing Technology 139:71–76.

Lu J, Xu RQ, Chen X, Shen ZH, Ni XW, Zhang SY, Gao CM (2004) Mechanisms of laser drilling of metal plates underwater. Journal of Applied Physics 95:3890–3894.

Mazumder J (1991) Overview of melt dynamics in laser processing. Optical Engineering 30:1208–1219.

Meijer J (2004) Laser beam machining (LBM), state of the art and new opportunities. Journal of Materials Processing Technology 149:2–17.

McNally CA, Folkes J, Pashby IR (2004) Laser drilling of cooling holes in aeroengines: state of the art and future challenges. Materials Science and Technology 20:805–813.

Nagano Y (1978) Laser drilling. Technocrat 11:19–25.

Paek UC, Gagliano FP (1972) Thermal analysis of laser drilling process. IEEE Journal of Quantum Electronics 8:112–119.

Patel RS, Brewster MQ (1991a) Gas-assisted laser-metal drilling: theoretical model. Journal of Thermophysics and Heat Transfer 5:32–39.

Patel RS, Brewster MQ (1991b) Gas-assisted laser-metal drilling: Experimental results. Journal of Thermophysics and Heat Transfer 5:26–31.

Ready JF (1997) Industrial Applications of Lasers. Academic Press, San Diego.

Rodden WS, Kudesia SS, Hand DP, Jones JD (2002) A comprehensive study of the long pulse Nd:YAG laser drilling of multi-layer carbon fibre composites. Optics Communications 210:319–328.

Rodden WS, Kudesia SS, Hand DP, Jones JD (2001) Use of "assist" gas in the laser drilling of titanium. Journal of Laser Applications 13:204–208.

Roos SO (1980) Laser drilling with different pulse shapes. Journal of Applied Physics 51:5061–5063.

Schoonderbeek A, Biesheuvel CA, Hofstra RM, Boller KJ, Meijer J (2004) The influence of the pulse length on the drilling of metals with an excimer laser. Journal of Laser Applications 16:85–91.

Tam SC, Yeo CY, Jana S, Lau MW (1994) A technical review of the laser drilling of aerospace materials. Journal of Materials Processing Technology 42:15–49.

Tokarev VN and Kaplan AF (1999) Suppression of melt flows in laser ablation: application to clean laser processing. Journal of Physics D: Applied Physics 32:1526–1538.

Verhoeven JC, Jansen JK, Mattheij RM (2003) Modelling laser induced melting. Mathematical and Computer Modelling 37:419–437.

Verhoeven J (2004) Modelling percission drilling, Ph.D. thesis, Eindhoven University of Technology, The Netherlands.

Voisey KT, Clyne TW (2004) Laser drilling of cooling holes through plasma sprayed thermal barrier coatings. Surface and Coatings Technology 176:296–306.

Voisey KT, Cheng CF, Clyne TW (2000) Quantification of melt ejection phenomena during laser drilling. Materials Research Society Symposium 617:J5.6.1–J5.6.7.

Voisey KT, Kudesia SS, Rodden WS, Hand DP, Jones JD, Clyne TW (2003) Melt ejection during laser drilling of metals. Materials Science and Engineering A 356:414–424.

von Allmen M (1976) Laser drilling velocity in metals. Journal of Applied Physics 47:5460–5463.

Wagner RE (1974) Laser drilling mechanics. Journal of Applied Physics 45:4631–4637.

Weaver I, Lewis CL (1996) Aspects of particulate production from metals exposed to pulsed laser radiation. Applied Surface Science 96–98:663–667.

Wu CY, Shu CW, Yeh ZC (2006) Effects of excimer laser illumination on microdrilling into an oblique polymer surface. Optics and Lasers in Engineering 44:842–857.

Yeo CY, Tam SC, Jana S, Lau MW (1994) A technical review of laser drilling of aerospace materials. Journal of Materials Processing Technology 42:15–49.

Yilbas BS, Sami M (1997) Liquid ejection and possible nucleate boiling mechanisms in relation to the laser drilling process. Journal of Physics D: Applied Physics 30:1996–2005.

Yilbas BS (1986) A study of affecting parameters in the laser hole-drilling of sheet metals. Journal of Mechanical Working Technology 13:303–315.

Yilbas BS (1987) Study of affecting parameters in laser hole drilling of sheet metals. Journal of Engineering Materials and Technology 109:282–287.

Yilbas BS (1997) Parametric study to improve laser hole drilling process. J. Materials Processing Technology 70:264–273.

Zacharia T, Eraslan AH, Aidun DK, David SA (1989) Three-dimensional transient model for arc welding process. Metallurgical Transactions B 20:645–659.

Zhang Y, Faghri A (1999) Vaporization, melting and heat conduction in the laser drilling process. International Journal of Heat and Mass Transfer 42:1775–1790.

Chapter 5
Laser Cutting

5.1 Introduction

Laser cutting is a two-dimensional machining process in which material removal is obtained by focusing a highly intense laser beam on the workpiece. The laser beam heat subsequently melts/vaporizes the workpiece throughout the thickness or depth of the material thus creating a cutting front. The molten material is expelled from the cutting front by a pressurized assist gas jet. The assist gas, in addition to facilitating the material removal by melt expulsion, may also help in enhanced material removal through chemical reactions such as oxidation of the material. The cutting of the material then proceeds by the movement of the cutting front across the surface of the material. This is carried out by the motion of either focused beam and/or the workpiece relative to each other.

Laser cutting is a high-speed, repeatable, and reliable method for a wide variety of material types and thicknesses producing very narrow and clean-cut width. The process is particularly suited as a fully or semiautomated cutting process for the high production volumes. One of the first industrial applications of laser cutting using 200 W laser is cutting of slots in die boards. The lasers are now capable of cutting a wide range of metallic materials such as steels, superalloys, copper, aluminum, and brass, and nonmetallic materials such as ceramic, quartz, plastic, rubber, wood, and cloth (Ready 1997).

Some of the advantages of laser cutting over the conventional machining techniques can be listed as:

1. Noncontact process: The workpiece need not to be clamped or centered on the precise fixtures as in conventional machining. Accurate positioning of the workpiece on the X–Y table with defined direction of cut can be easily obtained during laser cutting thus facilitating the machining of flimsy and flexible materials.
2. Ease of automation: Most of the laser-cutting processes are CNC-controlled giving accurate control over the dimensions of cut and faster cutting speeds.
3. High cutting speeds: Laser cutting is a fast process. The typical cutting speed for 4 mm carbon steels with a 1,250 W CO_2 lasers is 3 m/min.

N.B. Dahotre and S.P. Harimkar, *Laser Fabrication and Machining of Materials.*
© Springer 2008

4. Fine and precise cut dimensions: Laser cutting can be carried out with a very narrow kerf width (~0.1 mm) such that the process can be used for fine and profile cutting.
5. Better quality of cuts: Laser heats and melts the material in a localized fashion. This in combination with the melt expulsions during cutting minimizes the heat-affected zones and the thermal stresses.
6. Flexible process: Laser cutting can be used for a variety of materials ranging from ferrous and nonferrous to nonmetallic materials.

5.2 Laser Cutting Approaches

Based on interaction of the laser beam with the workpiece and the role of assist gas in the material removal process, lasers can be used in several ways in the material removal processes during cutting. The four main approaches to cut the material using laser are evaporative laser cutting, fusion cutting, reactive fusion cutting, and controlled fracture technique. The selection of optimum technique and operation condition depends on the thermo-physical properties of the material, the thickness of the workpiece, and the type of laser employed. General schematic of the laser-cutting process is shown in Fig. 5.1 (Powell 1998). Table 5.1 presents the laser-cutting approaches for the various materials.

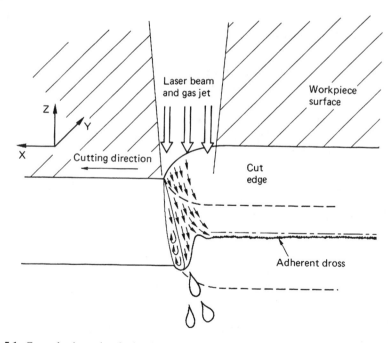

Fig. 5.1 General schematic of a laser-cutting process with a coaxial gas jet to blow the molten material. (Reprinted from Powell 1998. With permission. Copyright Springer.)

Table 5.1 Various approaches of laser cutting

Laser cutting approach	Principle	Material properties	Example
Vaporization	Direct vaporization of material by laser energy with inert assist gas	Low conductivity, low heat of vaporization	Cloth, wood, paper, etc.
Fusion	Laser energy melts the material which is subsequently expelled by impinging inert gas jet	High conductivity materials	Nonferrous materials (titanium, aluminum, etc.)*
Reactive fusion	Exothermic reaction creates additional source of energy. Molten materials is removed by reactive gas jet in the form of mixture of oxide and metal	High conductivity, reactive materials	Mild steel, titanium, stainless steel, etc.
Controlled fracture	Laser energy introduces stresses in localized area followed by mechanical or laser induced severing	Brittle materials	Alumina and other ceramics

*Titanium can also be cut by reactive fusion cutting; however, the quality of the cutting edges is better in fusion cutting. In addition, in some material (such as aluminum) the exothermic may be self–extinguishing hence efficient cutting results can be obtained with fusion cutting

5.2.1 Evaporative Laser Cutting

In evaporative laser cutting, the laser provides the latent heat until the material reaches the vaporization point and ablate into vapor state. Since the materials removal is due to direct phase change to the vapor, the cut quality is extremely high with clean edges. The method is primarily suitable for the materials with low thermal conductivity and low heat of vaporization such as organic materials, cloth, paper, and polymers. Nonreactive gas jet may be used to reduce charring (Ready 1997).

During evaporative laser cutting, the rate of penetration of beam into the workpiece can be estimated from a lumped heat capacity calculation assuming one-dimensional heat flow (Steen 1991). All the heat energy might be used in evaporation process without any heat conduction in the workpiece. The penetration velocity defined by the volume of material removed per unit second per unit area is then given by:

$$V = F_0 / \rho[L + C_p(T_v - T_0)], \tag{5.1}$$

where
F_0 = absorbed power density (W/m^2); ρ = density of solid (kg/m^3); L = latent heat of vaporization (J/kg); C_p = heat capacity of solid (J/kg °C); T_v = vaporization temperature (°C); T_0 = temperature of the material at start (°C).

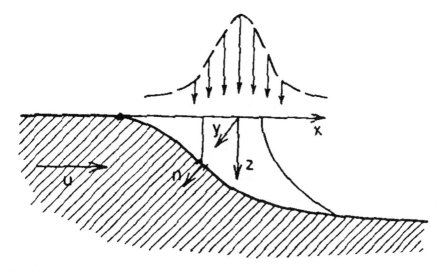

Fig. 5.2 Coordinate system for the groove formation during evaporative laser cutting. (Reprinted from Modest and Abakians 1986. With permission. Copyright American Society of Mechanical Engineers.)

Extensive studies have been carried out to study the formation of a deep groove by evaporation during laser cutting (Modest and Abakians 1986; Abakians and Modest 1988). The most important parameters during laser cutting such as groove width and shape can be calculated by numerically solving the relevant partial differential equations. The schematic of the laser-cutting process with the coordinate system is presented in Fig. 5.2. A semi-infinite body is irradiated with a Gaussian laser beam and moves with a constant velocity v.

Laser-cutting process is governed by a three-dimensional heat transfer problem with a moving source. The corresponding heat transfer equation is given by (Modest and Abakians 1986):

$$\rho C_p v \frac{\partial T}{\partial x} = k \left[\frac{\partial^2 T}{\partial x^2} + \frac{\partial^2 T}{\partial y^2} + \frac{\partial^2 T}{\partial z^2} \right]. \tag{5.2}$$

The boundary conditions are:

$$x \to \pm\infty, \quad y \to \pm\infty, \quad z \to \infty: \quad T \to T_\infty \tag{5.3}$$

Evaporative laser cutting can be approximated as a single stage solid to vapor phase transformation. The three different regions on the surface of the workpiece during cutting can be identified: Region I, the part of the workpiece which is away from the evaporation front and still too far to undergo the evaporation ($x << 0$); Region II, the region of evaporation front which is close to the center of laser beam; and Region III, where the evaporation has already been taken place with a formation of established groove (Fig. 5.3).

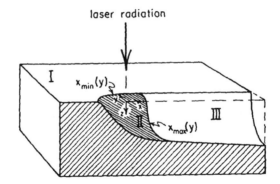

Fig. 5.3 Locations of three regions of the workpiece during evaporative laser cutting: I, region away from evaporation front; II, region of evaporation front, and III, region where evaporation has already been taken place establishing a well-defined groove. (Reprinted from Modest and Abakians 1986. With permission. Copyright American Society of Mechanical Engineers.)

The energy balance on the surface element on or away from the vaporization front gives the following boundary condition at the surface (Modest and Abakians 1986):

$$\alpha J_0 (n \bullet \hat{k}) e^{-(x^2 + y^2)/R^2} = h(T - T_\infty) - k(n \bullet \nabla T) - \rho L_s v(i \bullet \hat{n}) \qquad (5.4)$$

where α is the absorptivity; J_0 is the laser intensity at the center of beam; $\hat{i}$ and $\hat{k}$ are unit vectors in x- and y-directions respectively; $\hat{n}$ is the surface normal on evaporation front; R is laser beam radius; h is convective heat transfer coefficient; $T\infty$ is the ambient temperature; k is thermal conductivity; ρ is density; and L_s is the heat of sublimation. Equation (5.4) can be modified for the three different regions as follows (Modest and Abakians 1986):

Region I: In this region,

$$\hat{i} \bullet \hat{n} = 0 \quad \text{and} \quad \hat{n} = \hat{k}$$

such that

$$\alpha J_0 e^{-(x^2 + y^2)/R^2} = h(T - T_\infty) - k \frac{\partial T}{\partial z}, \quad x < x_{\min}(y) \qquad (5.4a)$$

Region II: In this evaporation region which is close to the center of the laser beam,

$$\hat{i} \bullet \hat{n} < 0$$

such that Eq. (5.4) becomes

$$\alpha J_0 e^{-(x^2 + y^2)/R^2} = (h(T - T_\infty) - k\,\hat{n} \bullet \nabla T)\left(1 + \left(\frac{\partial s}{\partial x}\right)^2 + \left(\frac{\partial s}{\partial y}\right)^2\right)^{1/2} + \rho L_s v \frac{\partial s}{\partial x};$$

$$z = s(x,y), \quad x_{\min}(y) < x < x_{\max}(y) \qquad (5.4b)$$

Additional boundary condition for evaporation region is

$$T = T_{ev}, \quad z = s(x,y), \quad x_{min}(y) < x < x_{min}(y) \tag{5.5}$$

where $s(x, y)$ is the local depth of groove and T_{ev} is the evaporation temperature.

Region III: This is the region where the established groove is formed and the region is moved too far from the evaporation front such that the temperature is below the evaporation temperature T_{ev}. In this region

$$\hat{i} \bullet \hat{n} = 0 \quad \text{and} \quad \hat{n} \bullet \hat{k} \neq 0$$

such that

$$\alpha J_0 e^{-(x^2 + y^2)/R^2} = (h(T - T_\infty) - k\hat{n} \bullet \nabla T)\left(1 + \left(\frac{\partial s_\infty}{\partial y}\right)^2\right)^{1/2} \, ;$$

$$z = s_\infty(y), \quad x > x_{max}(y) \tag{5.4c}$$

where $s_\infty(y)$ is the maximum established depth of the groove.

The following nondimensional parameters can be introduced to arrive at the solution of the above heat transfer equations:

$$\xi = x/R, \quad \eta = y/R, \quad \zeta = z/R;$$
$$S = s(x, y)/R, \quad \Theta = (T - T_\infty)/(T_{ev} - T_\infty);$$
$$N_e = \frac{\rho v L_s}{\alpha J_0}, \quad N_k = \frac{k(T_{ev} - T_\infty)}{R\alpha J_0}, \quad Bi = \frac{hR}{k}, \quad U = \frac{\rho C_p vR}{k} \tag{5.6}$$

Introducing the above Eqs in (5.2)–(5.5) gives

$$U\frac{\partial \Theta}{\partial \xi} = \frac{\partial^2 \Theta}{\partial \xi^2} + \frac{\partial^2 \Theta}{\partial \eta^2} + \frac{\partial^2 \Theta}{\partial \zeta^2}; \quad -\infty < \xi, \eta < +\infty, S \leq \zeta < +\infty \tag{5.7}$$

subject to

$$\xi \rightarrow \pm\infty, \quad \eta \rightarrow \pm\infty, \quad \zeta \rightarrow +\infty : \quad \Theta \rightarrow 0 \tag{5.8}$$

Region I:

$$\zeta = 0: \quad e^{-(\zeta^2 + \eta^2)} - N_k\left(Bi\Theta - \frac{\partial \Theta}{\partial \zeta}\right) = 0;$$

$$-\infty < \xi < \xi_{min}(\eta), \quad -\infty < \eta < +\infty \tag{5.9a}$$

Region II:

$$\zeta = S_\infty(\eta): \quad N_e\frac{\partial S}{\partial \xi} = e^{-(\xi^2 + \eta^2)} - N_k\left(Bi - \frac{\partial \Theta}{\partial n}\right)\left(1 + \left(\frac{\partial S}{\partial \xi}\right)^2 + \left(\frac{\partial S}{\partial \eta}\right)^2\right)^{1/2} \tag{5.9b}$$

$$\Theta = 1; \quad \xi_{min}(\eta) \le \xi \le \xi_{max}(\eta), \quad -\eta_{max} \le \eta \le +\eta_{max} \tag{5.10}$$

Region III:

$$\zeta = S_\infty(\eta): \quad e^{-(\xi^2+\eta^2)} - N_k \left(Bi\Theta - \frac{\partial\Theta}{\partial n} \right) \left(1 + \left(\frac{\partial S}{\partial \eta} \right)^2 \right)^{1/2}$$

$$\xi_{min}(\eta) < \xi < +\infty, \quad -\eta_{max} \le \eta \le \eta_{max} \tag{5.9c}$$

In order to solve Eq. (5.9b), the following assumption can be made: $\nabla^2\Theta \approx \partial^2\Theta/\partial n^2$, where n is distance along the vector $\hat{n}$.

Temperature profile normal to the surface can be approximated by a quadratic polynomial: $\Theta \approx \Theta_0[1 - n/\delta]^2$, where Θ_0 is the nondimensional surface temperature and δ is a penetration depth.

With these simplifications, the equations for three regions can be rewritten as:

Region I:

$$\frac{\partial}{\partial\xi}(\Theta_0\delta) = \frac{6\Theta_0}{U\delta} \tag{5.11}$$

$$N_k\Theta_0 \left(Bi + \frac{2}{\delta} \right) = e^{-(\xi^2+\eta^2)} \tag{5.12}$$

Region II:

$$\frac{\partial\delta}{\partial\xi} = \frac{6}{U\delta} - \frac{3\partial S}{\partial\xi} \bigg/ \left(1 + \left(\frac{\partial S}{\partial\xi} \right)^2 + \left(\frac{\partial S}{\partial\eta} \right)^2 \right)^{1/2} \tag{5.13}$$

$$N_e\frac{\partial S}{\partial\xi} = e^{-(\xi^2+\eta^2)} - N_k \left(Bi + \frac{2}{\delta} \right) \left(1 + \left(\frac{\partial S}{\partial\xi} \right)^2 + \left(\frac{\partial S}{\partial\eta} \right)^2 \right)^{1/2} \tag{5.14}$$

Region III:

$$\frac{\partial}{\partial\xi}(\Theta_0\delta) = \frac{6\Theta_0}{U\delta} \tag{5.15}$$

$$N_k\Theta_0 \left(Bi + \frac{2}{\delta} \right) \left(1 + \left(\frac{\partial S}{\partial\eta} \right)^2 \right)^{1/2} = e^{-(\xi^2+\eta^2)} \tag{5.16}$$

In order to determine the groove depth and shape given in region II, the boundary between regions I and II must be found by solving Eqs (5.11) and (5.12) for region I. The solution for the region I is given by:

$$\frac{d\phi}{d\xi} = \frac{\frac{3Bi^2}{2U}\left(e^{-\xi^2} - \phi\right)^3 - 2\xi e^{-\xi^2}\phi^2}{\phi\left(2e^{-\xi^2} - \phi\right)}, \tag{5.17}$$

where $\phi = N_k Bi\Theta_0 e^{\eta^2}$.

Equation (5.17) can be readily solved by the standard Runge-Kutta method. To determine $\xi_{min}(\eta)$, Eq. (5.17) can be inverted and solved for ξ as a function of φ up to $\varphi = N_k Bie^{-\eta^2}$, i.e., to the point where $\Theta_0 = 1$. Also, the local $\eta_{max}(\xi)$ can be found by integrating to the given ξ, such that $\eta_{max}(\xi) = (\ln[\varphi(\xi)/N_k Bi])^{1/2}$. For some $\xi > 0$, $d\varphi/d\xi = 0$, indicating the maximum groove width has been reached. The results of the evaporative laser cutting are given in Figs 5.4 and 5.5. Figure 5.4 shows that, in the case of low conduction losses (large values of N_k) the evaporating region is bounded by a semicircle for negative values of ξ. The value of $\eta(\xi)$ increases gradually with ξ and reaches its maximum at $\xi = 0$. As the conduction losses increase (N_k decreases) the evaporating region may be approximated by an elliptic shape. Typical cross section of the groove developed along ξ direction during evaporative laser cutting is shown in Fig. 5.5, which indicates that the maximum groove width reaches around $\xi = 0.3$ given by η_{max}. Beyond this ξ, no additional evaporation occurs because the part of the groove has moved far away from the center of laser beam such that the laser energy is lost by conduction and convection (Modest and Abakians 1986).

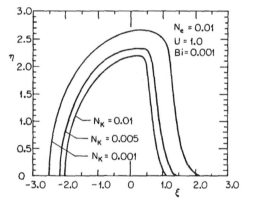

Fig. 5.4 Top view of evaporation front for different conduction-loss levels. (Reprinted from Modest and Abakians 1986. With permission. Copyright American Society of Mechanical Engineers.)

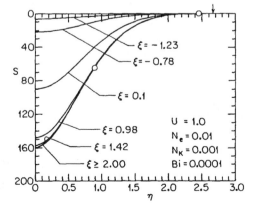

Fig. 5.5 Cross sections of the grooves formed during evaporative laser cutting at different locations in the evaporation region. (Reprinted from Modest and Abakians 1986. With permission. Copyright American Society of Mechanical Engineers.)

The above analysis was further extended by considering the effects of multiple reflection and beam guiding for highly reflective materials. The multiple reflection results in enhanced material removal rates and deeper grooves. The grooves obtained in evaporative laser cutting with multiple reflection effects have a flatter profile at the centerline and steeper slope in other parts of the grove cross section (Biyikli and Modest 1988, Bang and Modest 1991).

Recently, a computational model for the simulation of a transient laser-cutting process using time-dependant boundary element method (BEM) has been developed. Boundary element method is a numerical method that utilizes the picking-out property of the fundamental singular solution of a differential equation. In BEM, the solution of the domain boundary problems is expressed in terms of the boundary integrals of the fundamental solution and its derivatives. This feature of BEM makes typical discretization and formulation simpler than other methods in which the domain is discretized (Kim and Majumdar 1996, Kim 2000). Numerical analysis was carried out on the amount of material removal and groove smoothness with laser power and number of pulses and it is shown that there exist threshold values in a number of pulses and laser power in order to achieve predetermined amount of material removal and smoothness of groove shape (Kim et al. 1993; Kim and Majumdar 1995; Majumdar et al. 1995; Kim and Zhang 2001).

5.2.2   Laser Fusion Cutting

In fusion cutting, a laser beam moves relative to workpiece and follows a straight or curved path (straight cut or profile cut). Fresnel absorption of energy from the impinging high-intensity beam melts the metal throughout the thickness (through cut) of material thus creating a cutting front. In case of blind cuts the height of cutting front is smaller than the thickness of workpiece. The cutting front is in the form of a thin film of molten material. A conical nozzle with assist gas is usually employed coaxial with the laser beam. In fusion cutting, only nonreactive gases such as argon or nitrogen are used as against the oxygen used in reactive laser cutting. The process gas is at high pressure and transforms momentum to the melt film. If the momentum of the film exceeds the surface tension forces, melt is vertically accelerated and ejected from the bottom of the kerf in the form of droplets, otherwise dross attachment occurs. Since the primary phase transformation is melting, the energy requirements for fusion cutting are lower compared to evaporative laser cutting.

The modeling procedures for the fusion cutting are complex and various phenomena such as heat transfer, fluid flow, and gas dynamics needs to be considered. Extensive literature is available on the modeling efforts of fusion cutting. The most general procedure is to proceed with the solutions of mass, momentum, and energy balance equations in a controlled volume enclosed by cutting front and melting front (Kaplan 1996). Figure 5.6 shows cutting model geometry consisting of a control volume enclosed by a cutting front (gas/liquid interface) and melting front (solid/liquid interface).

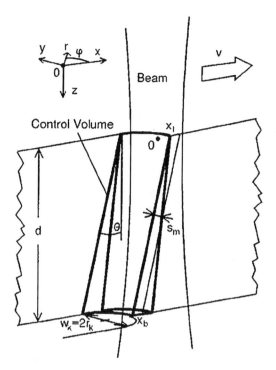

Fig. 5.6 Model geometry for laser cutting showing a control volume bounded by two inclined semicircular cylinders. (Reprinted from Kaplan 1996. With permission. Copyright American Institute of Physics.)

The heat flow at the top surface of the workpiece is balanced by absorbed beam intensity (Fresnel absorption):

$$q_{mcs} = \alpha \tan(\theta) I , \tag{5.18}$$

where θ is the angle of inclination of the front. The heat flow is given by the heat flux q_{mcs} of a cylindrical moving source of heat. Thus, the heat flux at the front ($\varphi = 0$) and side of the cylinder ($\varphi = \pi/2$) can be given as (Kaplan 1996):

$$q_{mcs}(r_k, v, \varphi = 0) = \alpha(2\theta_{max}) \tan(2\theta_{max}) \times I(x = x_t, y = 0, z = 0), \tag{5.19}$$

$$q_{mcs}(r_k, v, \varphi = \pi/2) = \alpha \tan(\theta_{max}) \times I(x = x_t - r_k, y = r_k, z = 0), \tag{5.20}$$

where $q_{mcs}(r_k, v, \varphi)$ is the local heat flow of a cylinder of diameter $w_k = 2r_k$; v is the processing speed; $\alpha(\theta)$ is the angle-dependent Fresnel absorption; θ is the angle of incidence; $I(x, y, z)$ is the spatial distribution of laser energy described by Gaussian beam; $\hat{\alpha}$ is the maximum Fresnel absorption at angle θ_{max}; and x_t is the horizontal location of the cutting front at the top surface. The horizontal location of the cutting front at the bottom surface of the workpiece x_b is given as (Kaplan 1996):

$$x_b = x_t - d\tan(\theta),\tag{5.21}$$

where d is thickness of the workpiece. Absorption front of the control volume is thus described by the parameters x_b, x_t, w_k, and θ, which along with the mean melt film thickness (s_m) describe the whole control volume.

5.2.2.1 Mass Balance

Mass balance equation is given by setting the rate of molten material, N_m (m^3/s), equal to the rate of ejected material, N_e(m^3/s). For fusion cutting, the evaporative losses are generally neglected. In addition, neglecting the dross attachment at the bottom of the front, the corresponding mass balance equation is (Kaplan 1996):

$$N_m = N_e.\tag{5.22}$$

The equation can be written in terms of workpiece thickness (d), translation velocity (v), melt film velocity (v_m), melt film thickness (s_m), and kerf width (w_k):

$$vdw_k = v_m s_m w_k,\tag{5.23}$$

such that

$$s_m = \frac{vd}{v_m}.\tag{5.24}$$

If the fusion cutting is accompanied with the oxidizing assist gases (Section 5.2.3), the ejected material consists of oxides in addition to the bulk metal. For mild steel, the preferred oxidation reaction is the formation of FeO. The melt film thickness is thus given by the addition of thickness of molten bulk metal (s_{Fe}) and thickness of oxide layer (s_{FeO}):

$$s_m = s_{Fe} + s_{FeO}.\tag{5.25}$$

From the laws of diffusion, s_{FeO} is given by:

$$s_{FeO} = \sqrt{2D_{eff}t_{eff}} \le s_m.\tag{5.26}$$

The effective reaction time (t_{eff}), is given by the mean duration taken by an atom from melting to ejection when averaged over the depth.

$$t_{eff} = \frac{d}{v_m}.\tag{5.27}$$

The fraction of oxide formed during oxidation can be given by s_{FeO}/s_m and the absolute reaction rate (mol/s) is:

$$N_{e,FeO} = v_m w_k s_{FeO}\frac{\rho}{A_{r,Fe}},\tag{5.28}$$

where $A_{r,Fe}$ is the relative atomic mass of iron.

5.2.2.2 Momentum Balance

Momentum equation is given by balancing the effective forces of the gas flow on the left-hand side and the forces acting on the melt on the right-hand side of the cutting front (Kaplan 1996):

$$F_0 + F_n + F_t = F_a + F_{st} + F_d + F_m,$$ (5.29)

where F_0 is the force due to static pressure; and F_n and F_t are the normal and tangential components of dynamic gas force, respectively, acting on left-hand side. In addition, F_a is the force due to ambient pressure; F_{st} is the surface tension force; F_d is the dynamic force; and F_m is the force describing frictional losses in the melt. When the supersonic gas jet reaches the surface of workpiece, a fraction of it flows radially outward along the workpiece and the remaining gas actually enters into the kerf. Moreover, a fraction of gas which enters into the kerf oxidizes the molten surface layer of metal and the remaining gas passes through the kerf after transferring some of its momentum to the melt film. The geometrical areas of these two flows are given by:

$$A_r = d_n \pi z_n,$$ (5.30)

$$A_k = w_k \left(\frac{d_n}{2} + x_t \right),$$ (5.31)

where A_r and A_k are the cross section of the flow expanding radially outward and effective area of the flow entering the kerf, respectively; d_n is the nozzle diameter; and z_n is the nozzle–workpiece standoff distance.

At the kerf entrance, the various parameters can be expressed as:

$$p_g \equiv p * \frac{A_k}{A_r + A_k},$$ (5.32)

$$\rho_g \equiv \rho*,$$ (5.33)

$$v_g = c*,$$ (5.34)

$$T_g = T*,$$ (5.35)

where p is pressure; v is velocity; and T is temperature of the processing gas. Indices g and $*$ denotes the state of gas at the entrance of the kerf and at the nozzle, respectively. Various forces can be expressed as:

$$F_0 = dw_k \frac{\pi}{2} p_g,$$ (5.36)

$$F_n = dw_k \frac{\pi}{2} p_g v_g^2 \tan(\theta),$$ (5.37)

$$F_t = dw_k \frac{\pi}{2} \sqrt{\rho_g \mu_g} \int_0^d \left(\frac{v_g}{z} \right)^{3/2} dz = \sqrt{d} w_k \frac{\pi}{2} \sqrt{\rho_g \mu_g} 2 v_g^{3/2},$$ (5.38)

$$F_a = s_m w_k p_a, \tag{5.39}$$

$$F_{st} = dw_k \frac{\pi}{2} \frac{2\sigma}{w_k}, \tag{5.40}$$

$$F_d = dw_k \frac{\pi}{2} \rho \frac{v_m^2}{3} \frac{s_m}{d}, \tag{5.41}$$

$$F_m = dw_k \frac{\pi}{2} \mu \frac{v_m}{s_m}, \tag{5.42}$$

where ρ_g and μ_g are the density and dynamic viscosity of the gas; p_a is the ambient pressure; and σ is the surface tension.

5.2.2.3 Energy Balance

Energy balance is given by setting the energy gain equal to the energy losses. The terms corresponding to the energy gains are the absorbed laser power (P_a) and the power generated due to exothermic reaction (P_r), whereas the terms corresponding to the energy losses are powers due to conduction heat losses (P_s), heat of melting (P_m), and heating of liquid material (P_l). The corresponding equations can be given as (Kaplan 1996):

$$P_a = \hat{\alpha} \int_0^{x_t} I(r, z = 0) r\pi dr + \hat{\alpha} \int_{x_b}^{0} I(r, z = d) r\pi dr, \tag{5.43}$$

$$P_r = N_{e,FeO} H_{FeO}, \tag{5.44}$$

$$P_s = P'_{mess}(r_k, v) d, \tag{5.45}$$

$$P_m = \rho v d w_k H_m, \tag{5.46}$$

$$P_l = \rho v d w_k c_p \frac{(T_l - T_m)}{2}, \tag{5.47}$$

where c_p is the heat capacity of metal; ρ is the density of metal; H_{FeO} is the heat of the oxidation reaction; H_m is the latent heat of phase transformation; T_m is the melting point; and T_l is the surface temperature of liquid.

Fourier's second law can be used to balance the heat flux at the front of the moving cylinder $q_{mcs}(\varphi = 0)$ with the temperature gradient in the melt film:

$$q_{mcs}(r_k, v, \varphi = 0) \equiv -k\nabla T = k \frac{T_l - T_m}{s_m}, \tag{5.48}$$

which gives an explicit relation between surface temperature (T_l) and melt film thickness (S_m):

$$T_l = T_m + \frac{s_m}{k} q_{mcs}(r_k, v, \varphi = 0), \tag{5.49}$$

where k is the heat conductivity of the material.

The results of the numerical solutions of the above model for fusion cutting of mild steel of various thicknesses (1–3 mm) with 1,500 W CO_2 laser are presented in Figs 5.7 and 5.8. Figure 5.7 shows that the kerf width decreases with increasing velocity because

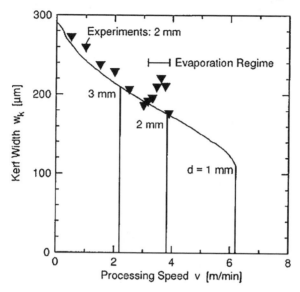

Fig. 5.7 Variation of kerf width with processing (cutting) speed during laser fusion cutting of 1, 2, and 3 mm thick mild steel. Experimental data for 2 mm thick mild steel is overlapped showing good agreement with the calculated values. (Reprinted from Kaplan 1996. With permission. Copyright American Institute of Physics.)

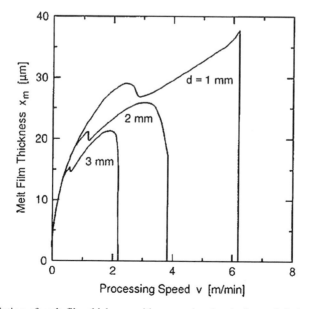

Fig. 5.8 Variation of melt film thickness with processing (cutting) speed during laser fusion cutting of 1, 2, and 3 mm thick mild steel. (Reprinted from Kaplan 1996. With permission. Copyright American Institute of Physics.)

heat flow increases the threshold values which are responsible for ignition of the front. The experimental data for the fusion cutting of 2 mm steel sheet matches closely with the calculated data except in the maximum velocity region. This small discrepancy near the maximum velocity region is due to the evaporation losses which are neglected in the present calculations. In addition, Fig. 5.8 indicates that the melt film thickness increases for higher speeds due to higher melting rate. The model also gives the results for the temperature of the cutting front at various velocities (Kaplan 1996).

5.2.3 Reactive or Oxygen-Assisted Laser Cutting

Reactive laser cutting is a variation of fusion cutting in which reactive gas is used instead of inert gas. In reactive fusion cutting, the material is heated to the point where an exothermic reaction with an oxidizing coaxial gas jet is triggered thus adding another source of heat to the process. Generally, the temperature at which the oxidation becomes dominant is much less than the evaporation temperature. Due to combined effects of the absorbed laser radiation and the exothermic reaction, the molten layer at the cutting front reaches the evaporation temperature thus facilitating the material removal by evaporation from the surface. In addition, the frictional forces between the impinging gas jet and the molten film at the cutting front causes the molten layer (oxide and the melt) to be ejected from the bottom surface of the workpiece (Schuöcker and Müller 1987). The amount of material lost by the material is then compensated by the further melting of the workpiece at the solid/liquid boundary below the cutting front, thus accomplishing the cutting by movement of solid/liquid interface into the uncut region of the workpiece. The method is mainly used for thick section cutting of stainless steels, titanium, and aluminum alloys.

The role of oxygen in the laser cutting of mild and stainless steel have been extensively investigated. During oxygen-assisted laser cutting, the dominant mechanism of material removal is the oxidation of materials, whereas the evaporation of the material is negligible. This was confirmed by the direct optical observations of the oxygen-assisted laser-cutting zone which showed the temperature in the pale yellow band (~2,000 K) as against the blue–violet band (temperature higher than 2,000 K) observed in evaporative laser cutting. In general, the exothermic reactions for mild and stainless steels during the oxygen-assisted laser cutting can be given as:
Mild steel:

$$Fe + O \rightarrow FeO + heat\ (257.58\ kJ/mol)$$

$$2Fe + 1.5O_2 \rightarrow Fe_2O_3 + heat\ (826.72\ kJ/mol)$$

Stainless steel:

$$2Fe + 1.5O_2 \rightarrow Fe_2O_3 + heat\ (826.72\ kJ/mol)$$

$$2Cr + 1.5O_2 \rightarrow Cr_2O_3 + heat\ (1,163.67\ kJ/mol)$$

$$Ni + 0.5O_2 \rightarrow NiO + (248.23 kJ/mol)$$

The extent of oxidation during the reactive laser cutting can be evaluated by the detailed chemical analysis of the material ejected from the bottom of cut zone. Typical average values of amount of unoxidized iron, oxidized iron, and oxygen present in the molten droplets ejected from the cut zone during laser cutting of mild and stainless steel with 900 W CO_2 laser are given in Table 5.2 (Ivarson et al. 1991). A typical cross section of an ejected droplet is shown in Fig. 5.9. The ejected droplets are made of a mixture of iron and oxidized iron (Powell 1998).

The role of oxygen during laser cutting can be quantified in terms of the oxidation energy generated which in combination with the irradiated laser energy

Table 5.2 Average compositions of the ejected droplets during oxygen assisted laser cutting of 1–4 mm thick mild and stainless steel using 900 W CO_2 laser (Reprinted from Ivarson et al. 1991. With permission. Copyright Laser Institute of America.)

Material	Free Fe (wt. %)	Oxidized Fe (wt. %)	Free Ni (wt. %)	Oxidized Ni (wt. %)	Total Cr (wt. %)	Total O (wt. %)
Mild steel	43.1	44.7	–	–	–	13.1
Stainless steel	43.0	19.5	10.9	0.67	15.2	10.7

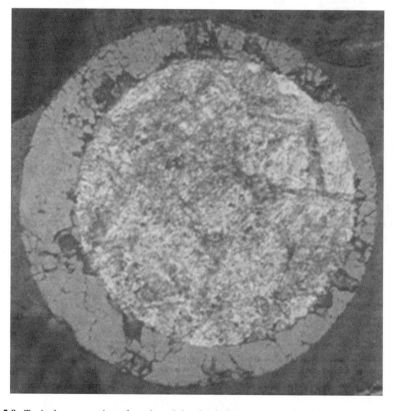

Fig. 5.9 Typical cross section of an ejected droplet during oxygen-assisted laser cutting showing a layer of oxidized metal. (Reprinted from Powell 1998. With permission. Copyright Springer.)

Table 5.3 Oxidation heat generated during reactive laser cutting of mild and stainless steel (Reprinted from Ivarson et al. 1991. With permission. Copyright Laser Institute of America.)

Materials	Thickness (mm)	Kerf width (mm)	Material removal rate (mm³/s)	Oxidation heat evolved (W)
Mild steel	1	0.23	28.7	515.2
	2	0.21	35.7	643.9
	3	0.22	35.0	618.2
Stainless steel	1	0.27	35.9	1156
	2	0.29	37.7	1215
	3	0.32	40.3	1297

determines the total energy input to the laser-cutting process. Table 5.3 gives the details of oxidation heat generated during the laser cutting of mild and stainless steel with 900 W CO_2 laser for various workpiece thicknesses. As seen from the table, the average heat of oxidation for mild steel is around 579 W, which in combination with 900 W of laser energy, determines the 40% of the total energy input to the cut zone. Moreover, for stainless steel, the average oxidation heat evolved is 1,220 W which is significantly greater than the input laser energy (Ivarson et al. 1991). In view of the enormous conductive thermal losses to the base material during laser cutting, the role of oxidation in laser cutting is significant. Higher cutting speeds are achieved in the oxygen-assisted cutting due to release of a large amount of exothermic heat, higher absorptivities of hot oxide films, and the high fluidity of oxide slag on the cutting front. However, it is important to deliver the sufficient oxygen at high pressure to sustain the exothermic reaction during laser cutting (O'Neill and Gabzdyl 2000).

Early attempts for the modeling of reactive gas cutting are done by Schuöcker. The various physical processes during reactive laser cutting are shown in Fig. 5.10 (Schuöcker 1986). When the laser beam interacts with the workpiece, a nearly vertical cutting front (or erosion front) is created. The erosion front is covered with a thin layer of molten liquid and the material removal takes place at the erosion front. The material removal takes place by evaporation from the erosion front and also by the ejection of molten liquid from the bottom surface of the workpiece due to frictional forces between the impinging gas and the molten film of material. The heat balance at the molten layer is given by setting the heat gain by the molten film by laser radiation and oxidation reaction equal to the heat losses by conduction, melting, vaporization, and ejection of liquid material by from the bottom surface of the workpiece (Schuöcker 1986).

The mass balance equation is given by equating the mass gained by the reactive gas particles impinging on the molten layer with the losses by reaction between gas and metal particles with rate k_R, losses by ejection of liquid material from the bottom surface with velocity v_s, and losses by evaporation described by temperature-dependent quantity γ_R. The resulting equation is (Schuöcker 1986):

$$wdq_R = swdk_R n_R n_A + swn_R v_s + wd\gamma_R n_R , \qquad (5.50)$$

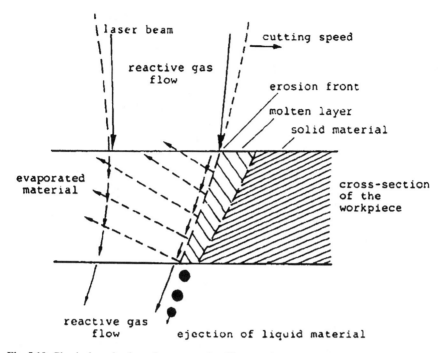

Fig. 5.10 Physical mechanism of oxygen assisted laser cutting. (Reprinted from Schuöcker 1986. With permission. Copyright The International Society for Optical Engineering.)

where w is kerf width; d is thickness of workpiece; S is thickness of molten layer; and n_A and n_R are the densities of pure metal particle in the molten layer and reactive gas, respectively. The quantity q_R equals βq_0, where β is fraction of reactive gas particles absorbed by the erosion front and q_0 is the density of reactive gas particles impinging on the erosion front. The equation for mass balance for pure metal particles is similar to Eq. (5.50) and is obtained by equating the mass gain by melting of solid material with the mass losses by reaction with rate k_R, ejection of liquid material from the bottom surface with velocity v_s, and losses by evaporation described by temperature-dependent quantity γ_A. Thus,

$$wdvn_0 = swdk_R n_R n_A + swn_A v_s + wd\gamma_A n_A , \qquad (5.51)$$

where v is the cutting speed and n_0 is atomic density of solid material.

Equation (5.51) can be solved to calculate the number of reaction events per unit time in the molten layer and the heat produced by reaction in the molten layer (Schuöcker 1986).

$$\left(\frac{dE}{dt}\right)_R = \varepsilon_R wdvn_0 \frac{k_R n_R}{k_R n_R + \dfrac{v_s}{d} + \dfrac{\gamma_A}{s}} , \qquad (5.52)$$

where ε_R is the reaction energy per particle. Since, a large fraction of reactive gas particles impinging on the surface of liquid layer gets reflected or diffused; it carries an average energy ε_{th} away from the melt. Thus the net energy gain by the molten layer due to reactive gas flow can be given as (Schuöcker 1986):

$$\left(\frac{dE}{dt}\right)_{netR} = \frac{\varepsilon_R wdvn_0 q_R}{q_R + \dfrac{s}{k_R} + \left(\dfrac{v_s}{d} + \dfrac{\gamma_A}{s}\right)\left(\dfrac{v_s}{d} + \dfrac{\gamma_R}{s}\right)} - \varepsilon_{th}\left(\frac{1-\beta}{\beta}\right)q_R. \tag{5.53}$$

The net energy gain due to reactive gas flow is thus composed of energy gain due to reaction and energy lost by convective cooling. The cooling effects of the reactive gas increase with increasing the strength of the gas flow. The energy gain given by Eq. (5.53) can be maximized if all the metal atoms entering the molten region are burned. Thus, the maximum energy gain is given by the number of metal atoms entering the melt per unit time and by the reaction energy:

$$\left(\frac{dE}{dt}\right)_{Rmax} = \varepsilon_R wdvn_0 . \tag{5.54}$$

The momentum balance at the molten layer in the vertical direction between the momentum gain by friction with gas flow and momentum loss by the ejection of melt from bottom gives the expression of velocity of ejected molten material:

$$v_s = \sqrt{\frac{\eta_R}{\rho_s}\frac{d}{sw}}v_R , \tag{5.55}$$

where η_R is the dynamic viscosity of the reactive gas and ρ_s is the density of melt. Assuming a rotational symmetry at the erosion front, the solution of heat conduction equation for a moving source gives the expression for energy lost from the erosion front into the workpiece by conduction:

$$\left(\frac{dE}{dt}\right)_K = 2\pi dkT \frac{\exp(-vw/4K)}{K_0(vw/4K)}, \tag{5.56}$$

where k is thermal conductivity; K is thermal diffusivity; and K_0 is Bessel function. The energy balance equation with the energy gains by the laser radiation and reaction along with energy losses due to heat conduction, evaporation can be written as:

$$P_L(1-e^{-\alpha_L d}) + \varepsilon_R swdk_R n_R n_A = 2\pi dkT \frac{\exp(-vw/4K)}{K_0(vw/4K)}$$
$$+ \varepsilon_v dw\left(\frac{133.3}{\sqrt{2\pi k_B Tm_s}}10^B T^C 10^{A/T}\right) \tag{5.57}$$

where P_L is the laser power; α_L is the waveguide attenuation of the cut; ε_v is the evaporation energy per atom; m_s is atomic mass; k_B is Boltzmann constant; and A, B, and C are the evaporation constants.

Combining the above equations, the mass balance equation can be written as (Schuöcker 1986):

$$\frac{wv}{4\kappa} = \sqrt{\frac{\eta_R}{2\rho_s}} \frac{w}{4\kappa} \sqrt{\frac{v_R}{d}} \sqrt{1 - \frac{T_s}{T}} \frac{1}{\sqrt{vw/4\kappa}} \sqrt{\frac{K_0(vw/4\kappa)}{K_1(vw/4\kappa)}}$$
$$+ \frac{1}{n_0}\left(\frac{133.3}{\sqrt{2\pi k_B T m_s}} 10^B T^C 10^{A/T} \right) \tag{5.58}$$

The above two equations correspond to the equilibrium energy and mass balance and establish the interrelationships between temperature of the molten layer, cutting speed, thickness of the workpiece, kerf width, and absorbed laser power. The mass and energy balance relationships are presented in Fig. 5.11 for different thicknesses of molten layer using temperature of the molten layer and cutting speed are as axes. The limiting conditions are $S = 0$ and $S = S_{max}$. When, $S = 0$, i.e., when there is no molten layer, the cutting is mainly due to evaporation of material resulting in correspondingly higher temperature given by mass balance (evaporative or sublimation laser cutting). The intersection of the mass balance with the energy balance at $S = 0$ and $S = S_{max}$ gives the minimum and maximum cutting speeds that can be obtained for a given laser power and thickness of the workpiece. Consequently, the thickness of molten layer is given by (Schuöcker 1986):

$$s = \frac{2\kappa}{v}\left(1 - \frac{T_s}{T} \right) \frac{K_0(vw/4\kappa)}{K_1(vw/4\kappa)}. \tag{5.59}$$

The results of numerical calculations of the total cutting speed and thickness of the workpiece during laser cutting of mild steel with two different values of waveguide attenuation are presented in Fig. 5.12. The figure indicates that cutting speed decreases with the increasing thickness of the workpiece. In addition, the figure clearly shows that the contribution of evaporative material removal is predominant at the moderate thicknesses of the workpiece supporting the experimental observation of the better cut quality at smaller thicknesses (Schuöcker 1986).

Extensive modeling efforts have been done for the evaporative, fusion, and reactive laser-cutting processes. Table 5.4 gives a comprehensive summary of the various laser-cutting models (O'Neill and Steen 1994).

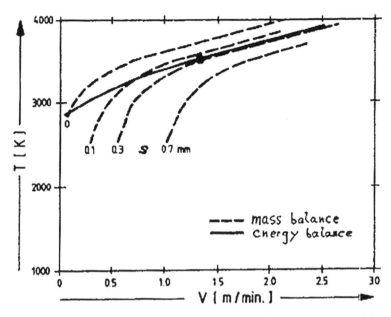

Fig. 5.11 Energy and mass balance relationships plotted between the temperature of the molten layer and the cutting speed during the reactive fusion cutting of mild steel for various thicknesses. (Reprinted from Schuöcker 1986. With permission. Copyright The International Society for Optical Engineering.)

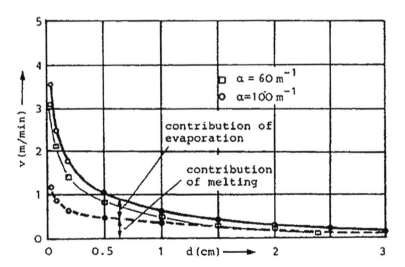

Fig. 5.12 Total cutting speed verses thickness of the workpiece showing the contribution of evaporation and melting during fusion cutting of mild steel. (Reprinted from Schuöcker 1986. With permission. Copyright The International Society for Optical Engineering.)

Table 5.4 Summary of modeling of laser-cutting process (O'Neill and Steen 1994)

Cutting process	Geometry	Analysis	Results	Reference
Evaporative laser	Semi-infinite body, moving	Partial differential equations and numerical solution	Groove shape	Modest and Abakians 1986
Cutting	Gaussian beam	Partial differential equations and numerical solution	Groove depth and shape	Abakians and Modest 1988
	Semi-infinite and Semitransparent body, moving	3D conduction equations, Boundary element method	Groove depth and shape	Roy and Modest 1990
	Gaussian beam	Quasi-1D conduction equations, multiple reflections and beam-guiding effects	Materials removal rates, groove depth and shape	Bang and Modest 1991
	Semi-infinite body, moving	2D steady-state conduction, Boundary element method	Groove shape and surface temperature	Kim and Mazumdar 1996
	Gaussian beam	2D unsteady state conduction, Finite element method	Materials removed and groove shape	Kim and Zhang 2001
	Semi-infinite body, moving Gaussian beam Semi-infinite body, moving Gaussian beam Finite body, Gaussian beam			
Laser fusion cutting	3D plane geometry, moving line source and diffuse source	Heat conduction equations	Cutting speed, kerf width, power density/unit thickness	Bunting and Cornfield 1975
	3D plane geometry	Energy balance equations	Cutting depth, speed, absorption efficiency	Schulz et al. 1987
	2D plane geometry	Melt ejection, stability, boundary layer equations	Molten layer thickness, flow velocity, stability	Vicanek et al. 1987

(continued)

Table 5.4 (continued)

Cutting process	Geometry	Analysis	Results	Reference
	2D plane geometry	Melt ejection, boundary layer equations	Pressure and shear stress distribution	Vicanek and Simon 1987
	3D plane geometry	Fresnel absorption, energy balance	Nomogram for fusion cutting, front contours	Petring et al. 1988
	3D geometry	Mass, momentum and energy balance	Temperature, Melt film thickness and velocity	Kaplan 1996
Reactive laser cutting	3D plane geometry	2D heat conduction equations, numerical	Nomogram, cutting speed	Babenko and Tychinsckii 1973
	Moving sheet, fixed point source	2D heat conduction equations	Cut width, fraction of incident energy used for melting	Gonsalves and Duley 1972
	Simple geometry, moving point source	Differential mass, energy, and energy balance	Surface temperature, film thickness, cutting speed	Schuöcker 1983
	Simple geometry, moving point source	Differential mass, energy, and energy balance	Cutting front oscillation frequencies	Schuöcker 1986

5.2.4 Controlled Fracture Technique

In controlled fracture technique, incident laser energy produces mechanical stresses in a localized area of the workpiece which cause the material to separate controllably along the path of the laser beam. The material separation takes place by the crack extension with controllable fracture growth. Since the material separation is due to propagation of crack and not due to evaporation and/or melting of the substrate throughout the thickness, the laser energy required in the controlled fracture technique is less than conventional evaporative laser cutting or laser scribing. Moreover, fast crack propagation characteristics of brittle materials give the cutting speeds in controlled fracture technique much higher than the conventional laser-cutting techniques. This method has been successfully applied to the brittle ceramic materials such as alumina and glass for simple and straight cuts with high cutting speeds. The cut quality is good because the cut edges are free of chipping and microcracks hence necessitating no further cleaning or finish grinding of the edges. The method of controlled fracture using lasers was first proposed by Lumley and successfully demonstrated for dicing brittle materials such as alumina ceramic substrate and glass using single CO_2 laser (Lumley 1969).

Most comprehensive work was carried out by Tsai and coworkers to understand the detailed mechanism of controlled fracture (Tsai and Liou 2001, 2003; Tsai and Chen 2003a, b, 2004). The experimental setup generally used for laser cutting using controlled fracture technique is shown in Fig. 5.13. It primarily consists of a Nd: YAG laser, a CO_2 laser, a XYZ positioning table, and a personal computer. The focused Nd:YAG laser with a focal plane on the surface of the substrate and the beam perpendicular to the surface is used to scribe a groove-crack on the

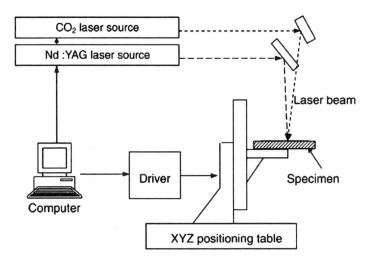

Fig. 5.13 Experimental setup for laser cutting using controlled fracture technique. (Reprinted from Ref. Tsai and Chen 2003b. With permission. Copyright Elsevier.)

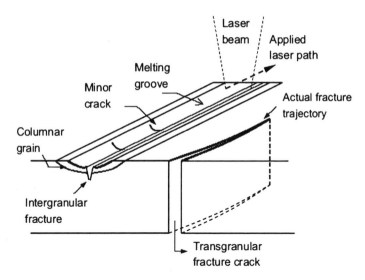

Fig. 5.14 Mechanism of laser cutting by controlled fracture for asymmetrical cutting. (Reprinted from Tsai and Chen 2004. With permission. Copyright Elsevier.)

surface of the ceramic substrate; whereas the defocused CO_2 laser, which is inclined to the Nd:YAG laser beam, induce thermal stresses in a localized area of the substrate material. Both the laser beams are applied synchronously on the substrate in a continuous mode of operation. The stress concentration at the tip of the groove-crack helps in extending the crack through the substrate followed by controlled separation along the moving path of laser beam (Tsai and Chen 2003b). Figure 5.14 illustrates the mechanism of fracture during asymmetric laser cutting by controlled fracture technique. In asymmetric cutting, the laser path is geometrically asymmetrical such that the state of stress at the crack tip is not pure. Since the stress is not symmetric to the crack, the actual fracture trajectory will deviate and would not exactly follow the applied laser path. The fracture is triggered by the solidification process with the main crack lying at the centre of the melting groove and the number of minor cracks branching from it (Tsai and Chen 2004).

The modeling of the laser controlled fracture technique involves the calculation of temperature and stress distribution around the laser spot along the thickness of the sample during cutting process. The typical model geometry for a straight line cutting by controlled fracture is shown in Fig. 5.15. The vertical separation surface is parallel to the laser path along the positive x-axis. For a three-dimensional case, the temperature distribution in a material during laser irradiation is given by the solution of heat transfer equation (Tsai and Liou 2003):

$$\rho c_{\mathrm{p}} \frac{\partial T}{\partial t} = k \left(\frac{\partial^2 T}{\partial x^2} + \frac{\partial^2 T}{\partial y^2} + \frac{\partial^2 T}{\partial z^2} \right) + q, \tag{5.60}$$

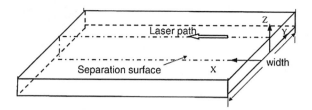

Fig. 5.15 Model geometry for calculating temperature and stress distribution during controlled fracture cutting. (Reprinted from Tsai and Chen 2003b. With permission. Copyright Elsevier.)

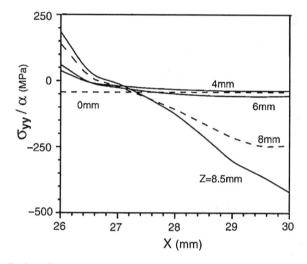

Fig. 5.16 Distribution of stresses during controlled fracture cutting of alumina at various locations in thickness (z). (Reprinted from Tsai and Chen 2003b. With permission. Copyright Elsevier.)

where q is laser power generation per unit volume. The initial condition can be written as:

$$T(x, y, z, 0) = T_0.$$ (5.61)

The convective conditions at the boundaries can be written as:

$$k \frac{\partial T}{\partial n} + h(T - T_\infty) = 0,$$ (5.62)

where h is the connective heat transfer coefficient and n is the direction cosine of the boundary. The temperature distribution can be used to calculate the distribution of stresses. The results of stress calculations for $108 \times 108 \times 10 \, mm^3$ alumina specimen irradiated with synchronous Nd:YAG (output power: 60 W) and CO_2 laser (output power: 80 W) are shown in Fig. 5.16. The energy of the focused Nd:YAG laser will be absorbed to induce a groove-crack of about 1 mm; whereas the energy of the defocused CO_2 laser will induce the thermal stresses. The stress calculations were done for traverse speed of 1 mm/s and time of 30 s for various values of thicknesses (z). After time $t = 30\,$s, the laser beams will reach $x = 30\,$mm and the separation front will be located at $x = 26\,$mm. The figure indicates that the stresses near the separation front ($x = 26$–$27\,$mm) are strongly tensile; whereas the stresses on the other side ($x > 27\,$mm) are strongly compressive. Thus, the stress near the laser spot is highly compressive and after the passage of laser beam, the plastic compressive stress will be relaxed, which then induces a residual tensile stress near the upper surface of the substrate. This residual tensile stress makes the fracture grow from the upper surface to the lower surface of the substrate. The stable separation of the brittle material is due to the local residual tensile stress; however, if the tensile stress is distributed throughout the thickness around the crack tip, the crack will extend unstably (Tsai and Chen 2003b).

Figure 5.17 presents a scanning electron microscopy (SEM) micrograph of the fracture surface of alumina ceramic cut by controlled fracture technique, showing the four distinct regions: evaporation, columnar grain, intergranular fracture, and transgranular fracture regions. The effects of cutting parameters such as the laser power, cutting speed, and laser spot diameter on the machining geometry has been obtained from the experimental analysis, and the phenomena have been also explained from the results of stress analysis (Tsai and Liou 2003; Tsai and Chen 2003a). Tsai and Liou also proposed an online crack detection and compensation technique for laser cutting by controlled fracture to eliminate the deviation of fracture trajectory (Tsai and Liou 2001; Tsai and Chen 2004).

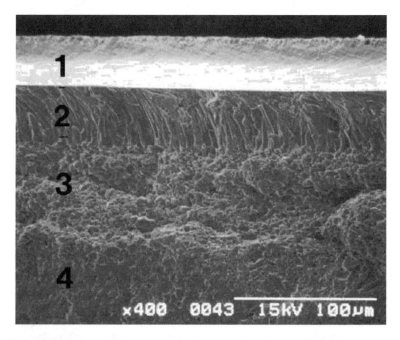

Fig. 5.17 SEM features of fracture surface of alumina ceramic cut by controlled fracture technique using laser power of 30 W and cutting speed of 4 mm/s. Four distinct regions 1–4 corresponds to evaporation, columnar grains, intergranular fracture, and transgranular fracture regions. (Reprinted from Tsai and Chen 2003a. With permission. Copyright Institution of Mechanical Engineers.)

A number of variations of the controlled fracture technique have been recorded. The process has been described for severing glass and vitro-crystalline materials with thickness greater than 5 mm (Lambert et al. 1976). The advantages of severing thicker vitreous sheet material is achieved by using two lasers, first of which has a wavelength such that at least 50% of the laser energy is used in melting of a 0.2 mm deep groove-crack, while the second laser beam generates thermal stress at the crack tip to make the material separate controllably. Recent studies using low power laser to separate the glass, employed water jet coolant to produce tensile stress along the cutting path. This method, in addition to increasing the cutting speed and accuracy, also control the depth, shape, and angle of the cut face formed by the crack (Kondratenko 1997; Unger and Wittenbecher 1998).

5.3 Quality Aspects

Laser cutting is finding increased utilization in the aircraft and automotive industry where the quality of cut is of the utmost importance. The various geometric and metallurgical quality factors sought in laser cutting include sharp corners at entry, narrow cut width, parallel sides, smooth cut surfaces, minimal thermal damage, and nonadherent dross. Due to the dynamic nature of the laser-cutting process, accurate prediction of cutting quality becomes difficult and is related primarily to the two factors: striations and dross.

5.3.1 Striations

Laser cutting is characterized by the formation of periodic striations along the cut edge. The presence of striations is undesirable since they may act as stress raisers in addition to the unpredictable geometric changes, necessitating the further finishing operations to achieve the smooth surface. In thin sections, these striations are generally clear and regular from the top of the cut edge to the bottom, whereas in thick sections these striations may be well defined at the top of the cut edge and become more random towards the bottom (Powell 1998). Typical striation pattern formed during laser cutting is shown in Fig. 5.18. A striation can be approximated by a partial elliptical geometry (Fig. 5.19). Hence the important parameters for characterizing the striation patterns in laser cutting are striation frequency (striation wavelength) and amplitude of striation (striation depth). Among various cutting parameters, these parameters are primarily affected by cutting speed. In general, the wavelength of the striation patterns increases with increasing the cutting speed (Fig. 5.20) (Ivarson et al. 1994).

Despite extensive research efforts directed to address this phenomenon, the mechanism of striation formation is still not completely understood. Several explanations have been put forward for the formation of periodic striations. The two possible

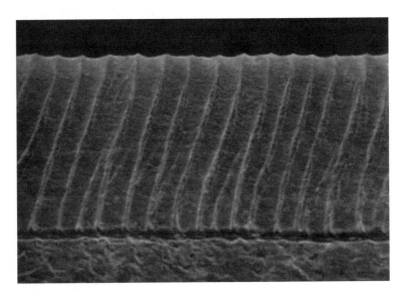

Fig. 5.18 Typical striation patterns formed during the laser cutting of 1.25 mm mild steel using 300 W CW laser using cutting speed of 1.8 m/min and oxygen pressure of 2.0 bar. (Reprinted from Powell 1998. With permission. Copyright Springer.)

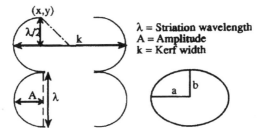

Fig. 5.19 Typical geometry of striations formed during laser cutting. (Reprinted from Ref. Ivarson et al. 1994. With permission. Copyright Elsevier.)

explanations for the formation of striations have been offered based on the pulsation in the molten layer and the side burning phenomenon. The first explanation for the formation of periodic striations is based on the fluctuations and oscillations of the liquid layer caused by instabilities associated with the dynamic nature of laser cutting (Shuöcker and Müller 1987). These fluctuations of the liquid layer induce perturbations on the cutting edges due to the movement of the liquid layer with the cutting front. The perturbed liquid layer subsequently solidifies into characteristic striation pattern. However, the most widely accepted mechanism for the striation formation is based on the side-burning effects associated with the laser cutting of mild steel (Arata et al. 1979; Miyamoto and Maruo 1991). The schematic of mechanism of striation formation by side burning is shown in Fig. 5.21. Various steps in the formation of striation can be written as (Powell 1998):

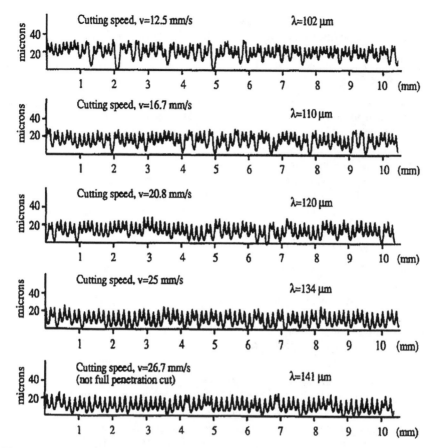

Fig. 5.20 Effect of cutting speed on the surface profile of striations (striation depth and frequency) during laser cutting of 2.5 mm thick mild steel with laser power of 400 W. (Reprinted from Ivarson et al. 1994. With permission. Copyright Elsevier.)

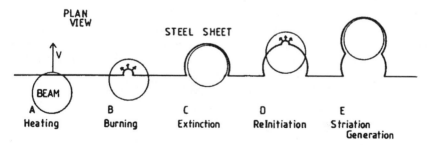

Fig. 5.21 Schematic mechanism of formation of striations during laser cutting. (Reprinted from Powell 1998. With permission. Copyright Springer.)

1. Ignition: The focused beam of laser in combination with the coaxial oxygen jet locally heats and subsequently oxidizes the edge of the steel, thus establishing a burning front.
2. Burning: Once the burning is initiated, the burning front receives energy from the oxidation reaction and also from the laser irradiation. This enhanced energy causes the burning front to move faster than the laser beam. This causes the burning front to be ahead of the center of laser beam in cutting direction. The burning front which propagates radially away from the center of the laser beam stops just outside the beam.
3. Extinction: The burning front eventually extinguishes after the laser beam is lagged behind the burning front thus creating a single striation.
4. Reintiation: As the laser beam moves forward in cutting direction, the re-ignition of the cutting front takes place which again goes through a cycle of burning and extinction. The repeated burning cycles as the cutting progresses results in the characteristic striation pattern on the cut edge.

Thus, the cyclic side burning theory suggests that at the cutting speed less than the speed of the reaction front caused by oxidation, sideways burning occur resulting in periodic striations. Excellent analysis of the striation generations by cyclic oxidation reaction of steels have been given based on dynamics of laser-cutting process (Ivarson et al. 1994).

Extensive modeling efforts have been directed towards understanding and controlling the striation formation. Assuming the hydrodynamic instabilities as a reason for the formation of striations, Vicanek has shown that the typical frequencies for the melt flow instabilities were 2.5 kHz. The simplistic calculations followed have predicted the striations on a 20 μm scale, which is about one order of magnitude too low compared to the actual observed structure suggesting the complexity of the process (Vicanek et al. 1987). Recently, based on heat transfer, transport, and chemical rate theory, the temperature fluctuations at the oxygen–melt interface and the associated molten layer fluctuation have been calculated for oxygen-assisted laser cutting of mild steel (Figs 5.22 and 5.23) (Chen et al. 1999). It is argued that during the oxygen-assisted laser cutting the oxide layer is formed which hinders the further oxidation due to resistance to oxygen diffusion resulting in temperature drop. The consequent removal of the oxide layer results in the temperature increase due to sudden increase of oxygen flux and reaction energy. Such temperature fluctuations result in the periodic striation. Chen et al. (1999) further presented an analytical analysis of two-dimensional melt using stability analysis in which melt dynamics were treated using the linearized Navier–Stokes equations and gas flow by power law of Schlichting. Based on a three-dimensional fully coupled model for the laser cutting, it was shown that the simplifications in modeling, such as reduction in the model dimension and neglect of gas dynamics, viscous shear, convection effects, and melt flow dynamics are not admissible if the accurate predictions of the cut surface profile is to be made and these small effects can induce identifiable differences in the cut surface profile (Gross et al. 2004). Current understanding of the striation formation in laser cutting has been reviewed in literature (Pietro and Yao 1994).

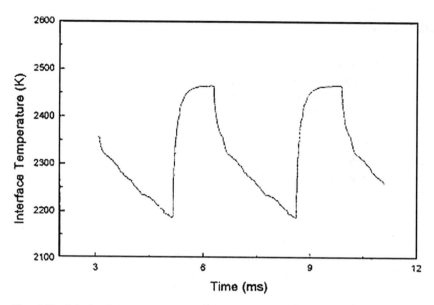

Fig. 5.22 Calculated temperature fluctuations at the oxygen-melt interface during oxygen-assisted laser cutting of mild steel using heat flux of 1.2×10^8 W/m², pressure of 2.5 bar, and cutting speed of 35 mm/s. (Reprinted from Chen et al. 1999. With permission. Copyright Springer.)

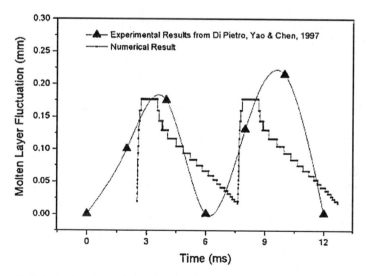

Fig. 5.23 Calculated and experimental molten layer fluctuation due to temperature fluctuations during oxygen-assisted laser cutting of mild steel using heat flux of 1.2×10^8 W/m², pressure of 2.5 bar, and cutting speed of 25 mm/s. (Reprinted from Ref. Chen et al. 1999. With permission. Copyright Springer.)

5.3.2   Dross

Dross is related with the incomplete expulsion of the melt from the bottom of the kerf. For precision applications where the clean cutting edges are important the formation of dross at the bottom of the cutting kerf must be controlled. This requires deep understanding of the mechanisms of dross formation and the various materials and laser-cutting parameters which potentially control the formation of dross. For example, the highly cleaned (dross-free) cutting edges can be obtained in mild steel at relatively higher cutting speed such as 7 m/min for thickness of 1 mm and laser power level of 1 kW; however, the stainless steels inadvertently tends to form dross during laser cutting.

The phenomenon of dross formation was first explained by Arata for the laser gas cutting of stainless steels (Arata et al. 1981). The melt flow from the bottom of the cutting kerf and the subsequent clinging of dross clinging during laser cutting was analyzed using the high speed camera. Figure 5.24 shows the high speed pictures and the corresponding schematics explaining the melt flow for the cutting of 2 mm thick stainless steel with laser power of 1 kW. At low cutting speed (1.4 m/min), melt flow at the bottom of the kerf gets stagnated and a very small amount of melt is expelled from the bottom in the form of a large amount of small spherical droplets. The large amount of melt flows backward and subsequently solidifies to form clinging dross. The characteristics melt flow and droplet ejection which in turn affect the nature and amount of dross are highly dependent on the cutting speed. At higher cutting speeds (3 m/min), the backward flow of melt becomes weak and the large amount of melt is expelled from the bottom of the kerf in the form of long pendent parts. This corresponds to the condition for the minimum amount of clinging dross. Further increase in the cutting speed (~5 m/min) causes the disturbances to the melt flow thus increasing the amount of clinging dross again at higher speeds. Hence, there exists an optimum cutting speed for which the cutting edges are highly cleaned with minimum associated clinging dross. These mechanisms are in agreement with the experimental observation which show the minimum roughness of the cutting edges at intermediate cutting speeds which corresponds to the minimum amount of clinging dross associated with it. In view of these mechanisms, two methods can be suggested to improve the cut quality: pile-up cutting and tandem nozzle cutting. In pile-up cutting stainless steel is cut with thin mild steel piled on it. Mild steel provides molten oxide which enhances the fluidity of the melt on cutting kerf of stainless steel and thus promoting the separation of molten oxide dross resulting in nearly dross-free cuts. An off-axis tandem nozzle can also improve the cut quality by effective removal of melt from the bottom of the curve due to enhanced dynamic force of the gas jet (Arata et al. 1981).

Extensive literature on the modeling of laser cut quality from the viewpoint of dross formation is available. The most important aspect of modeling the cut quality in laser cutting is the consideration of gas flow distribution. The modeling of laser fusion cutting to determine the geometry of the cutting kerf and gas flow distributions

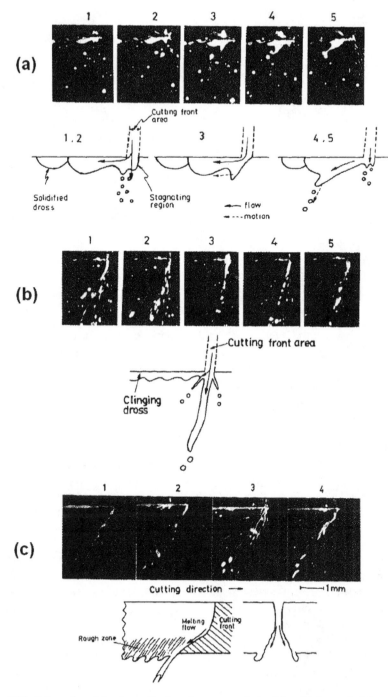

Fig. 5.24 Schematics of the dross-clinging phenomenon derived from high-speed pictures taken during the laser cutting of 2 mm stainless steel with laser power of 1 kW for cutting speeds of **(a)** 1.4 m/min, **(b)** 3 m/min, and **(c)** 5 m/min (Arata et al. 1981)

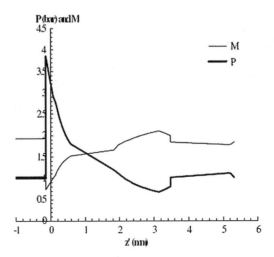

Fig. 5.25 Distribution of pressure and Mach number of the gas from the exit of the nozzle ($z = -1$) to the top of the cut kerf ($z = 0$) and along the cut kerf ($z = 0$ to 5). (Reprinted from Duan et al. 2001c. With permission. Copyright Institute of Physics.)

in the cut kerf which subsequently affect the quality of cut can be cited (Duan et al. 2001a, b, c). Figure 5.25 shows the gas flow field distribution (pressure P and Mach number M, i.e., the velocity of the gas) from the exit of the nozzle to the top of the cut curve and then along the cut kerf until exit at the bottom of the cut kerf for the laser fusion cutting of 5 mm thick stainless steel with laser power of 2 kW and cutting speed of 1.6 m/min. The figure indicates that a curved shock is formed in the flow field inside the cut kerf which ultimately influences the quality of the cut edge. As seen from the figure, the exit pressure of the gas (1.13 bar) is slightly greater than the ambient pressure (1.03 bar) which results in the slight expansion of the gas at the exit. The theory of gas dynamics suggests that the condition for maximum exit momentum thrust (hence the condition for achieving the dross-free cuts) corresponds to the exit gas pressure being equal to ambient pressure to achieve a cut edge without dross. Since the exit pressure at the bottom of the cutting front is higher than the ambient pressure the momentum thrust will be weaker due to decrease in velocity of the gas. This results in greater divergence of the ejected molten slag leading to excessive dross attachment at the bottom of the kerf. The flow fields at the entrance, along and exit of the cut kerf are also influenced by the laser power, cutting speed, focus position, inlet stagnation pressure, exit diameter of the nozzle, and the nozzle displacement range with the cut kerf (Duan et al. 2001c).

Recently, an analytical model of dross formation has been developed for evaluating the three-dimensional geometry of the cutting front, and the geometry and temperature fields of the melt film, by considering mass, force, and energy balances (Tani et al. 2003, 2004). The two different mechanisms responsible for dross formation were suggested: the former being dependent on the surface tension of the melt and directly linked to the ejection speed of the liquid phase from the kerf

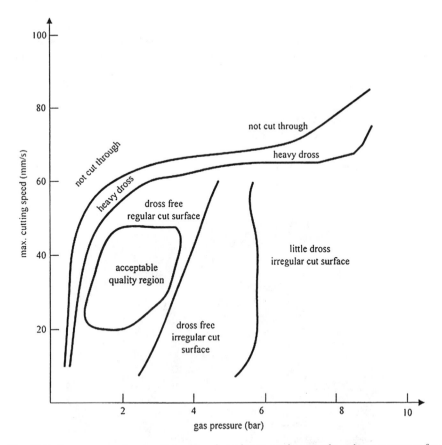

Fig. 5.26 Process map showing the range of maximum cutting speeds and gas pressures for achieving the dross-free acceptable quality cuts during oxygen-assisted laser cutting of 3 mm thick mild steel. (Reprinted from Chen 1999. With permission. Copyright Elsevier.)

bottom; the latter being linked to the occlusion of the kerf bottom, through an excess of liquid, which results in incomplete ejection of the molten material. The model predicted good quality of cut when the melt ejection speed exceeds 2,500 mm/s (Tani et al. 2003). The effect of high-pressure assistant gas flow on the dross formation in CO_2 laser cutting of 3 mm thick mild steel plate was studied by Chen (1998, 1999) and a dross-free feasibility area as a function of cutting speed and assistant gas pressure was experimentally investigated (Fig. 5.26).

5.3.3 Heat-Affected Zone

Laser cutting is often associated with thermal effects at the surface of the cutting kerf resulting in alteration of microstructure and/or mechanical properties. This results in the formation of distinct heat-affected zone (HAZ) at the surface of the cutting edge which can be revealed by appropriate polishing and etching techniques. HAZ is often associated with undesirable effects such as distortion, surface cracking, embrittlement, decrease in weldability, decrease in corrosion, fatigue resistance, etc. Hence laser-cutting parameters are selected so as to minimize the HAZ. Figure 5.27 shows a typical HAZ in titanium alloy sheet formed during laser cutting. The HAZ is characterized by the formation of metastable acicular martensite with hardness around HV 287–300. In general, the HAZ can be minimized by increasing the laser-cutting speed due to reduced transfer of heat to matrix at higher cutting speeds (Fig. 5.28). The figure also indicates thicker HAZ with air-assisted laser cutting compared to argon-assisted laser cutting. This is due to increased generation of heat

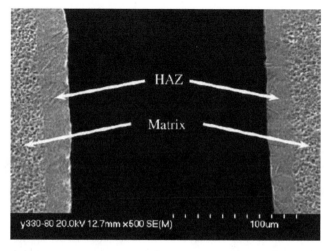

Fig. 5.27 Heat affected zone (HAZ) laser-cut titanium alloy. (Reprinted from Shanjin and Yang 2006. With permission. Copyright Elsevier.)

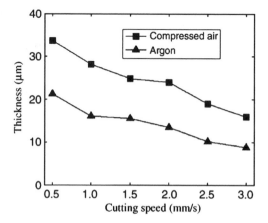

Fig. 5.28 Variation of thickness of HAZ with cutting speed during laser cutting of titanium alloy. (Reprinted from Shanjin and Yang 2006. With permission. Copyright Elsevier.)

corresponding to the reaction of titanium with oxygen and nitrogen and thus producing a thin layer of brittle oxides and nitrides (Shanjin and Yang, 2006).

Due to complexity of relationship between laser-cutting parameters and the cutting quality, statistical approaches are needed to arrive at the parameters of high significances. A parametric study by Rajaram et al. (2003) considered the combined effects of cutting speed and power on kerf width, surface roughness, striation frequency, and size of HAZ in CO_2 cutting of 4130 steel. The results are shown in Figs 5.29 and 5.30. It was shown that the laser power had major effect on kerf

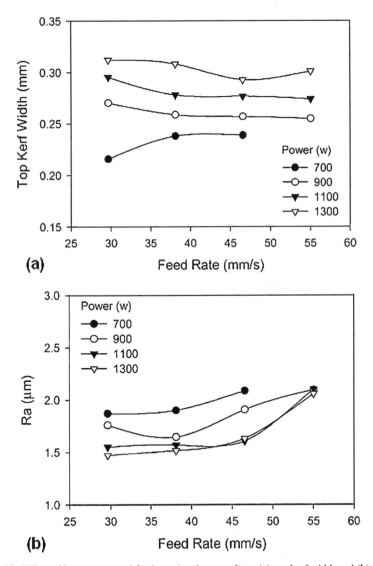

Fig. 5.29 Effect of laser power and feed rate (cutting speed) on **(a)** top kerf width and **(b)** surface roughness during CO_2 laser cutting of 4130 steel. (Reprinted from Rajaram et al. 2003. With permission. Copyright Elsevier.)

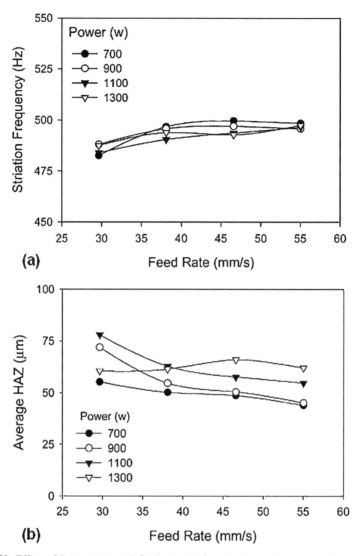

Fig. 5.30 Effect of laser power and feed rate (cutting speed) on **(a)** striation frequency and **(b)** average heat affected zone (HAZ) during CO_2 laser cutting of 4130 steel. (Reprinted from Rajaram et al. 2003. With permission. Copyright Elsevier.)

width, whereas the feed rate had major effect on striation frequency and roughness. Decreasing power and increasing feed rate led to a decrease in kerf width and HAZ, whereas increasing feed rate generally led to increasing roughness and striation frequency.

5.4 Practical Considerations

Laser cutting is a complex process involving parameters such as laser energy, pulse shape, focusing conditions, and assist gas. These parameters determine the geometric and metallurgical quality of the laser-cut workpieces. This section discusses the effects of some of the most important laser-processing parameters on the geometric and metallurgical quality of cuts.

5.4.1 Effect of Laser Type

Various types of lasers have been successfully used for cutting various materials for specific applications. For example, CO_2 and Nd:YAG lasers are primarily used in materials processing, while, ultraviolet lasers are used in medical cutting applications (such as eye surgery). However, laser cutting market is primarily dominated by the continuous wave (CW) CO_2 lasers particularly for sheet metal cutting applications. CO_2 lasers are capable of delivering high output powers, high beam quality, and highly controlled power mode. The popularity of CO_2 lasers for cutting is due to advantages such as high cutting speeds, excellent cut quality and processing flexibility, and widespread application possibility (Thawari et al. 2005; Tabata et al. 1996). Recently, increasing interests have been directed towards the use of Nd:YAG lasers for cutting applications. The most remarkable characteristics of the pulsed Nd:YAG lasers which are increasingly making them replace the CO_2 laser are short wavelength, good focusability, and high peak power. In addition, enhanced transmission through plasma, wider choice of optical material, and flexibility in handling due to fiber optically delivered beams further extends the applicability of pulsed Nd:YAG laser for laser cutting. Shorter wavelength (1.06 μm) of the Nd:YAG laser ensures the high absorption of laser energy thus making it suitable for cutting highly reflective materials (such as aluminum and copper); whereas very high peak powers associated with pulsed Nd:YAG lasers facilitate the cutting of thick metals. Laser cutting with pulsed Nd:YAG lasers is often useful for the precision cutting applications where narrow kerf widths, small HAZ, and intricate cut profiles are desired.

5.4.2 Effect of Laser Power

Laser power determines the direct the energy input to the cutting process. Both cutting quality and the cutting performance depend on the laser power. Figure 5.31 presents the effect of laser power on the kerf width and the roughness of cut edge for three different cutting speeds indicating that improved surface roughness and kerf width can be obtained at lower power and higher cutting speed (Ghany and Newishy 2005).

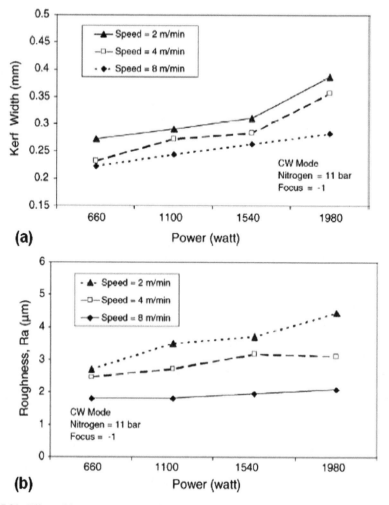

Fig. 5.31 Effect of laser power on the (**a**) kerf width and (**b**) roughness of cut edge for three different cutting speeds during laser cutting of 1.2 mm austenitic stainless steel by continuous Nd: YAG laser. (Reprinted from Ref. Ghany and Newishy 2005. With permission. Copyright Elsevier.)

Laser power also determines the maximum cutting speed which is defined as the minimum speed at which through cut is produced. There exist an optimum combination of cutting speed and the laser power which gives the maximum performance. This is illustrated in Fig. 5.32, for laser cutting of thin (0.7–0.8 mm) high strength steel sheets with a continuous CO_2 laser. The figure indicates that maximum cutting speed of 7,000 mm/min can be achieved at laser power of 300 W. At lower powers, the energy supplied to the cutting front may be insufficient to cut through the entire thickness of the material, whereas at higher powers, production of clean through cuts necessitates the reduction of cutting speed (Lamikiz et al. 2005).

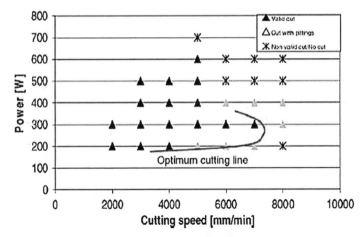

Fig. 5.32 Optimum laser-cutting area determined from the combination of laser power and cutting speed which gives clean thorough cuts. (Reprinted from Lamikiz et al. 2005. With permission. Copyright Elsevier.)

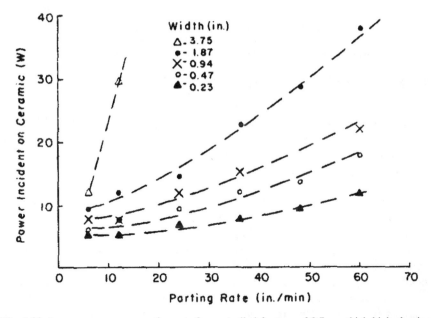

Fig. 5.33 Laser power verses parting rate for controlled fracture of 0.7 mm thick high alumina ceramic (>99%) (Reprinted from Lumley 1969. With permission. Copyright American Ceramic Society.)

In case of controlled fracture technique, the laser power determines the thickness of the ceramic that can be controllably fractured and the parting rate. Figure 5.33 shows that the parting rate of ceramics increases linearly with laser power for various widths of specimen. In addition, Fig. 5.34 indicates that laser power required for fracturing alumina ceramic increases with material thickness (Lumley 1969).

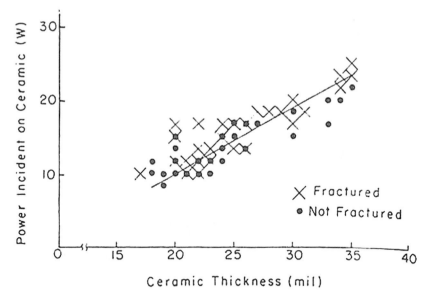

Fig. 5.34 Power required for separation verses thickness during controlled fracture of high alumina ceramic with a rate of separation 305 mm/min (Reprinted from Lumley 1969. With permission. Copyright American Ceramic Society.)

5.4.3 Effect of Optical System

Optical system determines the energy distribution in the beam (mode type), beam diameter, depth of focus, focus position relative to workpiece, and beam coupling with the workpiece. All of these factors affect the cutting performance in one or more ways (Steen and Kamalu 1983).

In general, symmetry of the laser mode is important to obtain the good cut quality. In asymmetric energy distribution (higher order modes), the local energy density differs substantially in different areas. Asymmetry in laser mode may lead to the increased roughness at the cutting edge, excessive dross, material burning at the corners, and reduced cutting speeds. The most preferred mode for laser cutting consists of Gaussian energy distribution. Gaussian beams can be focused to very small diameters thus resulting in higher power density. Moreover, axial symmetry of the Gaussian beam gives similar interaction with the workpiece in all directions resulting in the similar cutting edge profiles in all cutting directions with similar and low roughness. One more effect which is important in the context of laser cutting is the beam polarization. Laser beam consists of traveling photons which are characterized by oscillating electric and magnetic vectors perpendicular to each other. Polarization refers to the oscillations of electric and magnetic field vectors in a particular way. For most of the cutting applications, the beams are plane polarized, i.e., the electromagnetic waves oscillate in the same plane within the emergent beam. When the cutting direction is parallel to the electric field vector, the plane polarized beam offers advantages such as maximum absorption at the cutting front

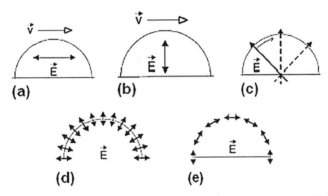

Fig. 5.35 Schematic of various laser beam polarization: (**a**) Plane polarization with electric vector parallel to the beam velocity vector; (**b**) plane polarization with electric vector perpendicular to the beam velocity vector; (**c**) circular polarization; (**d**) radial polarization; and (**e**) azimuthal polarization. (Reprinted from Niziev and Nesterov 1999. With permission. Copyright Institute of Physics.)

during the high speed cutting of thin sheets. It has been shown that for cutting metals with a large ratio of sheet thickness to width of cut, the laser-cutting efficiency of radially polarized beam is 1.5–2 times larger than for plane polarized and circularly polarized beams (Niziev and Nesterov 1999; Nesterov and Niziev 2000). Schematics of various laser beam polarization types are presented in Fig. 5.35.

Various focusing condition such as beam diameter, depth of focus, and the beam position relative to the workpiece primarily affect the laser-cut quality given by various geometrical factors. The most dominant effect of depth of focus is on the cutting front geometry described by kerf shape, kerf width, and taper of the cutting front.

5.4.4 Effect of Nozzle Parameters

Laser-cutting practices almost always use gas nozzle coaxial (sometimes off-axis) with the focused laser beam. The primary purpose of using gas nozzle is to effectively remove the molten material from the cutting kerf thus improving the cutting efficiency by enhancing cutting action. To achieve this, it is necessary to ensure the correct alignment of the nozzle with laser beam such that gas pressure at the cutting kerf is sufficient to expel the molten material effectively. In this context, various practical considerations are nozzle design, nozzle diameter, nozzle–material standoff, and nozzle gas pressure distribution.

A variety of nozzle geometries, based on designs derived from gas dynamics considerations, are used in industrial laser-cutting practices. Figure 5.36 shows a schematic of popular nozzle designs. Conical nozzles with minimum cross-sectional area at the exit diameter (0.8–2 mm) are the simplest and most commonly used nozzles. However, conical nozzles have some serious drawbacks. Above the

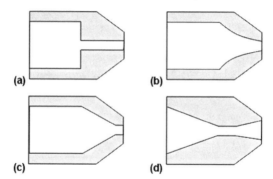

Fig. 5.36 Schematic of various nozzle designs: **(a)** parallel, **(b)** convergent, **(c)** conical, and **(d)** convergent–divergent (de Laval) nozzles

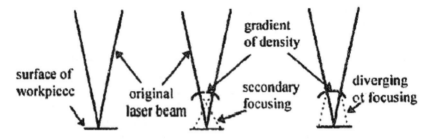

Fig. 5.37 Effects of interference of density gradient fields (DGFs) on focusing of laser beam. (Reprinted from Man et al. 1997. With permission. Copyright Elsevier.)

inlet gas pressure of 1.89 atm, gas flow increases but the gas velocity remains constant causing the transversal expansion of the jet in an explosive manner resulting in pressure reduction at the core of the jet. In addition, nonuniform jet flow produces density gradient fields (DGFs) leading to change in refractive index in the gas field. This phenomenon causes severe interference with the focusing of the laser beam resulting in secondary focusing or diverging of laser beam. Some of these possible interference patterns are shown in Fig. 5.37. This interference can result in poor cutting quality and low cutting speed (Man et al. 1997, 1999). This limits the placement of the nozzle very close to the workpiece to maintain the uniform pressure at the workpiece surface.

Supersonic nozzles such as convergent–divergent nozzle (de Laval nozzle) offer many advantages such as:

1. Higher cutting speed and smaller HAZ compared to the coaxial conical nozzle operating at same gas pressure.
2. Improved capability to remove the molten debris due to higher momentum of the jet.
3. Improved capability for precise cutting and cutting of thick plates due to long and sharp-pointed nature of jet boundary exiting from supersonic nozzle.

Another important nozzle parameter in the context of laser cutting is the nozzle diameter. The typical nozzle diameters used for laser-cutting applications range from 1 to 3 mm. Selection of nozzle diameter is primarily governed by the material being cut and the cutting speed desired. For example, smaller nozzles (~1 mm) are used for thin sections of cutting steel as they can supply the adequate pressure and results in minimum consumption of gas. Slightly larger diameter nozzles (~1.5 mm) are generally used for thicker section of steel due to insensitivity of such nozzles to minor misalignment and changes in supply pressure. Larger diameter nozzles are used for cutting polymers such as acrylic sheets where low velocity jets are desired to prevent the turbulence of molten material at the cutting edges (Powell 1998). In addition, the nozzle diameter influences the cutting speed. Figure 5.38 presents the variation of cutting speed with nozzle diameter during laser cutting of 2 mm

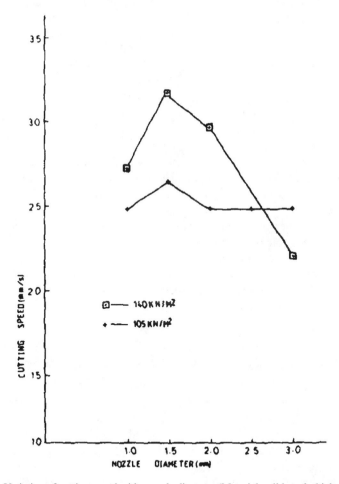

Fig. 5.38 Variation of cutting speed with muzzle diameter (Material: mild steel; thickness: 2 mm; power: 1,500 W; gas: Argon) (Reprinted from Steen and Kamalu 1983. With permission. Copyright M. Bass, Editor.)

mild steel at constant laser power and supply pressure of inert gas. For a given supply pressure, there exists an optimum nozzle diameter which gives maximum cutting speed. This is the direct consequence of improved pressure distribution for removal of molten material at intermediate value of nozzle diameter (Steen and Kamalu 1983).

Nozzle–material standoff is the distance between the surface of the workpiece and the exit of the nozzle and usually equals the nozzle–focal point standoff for the case where laser beam is focused on the surface of the workpiece. Nozzle–material standoff in the range of 2 ± 0.5 mm is usual. If the nozzle–material standoff is too less, severe back pressures will be created on the lens apart from mechanical wear and damage of nozzle due to splattered dross particle. On the other hand, if the nozzle–material standoff is too high, there is unnecessary loss of kinetic energy of the jet. Moreover, the nozzle–material standoffs significantly affect the pressure distribution of the gas on the surface of the workpiece which must be taken into consideration along with the other nozzle parameters such as nozzle geometry and nozzle design. In general, increasing the standoff distance increases the expansion of the jet flow on the kerf resulting in increased momentum and energy loss. In industrial laser-cutting practices, automatic height sensing devices are used to control the nozzle–material standoff (Man et al. 1999).

The most important effect of supply gas pressure is on the cutting speed. Increasing the supply gas pressure increases the cutting speed by expanding the pressure distribution curves (thus changing the position of low pressure areas) and also increasing the centreline pressure at the workpiece. However, the cutting speed reaches maximum for certain supply gas pressure beyond which the cutting speed decreases or remains constant. As explained before, this is the direct effect of interference of DGFs with the focusing of the laser beam. Figure 5.39 shows the variation of maximum cutting speed with assist gas pressure during laser cutting of 4.4 mm slate plate with laser power of 1,200 W using Laval nozzle. The maximum cutting speed increases up to a pressure of 5.5 bar in case of Laval nozzle, whereas in case of coaxial conical nozzles, the cutting speed increases only up to 2 bar due to the existence of DGF under the conical nozzles (Boutinguiza et al. 2002).

5.4.5   Effect of Assist Gas Type

Various oxidizing and inert gases such as oxygen, nitrogen, argon, and helium are commonly used for laser-cutting practices depending on the laser-cutting efficiency and the cut quality desired. For metallic and reactive material, the type of gas determines the amount of heat added for cutting action. For example, laser-cutting efficiency mild steel cutting is substantially improved by using oxygen as an assist gas. In this case, 60–70% of the energy required for cutting is provided by the energy released by oxidation reaction; whereas laser contributes only 30–40% of the energy. In addition, low viscosity of the metal oxides facilitates the easy ejection

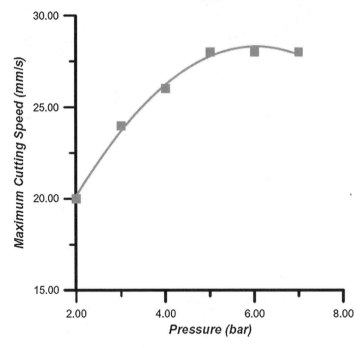

Fig. 5.39 Variation of maximum cutting speed with assist gas pressure using Laval nozzle (Laser power: 1200 W, plate thickness: 4.4 mm). (Reprinted from Boutinguiza et al. 2002. With permission. Copyright Elsevier.)

of molten material from the cutting kerf (Rao et al. 2005). Figure 5.40 indicates that for a given ratio of laser power to thickness (P/t), oxygen-assisted cutting gives higher cutting velocity compared to inert gas cutting of stainless steel. Similar relationships can be found for laser cutting of titanium (Steen and Kamalu 1983).

As mentioned earlier, oxygen-assisted cutting of titanium and its alloys also gives higher cutting speed than inert gas assisted cutting (Fig. 5.41). However, for such highly reactive materials, oxygen-assisted laser cutting is undesirable from the view point of desired cutting quality. During oxygen-assisted laser cutting of titanium, large amount of exothermic heat generated by the oxidation of titanium destabilizes melt from highly oxidized cut edges which produce relatively large kerf width. In addition, the absorption of oxygen causes the embrittlement of the cut edge thus drastically reducing the fatigue life of formed components. To retain the properties of the base material, the cut edge needs to be machined to remove the oxidized edge. In most of the cases inert gas cutting of titanium is desired. Inert gas cutting provide better quality of cutting edges, however, the cutting speeds are greatly reduced. It is necessary to protect the hot edges during inert gas cutting from atmospheric oxygen and nitrogen to prevent the embrittlement of cutting edges. Inert gas cutting of titanium is sometimes associated with considerable dross formation at the bottom of cut edge. This effect can be minimized or eliminated by

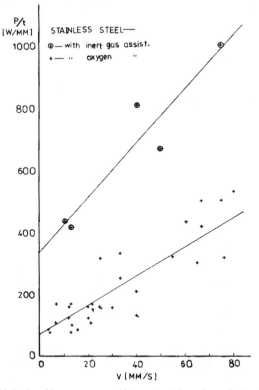

Fig. 5.40 Variation P/t (ratio of laser power to thickness) with cutting velocity for stainless steel (Reprinted from Steen and Kamalu 1983. With permission. Copyright M. Bass, Editor.)

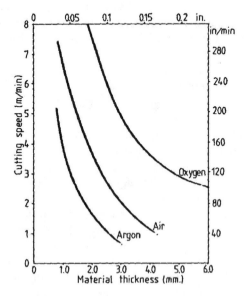

Fig. 5.41 Effect of assist gas on laser-cutting speeds of titanium alloys of various thicknesses. (Reprinted from Powell 1998. With permission. Copyright Springer.)

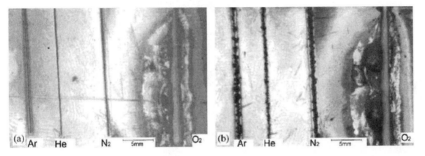

Fig. 5.42 Effect of assist gas on the quality of cutting edge for (**a**) top cut surface and (**b**) bottom cut surface of titanium. (Reprinted from Rao et al. 2005. With permission. Copyright Elsevier.)

employing the auxiliary inert gas jets or dross jets which are successful for inert gas cutting of titanium and its alloys (Powell 1998). Figure 5.42 shows the effect of assist gas on the quality of cutting edge of titanium clearly indicating that better quality cuts are obtained with inert gases compared to oxygen (Rao et al. 2005). Laser-cutting efficiency and cutting quality of nonmetallic materials is generally less influenced by the type of assist gas used (Caiazzo et al. 2005).

5.5 Laser Cutting of Various Materials

5.5.1 Metallic Materials

Laser cutting is successfully applied for cutting a range of ferrous and nonferrous metallic materials for various applications. The cutting mechanisms and the performance differ from material to material as a consequence of differing thermo-physical properties of the material. Among various industrially important ferrous materials, mild steel and stainless steels have been most extensively studied; whereas most commonly cut nonferrous materials are alloys based on but not limited to titanium, aluminum, copper, and nickel.

Laser cutting of mild and stainless steels is generally achieved by a combination of laser heating and oxidation reaction of iron with oxygen. The mechanisms of laser cutting for stainless steels are slightly different from that of mild steel due to presence of chromium. Chromium provides increased energy of oxidation compared to iron. However, during laser cutting of stainless steel, chromium preferentially reacts with oxygen forming a protective Cr_2O_3 layer on the cutting edge thus limiting the further oxidation of underlying melt. Even though the protective Cr_2O_3 layer ruptures continuously allowing further oxidation of the melt, the cut zone consists of substantial amount of unoxidized melt which is partly removed by shearing due to pressurized gas jet. The material removal during stainless steel cutting is generally inhibited due to higher surface tension of Cr_2O_3 covered melt than that of highly oxidized melt produced during mild steel cutting

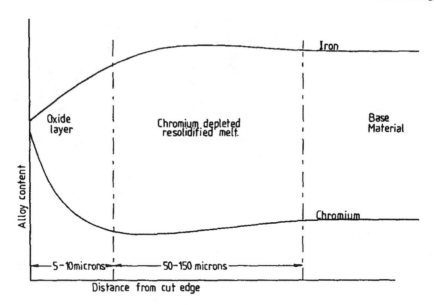

Fig. 5.43 Schematic of variation of chromium and iron content across the laser cut edge in stainless steel. (Reprinted from Powell 1998. With permission. Copyright, Springer.)

resulting in the formation of extensive dross. This may require the higher oxygen pressure for more complete and efficient removal of molten material from the cutting kerf. Figure 5.43 shows a schematic of changes in content of iron and chromium across the cut edge of laser cut stainless steel indicating the presence of higher chromium content in the form of chromium oxide at the surface of cut edge. Typical processing curve showing the relationship between cutting speed and material thickness for laser cutting of mild and stainless steel are presented in Fig. 5.44 (Powell 1998).

Most of the nonferrous metals and alloys can be cut using industrially popular CO_2 laser; however, the cutting of these materials is generally difficult than the ferrous materials. Industrially important nonferrous materials are characterized by high thermal conductivity and high reflectivity (less absorptivity) which causes inefficient absorption of laser energy resulting in less efficient oxidation reaction. Table 5.5 presents the absorptivity, thermal conductivity, and heat of oxidation for few nonferrous metals (absorptivity and the thermal conductivity are highly dependent of temperature; hence the values given here are for room temperature and should be considered for comparison purpose only). A combination of high conductivity and high reflectivity makes the high-power, short wavelength lasers such as Nd:YAG lasers ($\lambda = 1.06\,\mu m$) more suitable for cutting nonferrous materials due to higher absorptivity of these materials at shorter wavelengths. In general, the laser-cutting speeds for nonferrous materials are significantly lower (~an order of magnitude) than the ferrous materials due to inefficient oxidation reaction during laser dressing. Among the various nonferrous materials, titanium is

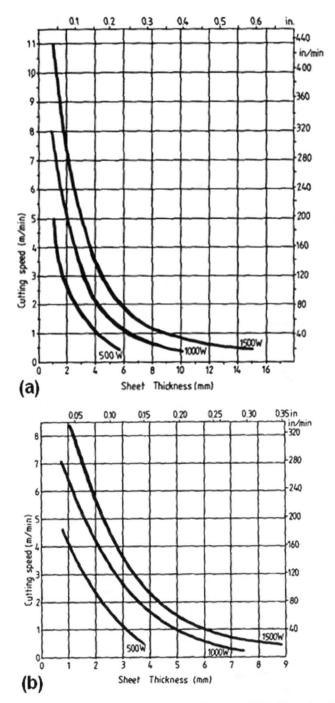

Fig. 5.44 Variation of laser-cutting speed with sheet thickness of (a) mild steel and (b) stainless steel. (Reprinted from Powell 1998. With permission. Copyright Springer.)

Table 5.5 Absorptivity, thermal conductivity, and heat of oxidation of selected metals (Reprinted from Powell 1998. With permission. Copyright Springer.)

Metal	Absorptivity (%)	Thermal conductivity (W/m K)	Heat of oxidation kJ/mol (oxidation product)
Iron	~5.0	80	822 (Fe_2O_3)
Nickel	~6.0	59	244 (NiO)
Titanium	~8.5	23	912 (TiO_2)
Copper	1.0–2.0	385	160 (CuO)
Aluminum	1.0–2.0	201	1670 (Al_2O_3)
Silver	0.5–1.0	419	–
Gold	0.5–1.0	296	–

Note: Absorptivity values are strongly dependent on temperature and surface conditions. The given values are at ambient temperature for unpolished metals and should be used for comparison purpose only. Conductivity values are also given at ambient temperature.

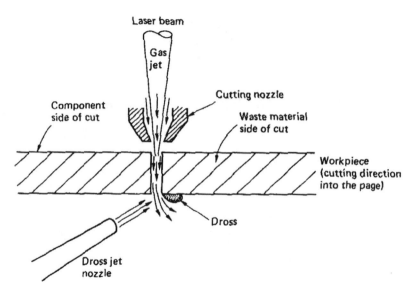

Fig. 5.45 Schematic showing operation of auxiliary inert gas jet (dross jet) in assisting the removal of dross from the bottom of cutting edge. (Reprinted from Powell 1998. With permission. Copyright Springer.)

most important due to its applications in aerospace industry. Oxygen-assisted cutting of titanium gives high cutting speed, but is often characterized by the presence of an undesirable oxidized layer on the cutting edge which has to be removed by mechanical means. This makes the oxygen-assisted cutting useful for rough machining applications and the application where oxidized cutting edge does not significantly affect the in-service performance. High-quality cutting edges in titanium can be produced by inert gas cutting where material removal is primarily due to shearing action of gas jets instead of exothermic oxidation of titanium. Inert gas cutting of titanium is often accompanied with the auxiliary inert gas jets to remove the dross from the bottom side of the cutting edge (Fig. 5.45) (Powell 1998).

5.5.2 Polymers

CO_2 laser radiation at 10.6 μm is efficiently absorbed by most of the natural and synthetic polymeric materials and products. Various thermoplastics, thermosets, and rubbers have been efficiently machined by laser cutting. Material removal mechanism during laser cutting of a polymer is a combination of melt shearing, vaporization, and chemical degradation, of which one or the other may dominate in any particular polymeric material. In most of the thermoplastic polymers (such as polyethylene, polypropylene, polystyrene, nylon, etc.), melt shearing dominates and the material is removed by blowing the molten material from the cutting kerf. The cut edges of polymers cut by melt shearing generally have better cut quality than the mechanical cuts. As with the metallic materials, the polymers too produce small amount of adherent dross due to incomplete ejection of molten material from the bottom of the cutting kerf. Vaporization is a dominant material removal mechanism in case of polymethyl methacrylate (PMMA) and polyacetal. During laser cutting, the acrylic sheets are dissociated into methyl methacrylate gas. Vaporization cutting of acrylics is generally associated with very high quality (polished edge quality) of cutting edges. The processing map showing the relationship between the cutting speed and the thickness for acrylic is shown in Fig. 5.46 (Chryssolouris 1991). Figure 5.47a illustrates the "polished" edge quality of the cutting edges of

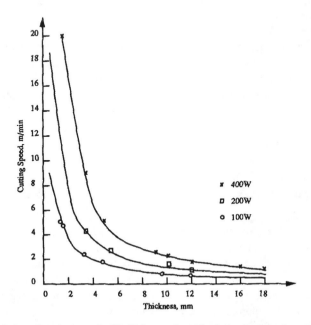

Fig. 5.46 Variation of cutting speed with thickness of acrylic during laser cutting at three different laser powers. (Reprinted from Chryssolouris 1991. With permission. Copyright Springer.)

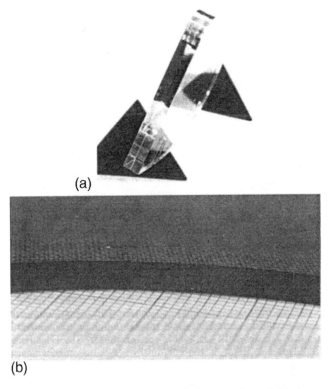

(a)

(b)

Fig. 5.47 Typical edge quality of laser cut polymers (a) acrylic and (b) thermoset polymer. (Reprinted from Powell 1998. With permission. Copyright Springer.)

laser cut acrylics. Chemical degradation of materials is a dominant mechanism of material removal for most of the thermosetting polymers such as phenolic or epoxy resins and rubber products. The chemical degradation is often associated with the generation of carbon-based smoke during cutting, and formation of carbon deposit on the cutting edge. Typical edge quality of thermoplastic and thermoset polymers is shown in Fig. 5.47. The figure shows a "polished" edge quality in thermoplastics and smooth-covered cutting edge with thermosets (Powell 1998).

5.5.3 Ceramics and Glasses

Ceramics and glasses are widely used in structural applications and considered as difficult-to-machine materials for the conventional machining techniques. Various industrially important ceramics include alumina, tungsten carbide, titanium nitride, and titanium carbide. In addition to cutting such difficult-to-machine materials, laser cutting is important from the viewpoint of the possibility of making profile cuts. Lasers offer various approaches for cutting ceramics such as evaporative laser cutting and controlled fracture technique. For through cutting of ceramics, evaporative cutting with pulsed lasers are preferred to achieve finer cuts and minimum damage to the surrounding materials. Often, the ceramics are cut by scribing a line or a row of blind holes along the cutting path followed by controlled breaking by applying the mechanical stress. The fracture can also be achieved by employing a second laser which introduces thermal stresses at the crack tip. Due to the brittle nature of ceramics, the fracture is fast resulting in cutting speeds as high as 15 m/min.

Representative data for laser cutting of various materials is given Table 5.6.

Table 5.6 Representative laser-cutting data for various materials

Material	Laser power (w)	Thickness (mm)	Cutting speed (mm/s)	Reference
Metallic				
Mild steel	2,000	2.0	90.0	Clarke and Steen
Low carbon	1,600	4.0	31.0	1979
Steel	200	4.75	10.6	Lunau and Paine
Stainless steel	100	1.0	26.6	1969
Titanium	850	2.2	30.0	Babenko and
Aluminum	100	1.0	15.6	Tychinsckii 1973
Nonmetallic	2,000	4.8	66.7	Clarke and Steen
Alumina	200	.03	37.5	1978
Porcelain tile	200	0.075	100	Duley and
Plywood	240	0.17	100	Gonsalves 1974
Wood	850	0.5	54	Harry and Lunau
PMMA	250	0.6	3.3	1972
Polyethylene	200	1.3	230	Babenko and
Polypropylene	100	0.75	25	Tychinsckii 1973
Polystyrene	150	0.635	8.3	Lee et al. 1985
Nylon	1,000	2.0	33.3	Harry and Lunau
ABS	950	8.0	1.6	1972
Polycarbonate	945	7.8	3.3	Powell et al. 1987
PVC	225	5.0	20	Powell 1998
Fiberglass	350	4.8	88.3	Pasucual-Cosp et al.
Glass	2,000	15.0	15.0	2002
	500	5.0	75	Harry and Lunau
	500	10.0	25	1972
	500	5.0	20	Clarke and Steen
	500	10.0	6.66	1978
	500	5.0	33.33	Powell 1998
	500	10.0	11.66	
	500	5.0	38.33	
	500	10.0	15.0	
	500	5.0	43.33	
	500	10.0	13.33	
	500	5.0	45.0	
	500	10.0	15.0	
	500	5.0	45.0	
	500	10.0	15.0	
	500	5.00	53.33	
	500	10.0	20	
	1,200	1.6	250	
	500	1.0	16.6	

5.6   Laser-Cutting Applications

As explained in the previous section, lasers are now used for cutting a variety of materials ranging from metals, plastics, ceramics, to composites. Due to high machining precision and processing speed, laser cutting offers tremendous economical benefits in the production lines. It is also suited for difficult-to-material, where conventional machining is either ineffective or uneconomical. Lasers are so flexible that the same laser can be used for a variety of other applications such as welding. Some of the applications of lasers are listed below:

1. Cutting of dieboards
2. Straight and profile cutting of metallic and nonmetallic sheets
3. Cutting of hard and brittle ceramics such as alumina, silicon nitride, etc.
4. Cutting of polymers and polymer matrix composites
5. Cutting of aerospace materials such as titanium- and aluminum-based alloys
6. Cutting of diamonds
7. Cloth and paper cutting
8. Cutting of decorative tiles
9. Cutting of wood
10. Cutting of polymers and polymer matrix composites

5.7   Advances in Laser Cutting

5.7.1   Laser Cutting Assisted by Additional Energy Sources

Extensive research efforts have been directed towards making laser cutting an efficient material-removal process by enhancing the energy inputs to the cutting front. Some of these are the use of coaxial oxygen assist gas, use of supersonic jets, use of multiple jets, and changes in gas compositions (Steen and Kamalu 1983). In addition, an alternate method for enhancing the energy input to the cutting process has been devised by employing an electric arc on the surface of the workpiece (Clarke and Steen 1978). Typical arrangement of arc-augmented laser-cutting process is shown in Fig. 5.48. Considerable improvements in the cutting speed have been reported when the arc is located on one side of the workpiece and the laser with its coaxial jet is located on the other side. When both arc and the laser with its coaxial jet are located on the same side of the workpiece, the arc wanders around the lip of the kerf thus deteriorating the quality of cutting edges (Steen 1980).

The important characteristic of the arc-augmented laser cutting is that the two sources of energy namely the laser and the arc act more in cooperation rather than merely adding to the total energy. It has been reported that during arc-augmented laser cutting of 2 mm thick mild steel with laser power of 1870 W and arc power up to 2 kW (i.e., total power around 3.8 kW) results in substantial increase in cutting speed with the cutting quality (parallel sides and no unusual side burning) comparable

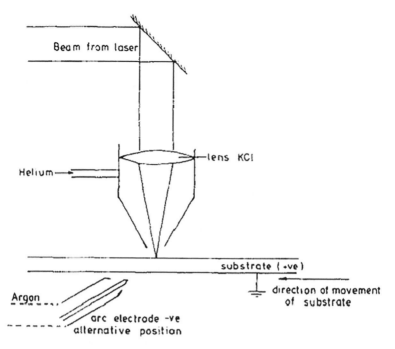

Fig. 5.48 Schematic of arc augmented laser-cutting process. (Reprinted from Steen 1980. With permission. Copyright American Institute of Physics.)

with that obtained using laser alone. Above 3.8 kW, there is no significant increase in cutting speed with the input power; however, the cutting quality deteriorates due to excessive side melting on the arc side of the workpiece (Steen 1980). A similar effect is presented in Fig. 5.49 for the arc-augmented laser cutting of 4 mm mild steel which indicates the initial increase in cutting speed with arc power followed by a region of almost constant cutting speed above arc power of around 6 kW. The figure also indicates the substantial extension of cutting speeds obtained with oxygen-assisted laser cutting by assisting it with the electric arc. Thus optimum laser-cutting speed and the quality of the cuts in arc-augmented laser cutting depend on the judicious selection of laser and arc powers (Steen and Kamalu 1983).

5.7.2 Underwater Cutting

Recently, underwater laser cutting have attracted considerable interests for the machining of heat sensitive materials while minimizing the thermal damages such as formation of cracks and/or chemical degradation. This concept is extended to laser cutting from the concept of laser etching in water. When underwater cutting is used with the CO_2 laser, a dry zone must be created at the workpiece surface because water does not transmit the radiation of CO_2 laser well. Various schemes of underwater and

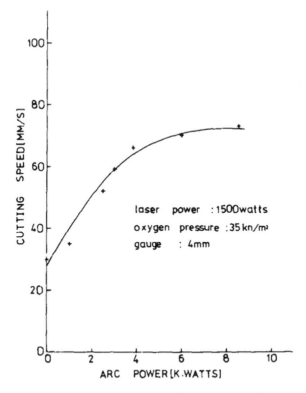

Fig. 5.49 Variation of cutting speed with arc power in arc-augmented laser cutting (Reprinted from Steen and Kamalu 1983. With permission. Copyright M. Bass, Editor.)

water-assisted laser cutting have been suggested; however, the most distinctive is the one in which laser light is transmitted along a water jet due to total internal reflection. Here, the 50 mm long water jet of diameter as small as 65–100 μm and jet speed of 200 m/s at 200 bar pressure transport up to 700 W light power corresponding to 21 MW/cm². Figure 5.50 illustrates that the cutting quality of underwater laser cutting with continuous wave laser is better than that obtained in dry air with continuous wave laser; however, pulsed laser beams with dry air ensures best cutting quality (Kruusing 2004a, b). Advantages, both in terms of enhanced cutting quality and speed, have been reported with underwater laser cutting.

5.7.3 Laser Cutting of Composites and Laminates

Conventional machining of composite materials is often difficult and result in poor cutting quality due to differing response of reinforcements and matrix to machining. Due to the thermal nature of the process, laser cutting offers an efficient alternative for cutting composite materials made of a variety of different reinforcements and

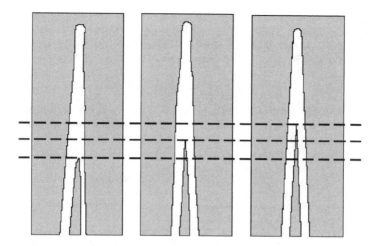

Fig. 5.50 Illustration of effect of water cooling on the V-cut contours in 1 mm thick X5CrNi1810 stainless steel sheet during laser cutting with **(a)** CW laser in atmosphere, **(b)** CW laser in water, and **(c)** pulsed laser in atmosphere (Laser: CO_2 laser; power: 1,000 W; gas: oxygen; cutting speed: 3.5 m/min). (Reprinted from Kruusing 2004b. With permission. Copyright Elsevier.)

matrices. Material removal mechanisms in composites are primarily melting and evaporation, however, when one of the constituent is a polymer, chemical degradation also contributes to the material removal. Laser-cutting quality of composites is highly dependent on the constituents of reinforcement and matrix. For example, aramid fiber reinforced plastic (AFRP) composites give better cut quality (due to polymeric nature of fiber and matrix) than carbon fiber reinforced plastic (CFRP) composites (due to large difference in thermal properties of fiber and matrix) (Cenna and Mathew 1997, 2002).

References

Abakians H, Modest MF (1988) Evaporative cutting of a semitransparent body with a moving CW laser. ASME Journal of Heat Transfer 110:924–930.

Arata Y, Maruo H, Miyamoto I, Takeuchi S (1979) Dynamic behavior in laser gas cutting of mild steel. Transactions of JWRI 8:15–26.

Arata Y, Maruo H, Miyamoto I, Takeuchi S (1981) Quality in laser-gas-cutting stainless steel and its improvement. Transactions of JWRI 10:129–139.

Babenko VP, Tychinsckii VP (1973) Gas-jet laser cutting (review). Soviet Journal of Quantum Electronics 2:399–410.

Bang SY, Modest MF (1991) Multiple reflection effects on evaporative cutting with a moving CW laser. ASME Journal of Heat Transfer 113:663–669.

Boutinguiza M, Pou J, Lusquinos F, Quintero F, Soto R, Perez-Amor M, Watkins K, Steen WM (2002) CO_2 laser cutting of slate. Optics and Lasers in Engineering 37:15–25.

Biyikli S, Modest MF (1988) Beam expansion and focusing effects on evaporative laser cutting. ASME Journal of Heat Transfer 110:529–532.

Bunting KA, Cornfield G (1975) Toward a general theory of cutting: a relationship between the incident power density and the cut speed. ASME Journal of Heat Transfer 97:116–121.

Caiazzo F, Curcio F, Daurelio G, Minutolo FM (2005) Laser cutting of different polymeric plastics (PE, PP and PC) by a CO_2 laser beam. Journal of Materials Processing Technology 159:279–285.

Cenna AA, Mathew P (1997) Evaluation of cut quality of fibre-reinforced plastics: a review. International Journal of Machine Tools and Manufacture 37:723–736.

Cenna AA, Mathew P (2002) Analysis and prediction of laser cutting parameters of fibre reinforced plastics (FRP) composite materials. International Journal of Machine Tools and Manufacture 42:105–113.

Chen K, Yao YL, Modi V (1999) Numerical simulation of oxidation effects in the laser cutting process. The International Journal of Advanced Manufacturing Technology 15:835–842.

Chen SL (1998) The effects of gas composition on the CO_2 laser cutting of mild steel. Journal of Materials Processing Technology 73:147–159.

Chen SL (1999) The effects of high-pressure assistant-gas flow on high-power CO_2 laser cutting. Journal of Materials Processing Technology 88:57–66.

Chryssolouris G (1991) Laser Machining: Theory and Practice. Springer, New York.

Clarke J, Steen WM (1978) Arc augmented laser cutting. Proceedings of the Laser'78 Conference, London.

Clarke J, Steen WM (1979) Arc augmented laser cutting. Proceedings of the Laser'79 Conference, Munich, p. 247.

Steen WM (1980) Arc augmented laser processing of materials. Journal of Applied Physics 51:5636–5641.

Duan J, Man HC, Yue TM (2001a) Modelling the laser fusion cutting process: I. Mathematical modelling of the cut kerf geometry for laser fusion cutting of thick Metal. Journal of Physics D: Applied Physics 34:2127–2134.

Duan J, Man HC, Yue TM (2001b) Modelling the laser fusion cutting process: II. Distribution of supersonic gas flow field inside the cut kerf. Journal of Physics D: Applied Physics 34:2135–2142.

Duan J, Man HC, Yue TM (2001c) Modelling the laser fusion cutting process: III. Effects of various process parameters on cut kerf quality. Journal of Physics D: Applied Physics 34:2143–2150.

Duley WW, Gonsalves JN (1974) CO2 laser cutting of thin metal sheets with gas jet assist. Optics and Laser Technology 6:78–81.

Ghany KA, Newishy M (2005) Cutting of 1.2 mm thick austenitic stainless steel sheet using pulsed and CW Nd:YAG laser. Journal of Materials Processing Technology 168:438–447.

Gonsalves JN, Duley WW (1972) Cutting thin metal sheets with the CW CO_2 laser. Journal of Applied Physics 43:4684–4687.

Gross MS, Black I, Müller WH (2004) Determination of the lower complexity limit for laser cut quality modeling. Modelling and Simulation in Mateials Science and Engineering 12:1237–1249.

Harry JE, Lunau FW (1972) Electrothermal cutting processes using a CO_2 laser. IEEE Transactions on Industry Applications 1A:418–424.

Ivarson A, Powell J, Magnusson C (1991) The role of oxidation in laser cutting stainless and mild steel. Journal of Laser Applications 3:41–45.

Ivarson A, Powell J, Kamalu J, Magnusson C (1994) The oxidation dynamics of laser cutting of mild steel and the generation of striations on the cut edge. Journal of Materials Processing and Technology 40:359–374.

Kaplan AF (1996) An analytical model of metal cutting with a laser beam. Journal of Applied Physics 79:2198–2208.

Kim MJ, Majumdar P (1996) Boundary element method in evaporative laser cutting. Applied Mathematical Modelling 20:274–282.

Kim MJ (2000) Transient evaporative laser-cutting with boundary element method. Applied Mathematical Modelling 25:25–29.

Kim MJ, Chen ZH, Majumdar P (1993) Finite element modelling of the laser-cutting process. Computers and Structures 49:231–241.

Kim MJ, Majumdar P (1995) A computational model for high energy laser-cutting process. Numerical Heat Transfer A 27:717–733.

Kim MJ, Zhang J (2001) Finite element analysis of evaporative cutting with a moving high energy pulsed laser. Applied Mathematical Modelling 25:203–220.

Kondratenko VS (1997) Method of splitting non-metallic materials. US Patent: 5,609,284.

Kruusing A (2004a) Underwater and water-assisted laser processing: Part 1-general features, steam cleaning and shock processing. Optics and Lasers in Engineering 41:307–327.

Kruusing A (2004b) Underwater and water-assisted laser processing: Part 2-Etching, cutting and rarely used methods. Optics and Lasers in Engineering 41:329–352.

Lambert E, Lambert JL, Longueville BD (1976) Severing of glass or vitrocrystalline bodies. US Patent: 3,935,419.

Lamikiz A, Lacalle LN, Sanchez JA, Pozo DD, Etayo JM, Lopez JM (2005) CO2 laser cutting of advanced high strength steels (AHSS). Applied Surface Science 242:362–368.

Lee CS, Goel A, Osada H (1985) Parametric studies of pulsed-laser cutting of thin metal plates. Journal of Applied Physics 58:1339–1343.

Lumley RM (1969) Controlled separation of brittle materials using a laser. American Ceramic Society Bulletin 48:850–854.

Lunau FW, Paine EW (1969) CO_2 laser cutting. Welding and Metal Fabrication 37:9–14.

Majumdar P, Chen ZH, Kim MJ (1995) Evaporative material removal process with a continuous wave laser. Computers and Structures 57:663–671.

Man HC, Duan J, Yue TM (1997) Design and characteristic analysis of supersonic nozzles for high gas pressure laser cutting. Journal of Materials Processing Technology 63:217–222.

Man HC, Duan J, Yue TM (1999) Analysis of the dynamic characteristics of gas flow inside a laser cut kerf under high cut-assist gas pressure. Journal of Physics D: Applied Physics 32:1469–1477.

Miyamoto I, Maruo H (1991) The mechanism of laser cutting. Welding in the World (UK) 29:283–294.

Modest MF, Abakians H (1986) Evaporative cutting of a semi-infinite body with a moving CW laser. ASME Journal of Heat Transfer 108:602–607.

Nesterov AV, Niziev VG (2000) Laser beams with axially symmetric polarization. Journal of Physics D: Applied Physics 33:1817–1822.

Niziev VG, Nesterov AV (1999) Influence of beam polarization on laser cutting efficiency. Journal of Physics D: Applied Physics 32:1455–1461.

O'Neill W, Gabzdyl JT (2000) New developments in laser-assisted oxygen cutting. Optics and Lasers in Engineering 34:355–367.

O'Neill W, Steen W (1994) Review of mathematical models of laser cutting of steels. Lasers in Engineering 3:281–299.

Pascual-Cosp J, Valle AJ, Garcia-Fortea J, Sanchez-Soto PJ (2002) Laser cutting of high-vitrified ceramic materials: development of a method using a Nd:YAG laser to avoid catastrophic breakdown. Materials Letters 55:274–280.

Petring D, Abels P, Beyer E (1988) The absorption distribution as a variable property during laser beam cutting. Proceedings of the International Congress on Applications of Lasers and Electro-Optics (ICALEO), Santa Clara, pp. 293–302.

Pietro PD, Yao YL (1994) A new technique to characterize and predict laser cut striations. International Journal of Machine Tools and Manufacture 35:993–1002.

Powell J, Ellis G, Young CD, Menzies PF, Scheyvaerts (1987) CO_2 laser cutting of non-metallic materials. In: Steen WM (ed.) Proceeding of 4th International Conference on Lasers in Manufacturing, Birmingham, UK, pp. 69–82.

Powell J (1998) CO_2 Laser Cutting. Springer, London.

Rajaram N, Sheikh-Ahmad J, Cheraghi SH (2003) CO_2 laser cut quality of 4130 steel. International Journal of Machine Tools and Manufacture 43:351–358.

Rao BT, Kaul R, Tiwari P, Nath AK (2005) Inert gas cutting of titanium sheet with pulsed mode CO_2 laser. Optics and Lasers in Engineering 43:1330–1348.

Ready JF (1997) Industrial Applications of Lasers. Academic Press, San Diego, CA.

Roy S, Modest MF (1990) Three-dimensional conduction effects during scribing with a CW laser. Journal of Thermophysics and Heat Transfer 4:199–203.

Shanjin L, Yang W (2006) An investigation of pulsed laser cutting of titanium alloy sheet. Optics and Lasers in Engineering 44:1067–1077.

Schulz W, Simon G, Urbassek HM, Decker I (1987) On laser fusion cutting of metals. Journal of Physics D: Applied Physics 20:481–488.

Schuöcker D, Müller P (1987) Dynamic effects in laser cutting and formation of periodic striations. Proceedings of SPIE (International Society for Optical Engineering) 801:258–264.

Schuöcker D (1986) Theoretical model of reactive gas assisted laser cutting including dynamic effects. Proceedings of SPIE (International Society for Optical Engineering) 650:210–219.

Schuöcker D (1983) Material removal mechanism of laser cutting. Proceedings of SPIE (International Society for Optical Engineering) 455:88–93.

Steen WM, Kamalu JN (1983) Laser cutting. In: Bass M (ed.) Laser Materials Processing. North-Holland, Amsterdam, The Netherlands, pp. 15–111.

Steen WM (1991) Laser Material Processing. Springer, London.

Tabata N, Yagi S, Hishii M (1996) Present and future of lasers for fine cutting of metal plate. Journal of Materials Processing Technology 62:309–314.

Tani G, Tomesani L, Campana G (2003) Prediction of melt geometry in laser cutting. Applied Surface Science 208–209:142–147.

Tani G, Tomesani L, Campana G, Fortunato A (2004) Quality factors assessed by analytical modelling in laser cutting. Thin Solid Films 453–454:486–491.

Thawari G, Sundar JK, Sundararajan G, Joshi SV (2005) Influence of process parameters during pulsed Nd:YAG laser cutting of nickel-base superalloys. Journal of Materials Processing Technology 170:229–239.

Tsai CH, Liou CS (2001) Applying an on-line crack detection technique for laser cutting by controlled fracture. International Journal of Advanced Manufacturing Technology 18:724–730.

Tsai CH, Liou CS (2003) Fracture mechanism of laser cutting with controlled fracture. Journal of Manufacturing Science and Technology 125:519–528.

Tsai CH, Chen CJ (2003a) Formation of the breaking surface of alumina in laser cutting with a controlled fracture technique. Proceedings of the Institution of Mechanical Engineers, Part B: Journal of Engineering Manufacture 217:489–497.

Tsai CH, Chen HW (2003b) Laser cutting of thick ceramic substrates by controlled fracture technique. Journal of Materials Processing Technology 136:166–173.

Tsai CH, Chen CJ (2004) Application of iterative path revision technique for laser cutting with controlled fracture. Optics and Lasers in Engineering 41:189–204.

Unger U, Wittenbecher W (1998) The cutting edge of laser technology. Glass 75:101–102.

Vicanek M, Simon G, Urbassek HM, Decker I (1987) Hydrodynamical instability of melt flow in laser cutting. Journal of Physics D: Applied Physics 20:140–145.

Vicanek M, Simon G (1987) Momentum and heat transfer of an inert gas jet to the melt in laser cutting. Journal of Physics D: Applied Physics 20:1191–1196.

Chapter 6
Three-Dimensional Laser Machining

6.1 Introduction

The previous two chapters dealt with the one- dimensional (drilling) and two-dimensional (cutting) laser machining techniques. Various laser machining approaches have been developed to facilitate the three-dimensional material removal during shaping of materials. Two most popular approaches for three-dimensional machining are laser-assisted machining (LAM) and laser machining (LM). LAM is a traditional machining approach assisted by laser energy, whereas LM is nontraditional machining approach which uses a laser beam as a direct tool for material removal. The detailed discussion of these approaches form the subject of this chapter.

6.2 Laser-Assisted Machining

LAM is a novel machining technique which combines the traditional machining methods such as turning, milling, grinding, etc. with the laser technology. The process is specially developed for the machining of difficult-to-machine materials such as ceramics and hard metals. Conventional machining of these materials is often costly due to slow machining speeds or is associated with machining defects such as surface and subsurface cracks, undesirable surface finish, etc. It has been reported that conventional grinding and diamond machining of structural ceramics represent 60–90% of the total cost of the final product. In this context, LAM offers an attractive alternative for the cost-effective machining of ceramics. In addition, due to recent developments in the high energy laser sources, energy delivery mechanisms, and process automation, LAM process can be configured to achieve higher material-removal rates, precise control over machined geometry, increased tool life, and significant reduction in man and machine time per part (Chrysolouris et al. 1997).

N.B. Dahotre and S.P. Harimkar, *Laser Fabrication and Machining of Materials.*
© Springer 2008

6.2.1 LAM Process

Schematic of the LAM approach for turning and grinding of workpiece is presented
in Fig. 6.1 (Chrysolouris et al. 1997). As indicated in the figure, in LAM, an intense
beam of laser is used to heat the workpiece surface ahead of the traditional cutting

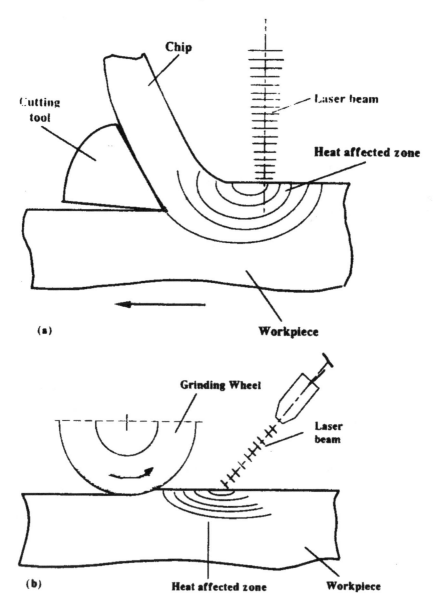

Fig. 6.1 Schematic of laser-assisted machining approaches in (**a**) turning and (**b**) grinding.
(Reprinted from Chryssolouris et al. 1997. With permission. Copyright American Society of
Mechanical Engineers.)

tool without phase changes (such as melting and evaporation) in the workpiece. The popular sources of lasers such as Nd:YAG or CO_2 lasers can be used for heating the workpiece. The traditional cutting tool may be a single point cutting tool as in laser-assisted turning (Fig. 6.1a) or a multipoint cutting tool such as a grinding wheel as in laser-assisted grinding (Fig. 6.1b). In both the configurations, laser heating causes the softening of material thereby changing the deformation mechanism from brittle to ductile during subsequent cutting. The softening of the material refers to the loss of bulk modulus and other mechanical properties of the workpiece upon laser heating such that cutting forces required for LAM are reduced. Another effect associated with the laser heating of workpiece ahead of the cutting tool is the heating of the shear plane as the chip is formed. Such heating of the shear plane may result in advantages such as decreased cutting forces, increased material removal rate, increased tool life, and improved surface conditions expressed in terms of surface roughness, residual stresses, and flaw distributions (Copley et al. 1983). In the context of machining of ceramic materials, the cost reductions in the range 60–80% are expected with LAM (Chrysolouris et al. 1997). The extent of these advantages may be influenced by the thermal aspects of laser heating, the important parameters being the distribution of temperature and thermal gradient. In addition, the distance between the position of laser spot on the workpiece and the position of cutting tool needs to be optimized by taking into consideration the speed of workpiece motion such that the laser-heated region of the workpiece is quickly machined without being cooled during its motion prior to the cutting action of the tool.

6.2.2 Analysis of LAM Process

During LAM, workpiece surface is heated to the temperature sufficient to soften the material such that subsequent ductile deformation facilitates the material removal during cutting. In addition, close control of the temperature is required to avoid the detrimental effects such as melting, heat treatment of the workpiece, and softening of the cutting tool. Thus, effectiveness of the LAM process is primarily determined by the temperature of the workpiece when irradiated with the laser beam and also at the onset of cutting. Hence, most of the modeling efforts for the LAM have been primarily directed towards predicting the temperature distribution in the workpiece during machining. Early efforts on the modeling for LAM started with the predictions of temperature distributions in the rotating workpiece heated with translating laser source without any material removal. These studies were based on transient, three-dimensional thermal models and resulted in good agreement between the predicted surface temperatures (using thermal model) and the measured surface temperature (using laser pyrometer) for silicon nitride workpieces for a wide range of operating conditions (Pfefferkorn et al. 1997; Rozzi et al. 1998a, b). Recently, these modeling efforts were extended to incorporate the physical processes associated with material removal during LAM to develop a comprehensive thermo-mechanical model for LAM (Rozzi et al. 2000a). This section is based on a transient, three-dimensional heat transfer model proposed by Rozzi et al. for the LAM of rotating silicon nitride

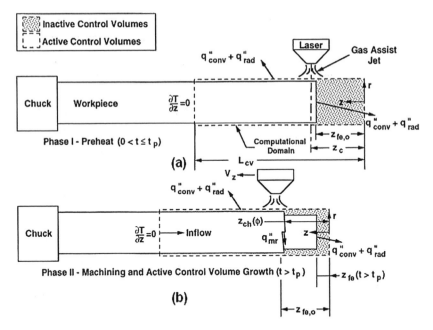

Fig. 6.2 Schematic of (**a**) preheating and (**b**) material removal geometries used for modeling the laser-assisted machining process. (Reprinted from Rozzi et al. 2000a. With permission. Copyright Elsevier.)

workpieces. The model is directed towards the prediction of surface temperature history in the workpiece while taking into account the laser heating and materials removal processes associated with LAM. Once such thermo-mechanical model is validated through comparisons with experimental results, it can help in establishing intelligent control systems capable of optimizing LAM process.

The schematic of the model formulation for LAM process is shown in Fig. 6.2. The cylindrical coordinate system given by (r, ϕ, z) is fixed relative to the laser. The transient numerical calculations were carried out in the following two phases (Rozzi et al. 2000a).

1. Phase I: This is a preheating phase where the laser is used to heat the free end of the workpiece. During this phase, there is no translational motion of the workpiece relative to the laser beam and also no material removal is initiated.
2. Phase II: This is a materials removal phase in which the free end of the workpiece is translated relative to the laser beam, and material removal is initiated.

The geometry of the workpiece after the initiation of phase II (material removal) with a cutting tool of radius $r_t = 0$ is shown in Fig. 6.3. The inset of Fig. 6.3 presents the materials removal plane corresponding to the actual tool geometry (with lead angle, $\Omega/_1 \neq 0$) and that corresponding to the approximated plane in modeling (with lead angle, $\Omega/_1 = 0$). At the actual material removal plane, the boundary between the

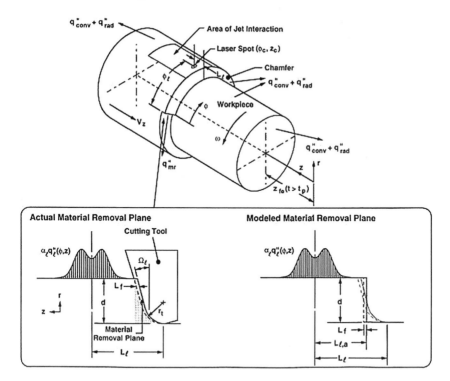

Fig. 6.3 Geometry of the laser-assisted machining process. Inset shows the actual and the approximated material removal planes. (Reprinted from Rozzi et al. 2000a. With permission. Copyright Elsevier.)

removed and the unremoved material is represented by the helical chamfer defined by the geometry of the cutting tool and feed rate. The important parameters which define the location of the material removal at $\phi = 0$ on the r–z plane and its position relative to the laser are ϕ_1 (circumferential distance between the laser center and material removal plane) and L_1 (laser–tool lead distance). The average distance $L_{1,a}$ between the laser center and the edge of the approximated material removal plane is given by:

$$L_{1,a} = L_1 - (d/2)\tan(\Omega_t) - r_t\left[1 - \sin(\Omega_1)\right]\left\{[\tan(\pi/4 - \Omega_1/2)]^{-1} - \tan(\Omega_1)\right\} \quad (6.1)$$

The governing equations for the transient heat transfer in the rotating workpiece with a nonrotating coordinate system attached to the translating laser and cutting tool is given by (Rozzi et al. 2000a):

$$\frac{1}{r}\frac{\partial}{\partial r}\left(kr\frac{\partial T}{\partial r}\right) + \frac{1}{r^2}\frac{\partial}{\partial \phi}\left(\frac{\partial T}{\partial \phi}\right) + \frac{\partial}{\partial z}\left(k\frac{\partial T}{\partial z}\right) + q''' = \rho c_p \omega \frac{\partial T}{\partial \phi}$$
$$+ \rho c_p V_z \frac{\partial T}{\partial z} + \rho c_p \frac{\partial T}{\partial t}, \quad (6.2)$$

where k is the thermal conductivity; ρ is the density; c_p is the specific heat; ω is the workpiece rotation speed; and q''' is the volumetric heat generation. Since the diameter of the laser beam (D_l) is smaller than the workpiece diameter, the following function (f_l) can be used to approximate the area of laser heating.

$$f_1(r,\phi,z) = \sqrt{[r(\phi - \phi_c)]2 + (z - z_c)2} / r_1 \leq 1, \tag{6.3}$$

where ϕ_c and z_c are the circumferential and axial locations of the laser center; and r is the radial location of laser heating for both the unmachined (r_w) and machined $(r_w = r_w - d)$ workpiece surfaces. The boundary conditions are given below.

Unmachined workpiece surface:

$$k\left(\frac{\partial T}{\partial r}\right)_{r=r_w} = \alpha_1\, q_1'' - q_{conv}'' - E(T), \tag{6.4}$$

for $z > z_{ch}$ (ϕ) and $f_1(r,z,\phi) \leq 1$ on the laser spot, and

$$k\left(\frac{\partial T}{\partial r}\right)_{r=r_w} = -q_{conv}'' - E(T), s \tag{6.5}$$

for $z > z_{ch}$ (ϕ) and $f_1(r,z,\phi) > 1$ on the laser spot.

Machined surface:

$$k\left(\frac{\partial T}{\partial r}\right)_{r=r_{r,m}} = \alpha_{1,m}\, q_1'' - q_{conv}'' - E(T), \tag{6.6}$$

for $z < z_{ch}$ (ϕ) and $f_1(r,z,\phi) \leq 1$ on the laser spot, and

$$k\left(\frac{\partial T}{\partial r}\right)_{r=r_{w,m}} = -q_{conv}'' - E(T), \tag{6.7}$$

for $z < z_{ch}$ (ϕ) and $f_1(r,z,\phi) > 1$ on the laser spot. Here, α_1 is absorptivity of the surface to the laser radiation; q'' is the heat flux; z_{ch} is the axial location of chamfer; E is the emissive power; and subscripts "1" and "conv" refers to laser beam and convection, respectively.

At the centerline of the workpiece $(r = 0)$, the conditions can be given as:

$$k\left(\frac{\partial T}{\partial r}\right)_{r=0} = 0 \tag{6.8}$$

At the interface between the machined and the unmachined material, the energy balance can be given as:

$$k \left(\frac{\partial T}{\partial r} \right)_{z=z_{\text{ch}}(\phi)} = q''_{\text{conv}} + E(T), \tag{6.9}$$

for $r_{\text{w,m}} \leq r \leq r_{\text{w}}$ and $0 < \phi \leq 2\pi - \phi_{\text{flank}}$, and

$$k \left(\frac{\partial T}{\partial r} \right)_{z=z_{\text{ch}}(\phi)} = -q''_{\text{flank}} \tag{6.10}$$

for $r_{\text{w,m}} \leq r \leq r_{\text{w}}$ and $2\pi = \phi_{\text{flank}} < \phi \leq 2\pi$. Here, subscript "flank" refers to the tool flank face. In addition, the condition at the end faces of the computational domain is given as:

$$k \left(\frac{\partial T}{\partial r} \right)_{\substack{z=z_{\text{Fe,o}}(t \leq t_p) \\ z=z_{\text{Fe}}(t > t_p)}} = q''_{\text{conv}} + E(T), \tag{6.11}$$

where, $z_{\text{Fe,o}}$ is the initial axial location of the workpiece free end; and t_p is the pre-heat time. In addition, $z_{\text{Fe}}(t > t_p) = z_{\text{Fe,o}} - V_z(t - t_p)$. At the inflow boundary, a zero-heat flux boundary condition can be described as:

$$\left(\frac{\partial T}{\partial r} \right)_{z=L_{\text{cv}}} = 0 \tag{6.12}$$

Energy advection out of the system at the material removal plane is given by:

$$-\frac{1}{r} \left(\frac{\partial T}{\partial \phi} \right)_{\phi=0_{\text{cv}}} = \rho c_p r \omega (T - T_{\text{ref}}), \tag{6.13}$$

for $r_{\text{w,m}} \leq r \leq r_{\text{w}}$, where T_{ref} is the reference temperature. Circumferential boundary condition away from the material removal location can be given by the continuity between temperature and temperature gradients at ϕ and $\phi + 2\pi$. Thus,

$$T(r,\phi,z) = T(r,\phi+2\pi,z), \tag{6.14}$$

$$\left(\frac{\partial T}{\partial \phi} \right)_{\phi} = \left(\frac{\partial T}{\partial \phi} \right)_{\phi+2\pi}. \tag{6.15}$$

The initial condition when the workpiece is in thermal equilibrium with the surrounding is expressed as:

$$T(r,\phi,z,0) = T_{\infty} \tag{6.16}$$

The detailed description of convection and radiation, and heat generation due to plastic deformation and workpiece–tool friction are given by Rozzi et al. (2000a).

To validate the numerical results, the experiments of LAM were carried out on silicon nitride workpieces using polycrystalline cubic boron nitride (PCBN)-tipped carbide tool insert assisted by a CO_2 laser heating. Figure 6.4 presents the numerical and the experimental results of the workpiece temperatures at three axial locations N1, N2, and N3 at distances 3.4, 6.4, and 9.4 mm from the laser center, respectively. The inset of the figure provides the details of the parameter used in the calculations. The figure indicates that the temperature of the workpiece at each measurement location increases as the laser approaches the measurement location. The initial temperature rise is also facilitated by the axial heat transfer due to conduction. Eventually, the temperature reaches maximum value when the laser passes over measurement location followed by the subsequent decay of temperature due to radial and axial conduction of heat away from the heated surface and the surface convection and radiation losses. The numerical results showed excellent agreement with the measured temperature histories obtained using pyrometer (Rozzi et al. 2000a). Rozzi et al. further extended their efforts to analyze the effect of laser translational velocity, laser–tool lead distance, laser diameter, and laser power on the numerical and the experimental temperature histories at a particular measure-

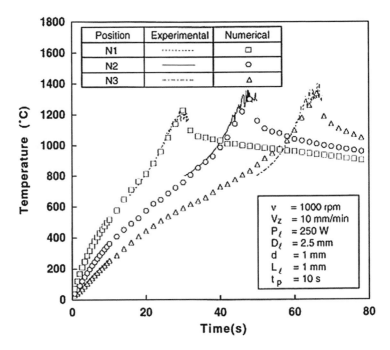

Fig. 6.4 Predicted and experimental surface temperature at three axial locations (N1, N2, and N3) during laser-assisted machining of silicon nitride. (Reprinted from Rozzi et al. 2000a. With permission. Copyright Elsevier.)

ment location. As the laser translational velocity increases and/or the laser–tool lead distance decreases, the amount of energy deposited in the workpiece decreases resulting in lower surface temperatures at each measurement location. In addition, as expected the decrease in laser beam diameter and increase in laser power increases the heat flux and energy deposition resulting in increase in surface temperature (Rozzi et al. 2000b). Similar modeling efforts have been conducted to predict the temperature histories in the mullite workpiece during LAM (Rebro et al. 2002, 2004). Figure 6.5 shows the temperature distribution in the mullite workpiece at the end of preheating phase and at the initiation of material removal phase during LAM. The figure indicates that as the material removal progresses, the quasi-steady state is approached and the temperature near the material removal zone reaches 1,100–1,300 °C corresponding to the desired range for material removal. Most of these modeling efforts were limited to the straight cylindrical workpieces with constant machined diameter such that the geometry of the material removal plane does not change during simulation. Recently, these modeling efforts have been extended to predict the temperature distributions in workpieces with geometrically complex features by means of partially deactivating the control volumes according to the geometry of the machined workpiece (Tian and Shin 2006).

6.2.3 LAM Process Results

6.2.3.1 Cutting Forces and Specific Cutting Energy

The three-dimensional machining geometry and the associated forces in the LAM are presented in Fig. 6.6. The three components of the machining forces are F_c (primary cutting force), F_r (radial force), and F_f (feed force). The geometry of the force balance can be simplified in two dimensions with two force components: F_c (primary cutting force) and F_t (tangential force) (Fig. 6.6b). These forces are measured using dynamometers in machining practices. The chip is formed by a localized shear process that takes place over a very narrow region referred to as the shear zone. The important parameter in this machining analysis is the shear angle(ϕ) defined as the angle the primary shear plane makes with the horizontal motion of the tool. In case of continuous chip formation, the shear angle is primarily determined by the chip thickness. In most of the LAM processes with ceramics, the chips are segmented and accurate measurement of chip thickness is difficult. Hence, an alternate equation is developed for calculating the shear angle, ϕ (Lei et al. 2000):

$$\frac{h}{\sin(\phi + \lambda - \alpha)} = \frac{t}{\sin(\phi)\cos(\lambda)}, \qquad (6.17)$$

Where h is the chip–tool contact length; t is the uncut chip thickness; λ is the friction angle; and α is the rake angle.

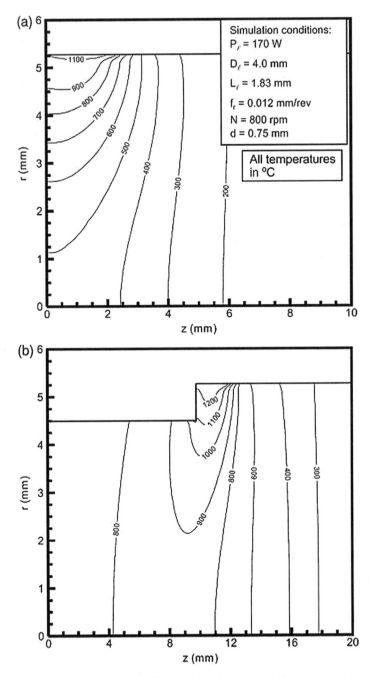

Fig. 6.5 Temperature distribution in mullite workpiece (**a**) at the end of preheating phase ($t = 10$ s) and (**b**) at the initiation of material removal during laser-assisted machining (depth of cut: 0.75 mm, cut length: 10 mm, $t = 72$ s). (Reprinted from Rebro et al. 2004. With permission. Copyright Elsevier.)

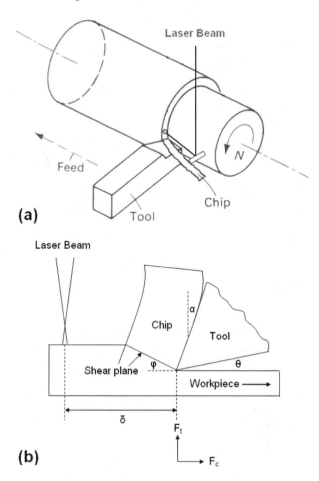

Fig. 6.6 (a) Geometry of laser-assisted machining and (b) geometry of force balance on chip in two dimensions

Figure 6.7 shows the variation of three components of machining forces during LAM of silicon nitride. Various parameters used for this study were: laser power: 280 W; cutting speed: 0.4 m/s; feed: 0.02 mm/rev; depth of cut: 1 mm; preheat time: 11 s; pyrometer–laser lead: 0.85 mm; laser beam diameter: 3.4 mm; laser–tool lead: 2.6 mm; and circumferential laser–tool angle: 55°. The results indicated similar variation for all the three force components without any significant changes (increase/decrease) during machining. The results suggest that tool wear during machining does not significantly influence the machining forces which may be due to the formation of thin glassy work-piece material, acting as lubricant, on the wear land (Lei et al. 2000).

One of the major advantages of the LAM process is the reduction in the cutting forces. This has been demonstrated for various materials such as magnesia-partially-stabilized zirconia (PSZ), Al_2O_3 particle reinforced aluminum matrix composites, etc. (Pfefferkorn et al. 2004; Wang et al. 2002). The results of the force ratio as a function of laser power during LAM of 15 mm diameter magnesia-PSZ workpieces are presented in Fig. 6.8.

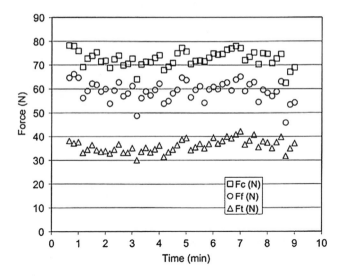

Fig. 6.7 Variation of machining forces with time during laser-assisted machining of silicon nitride. (Reprinted from Lei et al. 2000. With permission. Copyright Elsevier.)

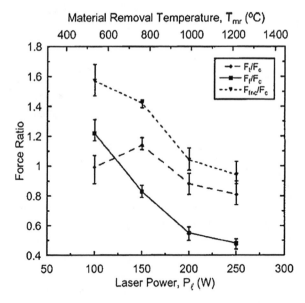

Fig. 6.8 Variation of force ratios with laser powers during laser-assisted machining of magnesia-partially-stabilized zirconia. (Reprinted from Pfefferkorn et al. 2004. With permission. Copyright American Society of Mechanical Engineers.)

The results were obtained using $1.5\,kW$ CO_2. The machining parameters employed in the tests were: depth of cut $= 0.5\,mm$, feed rate $= 0.02\,mm/rev$, laser-too lead $= 1.6\,mm$, spindle speed $= 800\,rpm$, and preheat time $= 6\,s$. The results indicated that the force ratios: F_t/F_c, F_f/F_c, F_{fric}/F_c decreases with increasing material removal temperature (corresponding to increasing laser power) (Pfefferkorn et al. 2004).

In addition to the cutting forces, the important parameter in machining is the specific energy for material removal. The specific energy (u_c) is given by the ratio of the total energy (given by product of cutting force, F_c and cutting velocity, V) and the material removal rate (given by product of depth of cut (t) width of cut (w), and cutting velocity (V)) (Cohen 1989):

$$u_c = \frac{F_c V}{t w V}. \tag{6.18}$$

Due to softening of the material during laser heating, its strength decreases resulting in correspondingly decreased energy for material removal in LAM. Figure 6.9 presents the results of specific cutting energy calculations during the LAM of mullite (reaction sintered from Al_2O_3–71.34 wt. % and SiO_2–25.83 wt. %) using 1.5 kW CO_2 laser. The results are normalized with the specific cutting energy (4.73 J/mm^3) corresponding to the laser power of 170 W. The figure clearly indicates that LAM is associated with marked decrease in the specific cutting energies compared to conventional machining without laser assist. Reduction in specific cutting energy from no-assist to the LAM case (with laser power of 170 W) is around 30%. Further decrease in specific energy is expected using higher laser powers during LAM. This is indicated by the decreasing trend of specific cutting energy with increasing laser power during LAM. The figure also indicates that the decreasing specific energy with laser powers is a direct consequence of higher material removal temperatures (Rebro et al. 2002).

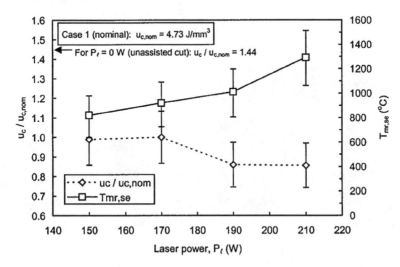

Fig. 6.9 Variation of normalized specific cutting energy and estimated material removal temperature with laser power during laser-assisted machining of reaction sintered mullite. (Reprinted from Rebro et al. 2002. With permission. Copyright American Society of Mechanical Engineers.)

6.2.3.2 Tool Wear

During machining, the cutting tools are subjected to wear due to various cutting and frictional forces acting at the contact surfaces of the tool with workpiece and chips. Due to higher temperatures associated with LAM, assessment and monitoring of tool wear becomes important. The wear behavior of the tool determines the overall life of the tool and thus ultimately influences the economics of the machining process. Typical wear profiles of the PCBN tool insert during LAM of magnesia- PSZ after 52 min of cutting are shown in Fig. 6.10. The LAM conditions for these wear profiles were: laser power: 200 W, depth of cut: 0.5 mm, feed: 0.02 mm/rev, and material removal temperature: 1,000 °C (Pfefferkorn et al. 2004). Similar wear profiles were also observed during the LAM of silicon nitride using PCBN tool inserts. As indicated in the figure, the grooves on the primary and secondary faces of the flank seem to be the result of abrasion of the tool. In addition, the adhesion of the tool with the workpiece or chip may result in failure of the tool (Lei et al. 2001).

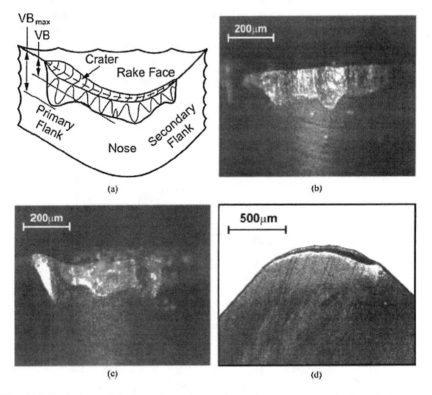

Fig. 6.10 Typical morphology of the tool wear during laser-assisted machining of PSZ using PCBN insert. (**a**) Schematic, images of (**b**) nose, (**c**) primary flank, and (**d**) rake face. (Reprinted from Pfefferkorn et al. 2004. With permission. Copyright American Society of Mechanical Engineers.)

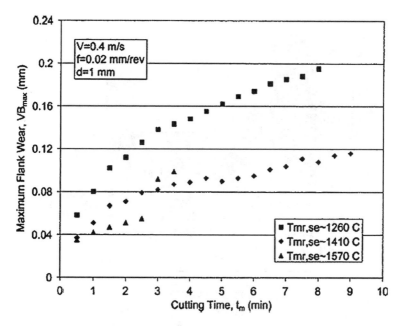

Fig. 6.11 Maximum tool flank wear (VB_{max}) as a function of cutting time during LAM of silicon nitride. (Reprinted from Lei et al. 2001. With permission. Copyright American Society of Mechanical Engineers.)

Tool wear during LAM can be characterized in terms of maximum flank wear (VB_{max}) and average flank wear (VB) measured from the geometry of tool after machining tests. Figure 6.11 presents the variation of maximum flank wear with the machining time during LAM of silicon nitride using PCBN tool insert at three different material removal temperatures (12,60 °C, 1,410 °C, and 1,570 °C). The results were obtained for a cutting speed of 0.4 m/s, feed rate of 0.02 mm/rev, and cutting depth of 1 mm. The results indicate that the tool wear increases progressively with the machining time. In addition, the tool wear decreases as the temperature of the workpiece increases as expected from the reduction in strength of the workpiece material at higher temperatures. It has been observed that for a typical tool wear limit ($VB = 0.3$ mm), life of the tool was 42 min at a material removal temperature of 1,410 °C. Such longer tool life in the LAM is comparable with that obtained in metal cutting. However, there is a maximum temperature limit beyond which the advantages of LAM ceases due to accelerated wear of the tool possibly because of the complex physical and chemical processes associated with high temperature wear of the tool (Lei et al. 2001).

LAM is expected to offer tremendous economic advantages compared with the machining with conventional ceramic and carbide tool. Figure 6.12 presents the total cost for machining a 1 m long workpiece of diameter 25 mm with a cut depth of 0.76 mm (corresponding to the total material removal volume of 58,000 mm³) using different machining processes. The results indicate that even though LAM is associated with laser operation cost, it decreases the total machining time (and thus

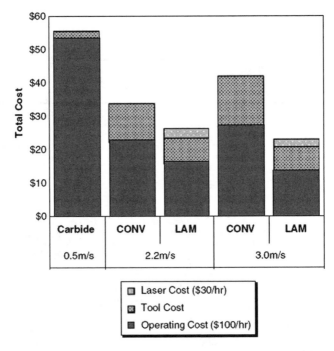

Fig. 6.12 Comparison of total cost for machining a workpiece 1 m long and 25 mm in diameter using conventional machining and laser-assisted machining. (Reprinted from Anderson et al. 2006. With permission. Copyright Elsevier.)

the total machining cost) by 30% at a cutting speed of 2.2 m/s. In addition, at high cutting speeds (~3.0 m/s) during LAM, the number of tool changes can be significantly minimized resulting in cost saving of almost 50% (Anderson et al. 2006).

6.2.3.3 Surface Finish and Integrity

One of the important concerns during the machining of materials is related with the surface condition of the machined workpiece. This is often expressed in terms of the surface finish and surface integrity of the workpiece. Surface finish is related with geometric irregularities at the surface and is primarily described in terms of the surface roughness and flaws such as cracks. Surface integrity is related with the metallurgical changes in the workpiece surface associated with the machining process (Field et al. 1989).

As in conventional machining, the surface finish of the workpiece machined during LAM may be influenced by the material removal rates and the morphology of the chip formation. The morphology of the chips generated during LAM is primarily determined by the temperature of the workpiece at the material removal plane. Figure 6.13 presents the various chip morphologies obtained in LAM of silicon

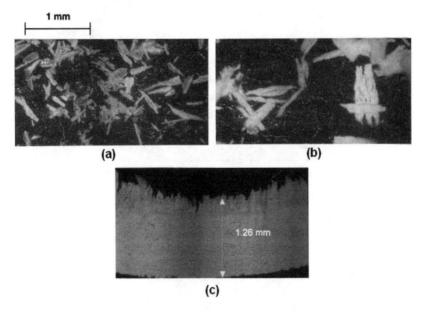

Fig. 6.13 Various chip morphologies during laser-assisted machining of silicon nitride **(a)** fragmented chips at $T < 1,151\,°C$, **(b)** semicontinuous chips at $1,151\,°C < T < 1,309\,°C$, and **(c)** continuous chips at $T > 1,329\,°C$ (T is temperature near material removal location). (Reprinted from Rozzi et al. 2000c. With permission. Copyright American Society of Mechanical Engineers.)

nitride depending on the temperature of the workpiece at the material removal plane. At temperatures less than $1,151\,°C$, the chips are small and fragmented possibly formed due to the brittle fracture of the workpiece. Further increase in the temperature causes a decrease in the viscosity of the glassy phase and resultant chips characteristic of visco-plastic flow associated with machining. This is evident in the formation of fragmented, large semicontinuous chips at the intermediate temperature ($1,151\,°C < T < 1,305\,°C$) and the long continuous chips at higher temperatures ($T > 1,329\,°C$) (Rozzi et al. 2000c).

Various different studies have reported differing trends of variations of surface roughness with machining conditions during LAM. This may be due to the differing fracture and deformation behavior of the various materials during the material removal at high temperature. During the LAM of aluminum/SiC metal matrix composites (MMCs), surface roughness decreased with increase in the preheating temperature. However, at very high temperature, the surface roughness increases. The results were based on the laser-assisted turning of aluminum/SiC MMCs after preheating to $200\,°C$, $300\,°C$, and $400\,°C$ (Barnes et al. 1996). However, it has been reported that the machining conditions during LAM of silicon nitride does not significantly influence the surface roughness of the finished workpiece. This suggests that the surface roughness of the silicon nitride machined by LAM is independent of the material removal temperature. The surface roughness values of silicon nitride subjected to LAM were found to be slightly higher than that obtained with surface grinding (Surface roughness (R_a): 0.3–$0.75\,\mu m$ with LAM and $0.2\,\mu m$ with surface grinding) (Rozzi et al. 2000c).

Similar results have been obtained during LAM of reaction-sintered mullite ceramics. Surface roughness of the mullite workpieces for most of the LAM conditions was around 4 μm, again suggesting that surface roughness is independent of the machining conditions. However, surface roughness was found to be influenced by the extreme machining conditions such as machining without laser-assist (R_a = 5.9 μm) and LAM at high power (R_a = 2.8 μm). In addition, comparison of the surface roughness of mullite machined by LAM and surface grinding showed that the surface roughness is comparable (~4 μm) (Fig. 6.14). However, surface grinding results in higher maximum valley depth (R_v) and peak to peak value (R_t) (Rebro et al. 2002).

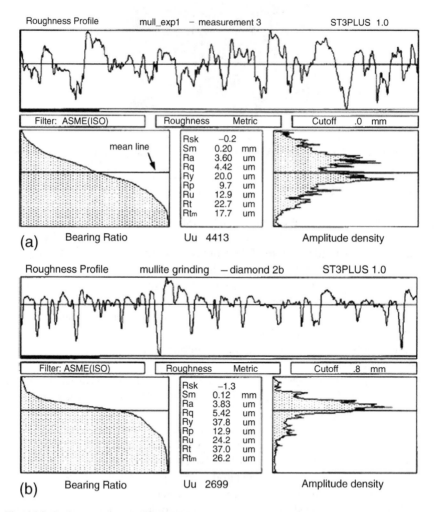

Fig. 6.14 Surface roughness of finished mullite workpiece by (**a**) laser-assisted machining (R_a = 3.60 μm, R_v = 12.9 μm, and R_t = 22.7 μm) and (**b**) surface grinding (R_a = 3.83 μm, R_v = 24.2 μm, and R_t = 37.0 μm). (Reprinted from Rebro et al. 2002. With permission. Copyright American Society of Mechanical Engineers.)

During machining, the surface of the workpiece is compressed and subsequently sheared by the tool such that a plastic deformation zone is formed near the cutting edge. As the machining continues, a part of the plastic deformation zone will be sheared off to form the chip keeping behind the other part of the deformation zone on the surface of the workpiece in highly stressed conditions. The machined surface of the workpiece is often associated with the machining-induced residual stresses due to thermal/mechanical misfit between the surface and the bulk material. The residual stresses induced into the workpiece during machining can be readily determined by x-ray diffraction analysis. It has been reported that LAM of Al_2O_3 particle-reinforced aluminum matrix composites introduces compressive stresses in the surface of the workpiece and the magnitude of these compressive stresses is almost three times of that obtained with conventional machining. The surface compressive stresses in the workpiece are expected to improve the fatigue strength of the workpiece (Wang et al. 2002).

6.3 Laser Machining

The previous section discussed the LAM of materials where the lasers can be used to facilitate the material removal processes in conventional machining setups. The laser intensities in such LAM techniques are generally selected such that regimes of laser–material interactions are limited to localized heating of the workpiece. The primary objectives in such processes were to reduce the tool wear and machining forces by softening the workpiece material ahead of the cutting tool, thus making the material removal relatively easy. Laser machining setups can also be configured in various ways to facilitate the shaping or three-dimensional machining of materials by efficient material-removal processes such as surface melting and/or evaporation. Two distinct machining approaches are explained: one is based on the use of single laser beam, while, the other is based on two intersecting laser beams. These approaches have been extensively demonstrated for the laser machining of hard ceramic materials such as Si_3N_4, SiC, SiAlON, and Al_2O_3.

6.3.1 Machining Using Single Laser Beam

This approach is based on the generation of overlapping shallow multiple grooves by scanning a single laser beam to systematically remove the surface layer of the workpiece material. Vaporization of the material during laser scanning is a primary material-removal mechanism for the formation of each groove. Overlapping of the grooves is realized by applying a continuous feed motion to the workpiece perpendicular to the laser scanning direction. This causes the removal of planar surface layer with orthogonal or cylindrical boundaries. The process is repeated to remove the additional layers of material by applying intermittent feed motion perpendicular to the previously machined surface to either the workpiece or the focusing lens (Wallace and Copley 1989).

One of the earliest arrangements of laser machining in milling configuration is presented in Fig. 6.15. The laser beam is directed along the z-axis, while, the workpiece is moved in x- and y-directions by linear translational stages. During machining, the workpiece is translated relative to the focal spot of the beam in $+x$-direction. The formation of continuous groove is facilitated by adjusting the workpiece translation speed and repetition rate of the laser. If the workpiece translation speed is very high, the material removal craters from the individual pulses can be separated resulting in a discontinuous groove. A series of overlapping continuous grooves, which forms the removal of a thin layer of workpiece material, can be produced by applying the intermittent feed motion to the focal spot of the laser in y'-direction. Subsequent layers of material are then removed by applying the intermittent feed motion in z'-direction to the surface of the previously machined layer. The final geometry of the machined surface can be given by stepwise approximation of the three-dimensional shapes (Hsu and Copley 1990). Figure 6.16 presents the laser-milled steps when the electric vector of the polarized beam was oriented parallel and perpendicular to the side walls of the groove. The results were obtained by using the laser power of 780 W with the laser beam focused either at the surface or 2 mm above the surface. In addition to the steps on planar surfaces, such an approach can also be used for milling the cylindrical surfaces (Fig. 6.17).

To design the laser machining process based on the overlapping multiple grooves (OMGs) approach, it is necessary to understand the formation of groove during laser

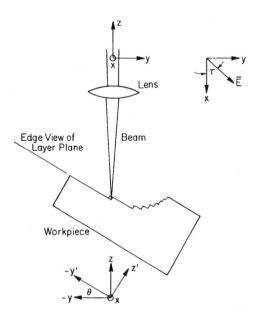

Fig. 6.15 Schematic of the laser machining setup in laser milling configuration. (Reprinted from Hsu and Copley 1990. With permission. Copyright American Society of Mechanical Engineers.)

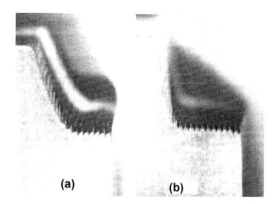

Fig. 6.16 Steps produced in graphite workpiece by laser milling with electric vector (**a**) parallel and (**b**) perpendicular to groove walls. (Reprinted from Hsu and Copley 1990. With permission. Copyright American Society of Mechanical Engineers.)

Fig. 6.17 Laser milled cylindrical surface of graphite workpiece. The figure shows a concave quadrant of 5 mm radius. (Reprinted from Hsu and Copley 1990. With permission. Copyright American Society of Mechanical Engineers.)

interaction with the workpiece. Extensive investigations have been carried out to study the formation of single and overlapping groove shapes during laser machining of silicon nitride ceramic (Wallace and Copley 1986, 1989; Wallace et al. 1986). The significant results obtained during these studies are related with dependence of single groove shape with the longitudinal position of the groove and the beam polarization. Figure 6.18 presents the cross sections of the laser machined groove in silicon nitride at the entrance and the exit surface. As indicated in the figure, the shape of the groove

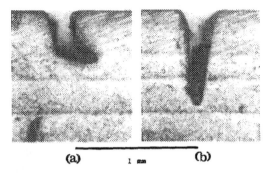

Fig. 6.18 Cross section of the laser machined grooves showing (**a**) curved shape at the entrance and (**b**) symmetric straight shape at the exit surface. Laser machining parameters: laser power = 560 W and sample velocity = 5 cm/s. (Reprinted from Wallace et al. 1986. With permission. Copyright American Institute of Physics.)

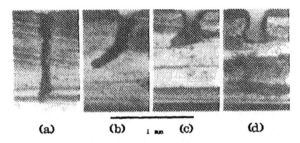

Fig. 6.19   Shape of the laser machined grooves for (**a**) $\theta = 0°$, (**b**) $\theta = 42°$, (**c**) $\theta = 72°$, and (**d**) $\theta = 90°$, where θ is the angle between the electric vector of the beam and the velocity vector of the sample. (Reprinted from Wallace et al. 1986. With permission. Copyright American Institute of Physics.)

is curved at the entrance, whereas, it becomes highly symmetric and straight at the exit surface. Furthermore, the transition from curved groove shape at the entrance to the highly symmetric, straight groove shape at the exit is progressive along the distance from entrance to exit surface. It was proposed that as the beam reaches the exit surface, there will not be any front surface with which beam can interact resulting in flattening of the groove. Once the groove becomes flat, the beam can be reflected only in the forward direction or down the exit surface leading to highly symmetric groove shape at the exit surface (Wallace et al. 1986).

The shape of the laser machined groove is greatly influenced by the beam polarization. Figure 6.19 presents the cross sections of the grooves for various angles (θ) between the electric vector of the polarized beam (E) and the velocity vector of the sample (V_s). As indicated in the figure, as the angle θ increases, the grooves transition from straight and deep cross sectional shape to significantly curved shape. In addition, with increasing angle θ, the depth decreases, while, width and curvature of the groove increases. The formation of the curved groove was related with the asymmetric spatial distribution of the intensity in the laser beam. This effect is illustrated in

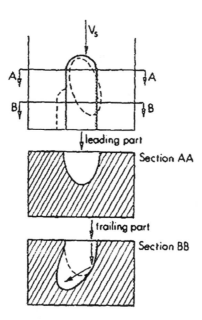

Fig. 6.20 Schematic of the mechanism of formation of curved groove during laser machining. (Reprinted from Wallace et al. 1986. With permission. Copyright American Institute of Physics.)

Fig. 6.20 using an elliptical laser beam. The major axis of the ellipse makes an angle with the translation direction. The shape of the groove due to interaction of the leading part of the beam is highly symmetric (section AA). As the workpiece is translated, the trailing part of the beam will encounter the walls of the groove formed by the leading part. Thus, the trailing part of the beam will be reflected from the walls of the groove resulting in curved shape (section BB). Due to interaction of the laser beam with the groove walls, the polarization of the laser beam is also expected to influence the absorption of radiation (Wallace et al. 1986).

As the shapes of laser machined grooves are influenced by the polarization effects, the material removal rates during laser machining are also expected to be dependent on the angle (θ) between the electric vector of the polarized beam, E and the velocity vector of the sample, V_s. The above analysis was further extended to estimate the material removal rates during single groove machining of silicon nitride for a wide range of incident powers and translation speeds (Wallace and Copley 1986). Figure 6.21 presents the material removal rates as a function of angle θ, using the laser power of 560 W and translation speed of 4.2 cm/s. The insets also show the typical shapes of the laser machined grooves corresponding to various angles θ. As indicated in the figure, the material removal rate is highest at $\theta = 0°$, decreases gradually and reaches its lowest value at $\theta = 90°$. Figure 6.22 shows the effect of laser power and scanning speed on the material removal rate during laser machining of silicon nitride. In addition, the figure indicates that the difference in the material removal rates at $\theta = 0°$ and $\theta = 90°$ decreases as the scanning speed is

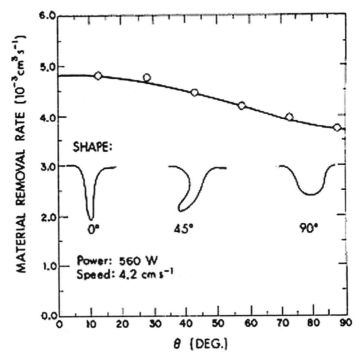

Fig. 6.21 Effect of beam polarization on the material removal rates during single groove laser machining of silicon nitride. θ is the angle between electric vector of the polarized beam and velocity vector of the sample (Reprinted from Wallace and Copley 1986. With permission. Copyright American Ceramic Society.)

increased. The differing trends of the material removal rates with laser scanning speed at various powers are intimately linked with the differing shapes of the laser machined grooves under various laser machining conditions. The material removal rates during laser machining are also influenced by the focusing conditions of the laser beam. Figure 6.23 indicates that material removal rates are slightly more when the focal plane was located at or just above the surface of the sample than when it was located below the surface of the sample. Such an observation is probably due to enhanced absorption due to plasma processes when the focal plane is above the sample surface (Wallace and Copley 1986).

The above analysis of single groove laser machining has been extended for the shaping of silicon nitride by OMGs using a continuous wave carbon dioxide laser. Schematic of the geometry of OMGs milled in a workpiece using a single point cutting tool is presented in Fig. 6.24. For the case of overlapping grooves, the feed (f) is less than the width of single groove (a). The materials removal rate (Z) for the OMGs, approximated by triangular geometries, is given by (Wallace and Copley 1989):

$$Z = D_0 \left(1 - \frac{f}{2a}\right) f V, \tag{6.19}$$

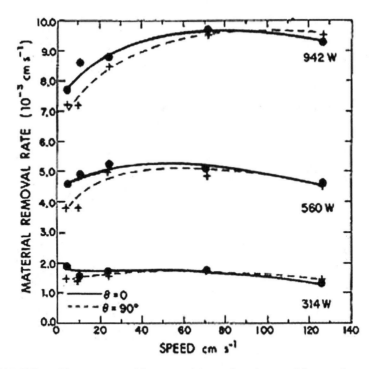

Fig. 6.22 Effect of laser power and laser scanning speed on the material removal rates during single grove laser machining of silicon nitride for $\theta = 0$ and $90°$ (Reprinted from Wallace and Copley 1986. With permission. Copyright American Ceramic Society.)

where D_0 is the single width depth and V is the laser scanning speed. Figure 6.25 presents the materials removal rate (Z) as a function of ratio of feed to groove width (f/a) for three different workpiece translation velocities (71, 128, and 238 cm/s) and incident laser power of 560 W. As indicated in the figure, the calculated values of material removal rates (dashed lines) predicted from the above equations applicable for milling with single-point cutting tool decreases with decreasing feed for $f < a$. However, material removal rates (solid lines) corresponding to laser milling with OMGs approach remains constant and equal to single groove values until a critical feed value is reached. If the feed is reduced further below the critical feed value, the material removal rate decreases. The difference in the observations is a direct consequence of the fundamental difference in the nature of material removal in single point milling laser milling. This is explained for the cases of $f = 0$ and $f = a/2$ (Fig. 6.26). For $f = 0$ and machining with multiple passes, the material is removed only during the first pass with single point cutting tool, whereas, material removal continues during the subsequent passes with laser machining as long as the surface is within the depth of focus (Fig. 6.26a). For $f = a/2$ and machining with two passes, additional material removal in the black and cross-hatched region is expected during the second pass due to reflected and direct exposure to laser radiation, respectively (Fig. 6.26b).

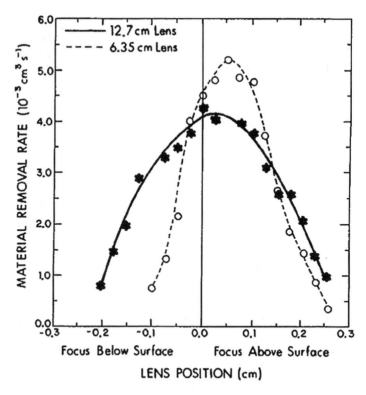

Fig. 6.23 Effect of focusing conditions on the material removal rates during single grove laser machining of silicon nitride with laser power of 520 W, traverse speed of 70.6 cm/s, and $\theta = 0$ (Reprinted from Wallace and Copley 1986. With permission. Copyright American Ceramic Society.)

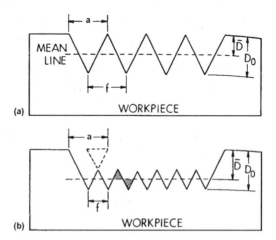

Fig. 6.24 Schematic of **(a)** nonoverlapping multiple grooves where feed, f = groove width, a and **(b)** overlapping multiple grooves where $f < a$. (Reprinted from Wallace and Copley 1989. With permission. Copyright American Society of Mechanical Engineers.)

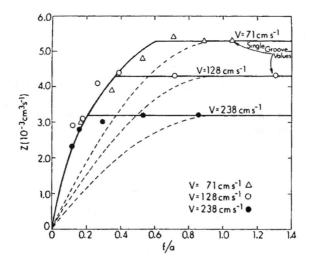

Fig. 6.25 Variation of material removal rates with a ratio of feed to groove width (*f/a*) for three sample translation velocities: 71, 128, and 238 cm/s. (Reprinted from Wallace and Copley 1989. With permission. Copyright American Society of Mechanical Engineers.)

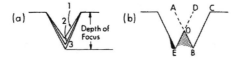

Fig. 6.26 Effect of feed on the material removal and groove formation during laser machining with multiple passes: (**a**) for $f = 0$ with three passes; and (**b**) for $f = a/2$ with two passes. (Reprinted from Wallace and Copley 1989. With permission. Copyright American Society of Mechanical Engineers.)

6.3.2 Machining Using Intersecting Laser Beams

In the laser machining concept using a single laser beam, the material removal mechanism is either melting or evaporation and the shaping of the workpiece is realized by making the multiple overlapping grooves. During machining all the material is removed in the molten or vaporized form. Hence, enormous energy of the laser beam is required for material removal in molten or vaporized form resulting in low overall energy efficiency of the process. In addition, the machining using single laser beam is limited to the production of simple three-dimensional shapes. Attempts have been made to overcome the limitations of the single-beam laser machining by employing multiple intersecting laser beams. One such concept of a three-dimensional laser machining is based on two intersecting laser beams (Chrysolouris et al. 1988). In addition to the three-dimensional machining capabilities, intersecting laser beams offer the significantly increased energy efficiency.

   The machining involves directing two laser beams on the workpiece such that the axis of the first laser intersects with the axis of the second laser. Each laser produces the corresponding blind cutting kerfs which converges and results in the solid stock removal. Since material removal is not completely due to melting and/or evaporation as in drilling and cutting, the machining with intersecting beams offer very high energy efficiency. The concepts of laser turning and milling have been introduced with the machining using intersecting laser beams. In laser turning, the two intersecting beams are directed on a rotating workpiece. When the laser beams are perpendicular to each other with one laser beam being parallel to the axis of the workpiece, the material stock removal is in the form of concentric rings (Fig. 6.27a). Moreover, when the laser beams intersect at an angle, the material stock removal is in the form of helix (Fig. 6.27b). In laser milling, the two laser beams are directed on the translating workpiece at an oblique angle such that material removed has a prismatic shape with triangular cross section (Fig. 6.27c) (Chrysolouris 1991).

   The process of laser machining using intersecting beams (e.g., laser turning) is based on the creation of blind kerfs where the material removal proceeds on the two nearly perpendicular planes: cutting front and drilling front. Hence, optimization of the laser machining process in terms of material removal rates, surface quality, and dimensional accuracy necessitates the understanding of the blind cutting. The early

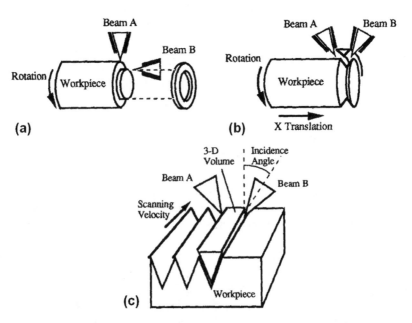

Fig. 6.27 Laser machining with intersecting beams: (**a**) laser turning with material removal in the form of rings, (**b**) laser turning with material removal in the form of helical elements, and (**c**) laser milling with material removal in the form of prismatic elements. (Reprinted from Chryssolouris 1991. With permission. Copyright Springer.)

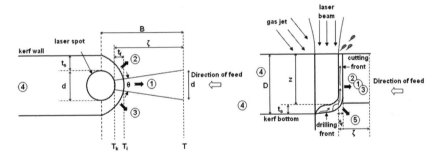

Fig. 6.28 Schematic of the various material removal regions during laser blind cutting. (Reprinted from Chryssolouris et al. 1988. With permission. Copyright American Society of Mechanical Engineers.)

efforts to model the laser machining process based on blind cutting were conducted by taking into account the energy balance approach for predicting depth of kerfs from the material properties and laser parameters (Chrysolouris et al. 1988). This analysis based on classical theory of quasistationary heat flow due to a moving source of heat is presented here (Rosenthal 1946).

Schematic of the machining front geometry during laser blind cutting is presented in Fig. 6.28. It also shows the various regions of material removal and heat-affected zones associated with blind cutting. The laser beam of spot diameter (d) is incident on the surface of workpiece moving with a velocity (v) under the laser beam. Laser irradiation causes the vaporization from the surface of the workpiece to the depth (z) thus forming a cylinder of diameter d and height z corresponding to the vaporized material. The vaporized cylinder with a temperature T_k forms a region of molten material surrounding it. During machining, the vaporized cylinder moves along the feed direction with velocity (v) corresponding to the feed rate of the workpiece. Based on the geometry of the cutting front shown in figure, it was assumed that the melting front is at distance t_f from the vaporized cylinder surface and that the bottom of the kerf is at distance t_s from the bottom of the cylinder (Chryssolouris et al. 1988).

The general differential equation of the heat flow expressed in rectangular coordinates (x, y, z) which are referred to a fixed origin in solid is given by:

$$\frac{\partial^2 T}{\partial x^2} + \frac{\partial^2 T}{\partial y^2} + \frac{\partial^2 T}{\partial z^2} = \frac{1}{\alpha}\frac{\partial T}{\partial t}, \tag{6.20}$$

where T is the temperature, t is the time, and α is the thermal diffusivity.

Assuming that the heat source of strength (rate of heat) q is moving with velocity v along x-axis, the above equation can be expressed by referring to the coordinate system moving with the laser beam. This requires replacing x by $\xi = x - vt$, where ξ is the point considered from heat source. This gives:

$$\frac{\partial^2 T}{\partial \xi^2} + \frac{\partial^2 T}{\partial y^2} + \frac{\partial^2 T}{\partial z^2} = -\frac{v}{\alpha}\frac{\partial T}{\partial \xi} + \frac{1}{\alpha}\frac{\partial T}{\partial t} \tag{6.21}$$

For quasistationary heat flow, the temperature distribution around the heat source becomes constant. The condition is expressed by:

$$\frac{\partial T}{\partial t} = 0. \tag{6.22}$$

In addition, for one-dimensional heat flow:

$$\frac{\partial^2 T}{\partial y^2} + \frac{\partial^2 T}{\partial z^2} = 0. \tag{6.23}$$

Substituting above equations, the governing heat transfer equation becomes:

$$\frac{\partial^2 T}{\partial \xi^2} = -\frac{v}{\alpha}\frac{\partial T}{\partial \xi} \tag{6.24}$$

The boundary conditions can be written as:

$$-\frac{\partial T}{\partial \xi} \times k = q'' \text{ at } \xi = 0, \tag{6.25}$$

and

$$T - T_0 = 0 \text{ at } \xi = \infty \text{ at.} \tag{6.26}$$

where k is the heat conductivity, q'' is the rate of heat per unit section, and T_0 is the initial temperature of solid.

The solution of the above equations gives the temperature distribution around the moving source of heat.

$$T - T_0 = \frac{q''}{c\rho v} \text{ for } \xi < 0, \tag{6.27}$$

and,

$$T - T_0 = \frac{q''}{c\rho v} e^{\frac{-v\xi}{\alpha}} \text{ for } \xi < 0, \tag{6.28}$$

where c is the specific heat and ρ is the density. In the above equation, $\frac{q''}{c\rho v}$

represent the rise in temperature at the location of the point source ($\xi = 0$).

Now, if we define $\zeta = \frac{\alpha}{v}$, the above equation can be rewritten as:

$$T - T_0 = \frac{q''}{c\rho v} e^{\frac{-\xi}{\zeta}}.$$ (6.29)

In addition,

$$T - T_0 = \frac{q''}{c\rho v} \frac{1}{e} \text{ for } \xi = \zeta.$$ (6.30)

Thus, ζ determines the position of a point along x-axis where the temperature rise is $\frac{1}{e} \times \frac{q''}{c\rho v}$.

During blind laser cutting, the heat flow takes place in five distinct regions due to characteristic geometry of the cutting front. These regions are

Region 1: Forward conduction
Regions 2 and 3: Side conduction
Region 4: Backward conduction
Region 5: Conduction beneath the laser spot

Region 1 is defined by a vertex angle θ which can be calculated from the laser beam diameter d and the distance ζ as:

$$\frac{\theta}{2} = \tan^{-1} \frac{d}{2B},$$ (6.31)

where $B = \zeta + d/2$. In addition, θ can be used to define regions 1–4 where the heat is dissipated radially.

The thickness of the cutting front in forward direction (t_f) and the thickness of the cutting front at the sides and the drilling front at the bottom (t_s) can be calculated by considering the heat balance between the laser source and the appropriate regions of heat conduction.

For region 1:

$$k\left(\frac{T_k - T_i}{t_f}\right) = \rho v \left(c(T_k - T_0) + L\right),$$ (6.32)

where T_k is the temperature at the laser spot; T_i is the melting temperature; and L is the latent heat.

For regions 2, 3, and 5:

$$kA_s\left(\frac{T_k - T_i}{t_s}\right) = \rho v t_s z \left(c\left(\frac{T_k - T_i}{2} - T_0\right) + L\right),$$ (6.33)

where z is the penetration depth corresponding to the height of the vaporized cylinder; and $T_k - T_i/2$ is the average temperature between the laser spot and the melting isotherm. A_s is given by:

$$A_s = (\pi - \theta)\frac{zd}{2} + \frac{\pi d^2}{4} \tag{6.34}$$

Based on above equations, the thickness of cutting front in forward direction (t_f) and the drilling front at the bottom of the laser beam (t_s) are given by:

$$t_f = \frac{\alpha}{v}\left[\frac{T_k - T_i}{(T_k - T_0) + (L/c)}\right], \tag{6.35}$$

and,

$$t_s = \sqrt{\frac{\alpha(\pi - \theta)d}{2v}\left(\frac{T_k - T_i}{\left(\frac{T_k + T_i}{2} - T_0\right) + \frac{L}{c}}\right)}. \tag{6.36}$$

Heat fluxes in forward direction $\dot{Q}_f$ and in the side and bottom of laser beam $\dot{Q}_s$ can then be expressed as:

$$\dot{Q}_f = kA_f\left(\frac{T_k - T_i}{t_f}\right), \tag{6.37}$$

and,

$$\dot{Q}_s = kA_s\left(\frac{T_k - T_i}{t_s}\right), \tag{6.38}$$

where $A_f = \theta \, dz / 2$.

Assuming that no other heat losses take place during machining, the energy supplied by the laser beam can be equated with the rate of heat flux given by the sum of $\dot{Q}_f$ and $\dot{Q}_s$:

$$P = \dot{Q}_f + \dot{Q}_s, \tag{6.39}$$

where P is average laser power.

Solving for z gives,

$$z = \frac{P - \frac{\pi}{4}d^2 k \left[\dfrac{T_k - T_i}{t_s}\right]}{k - \frac{\theta}{2}d \left[\dfrac{T_k - T_i}{t_f}\right] + k \left(\dfrac{\pi - \theta}{2}\right) d \left(\dfrac{T_k - T_i}{t_s}\right)} \tag{6.40}$$

Combining the depth of penetration (z) with the thickness of drilling front (t_s) the depth of the single-pass blind cut is given by:

$$D = z + t_s \tag{6.41}$$

Based on this theoretical analysis, Chrysolouris predicted the depth of kerfs as a function of laser power for several ceramic materials (porous and nonporous Al_2O_3, SiC). Figure 6.29 presents the predicted and experimental depths of cut during machining of the alumina ceramic. The figure indicates that the theoretical predictions overestimate the values of depth of cut probable due to consideration of complete absorption of the laser beam (i.e., absoptivity = 1).

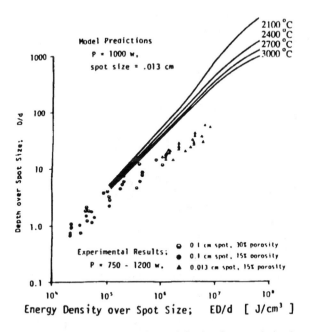

Fig. 6.29 Comparison of calculated and experimental depths of groove during laser blind cutting of alumina ceramic. (Reprinted from Chryssolouris et al. 1988. With permission. Copyright American Society of Mechanical Engineers.)

In addition, the comparisons of the predicted theoretical depths with the experimental results in alumina showed the better agreement in the lower energy density and higher cutting speed region. The model predicted that 10 kW laser with a spot size of 0.013 cm can make a 1 cm deep cut 60 times faster than a 1 kW laser provided a strong gas jet is applied to the workpiece (Chrysolouris et al. 1988).

Chrysolouris further extended the analysis of grooving process to the carbon/teflon and glass/polyester composites. Figure 6.30 presents the comparison of model predictions and experimental results of groove depths for carbon/teflon composites for the cases of continuous and pulsed beam machining. The figure indicates that for both the cases of continuous and pulsed beam machining, the

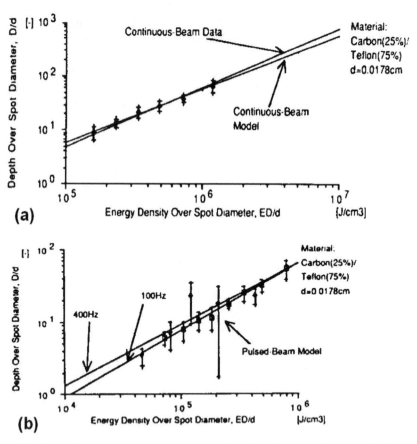

Fig. 6.30 Comparison of calculated and experimental depths of groove during laser blind cutting of carbon/teflon composites using (a) a continuous, and (b) a pulsed laser beam. (Reprinted from Chryssolouris et al. 1990. With permission. Copyright American Society of Mechanical Engineers.)

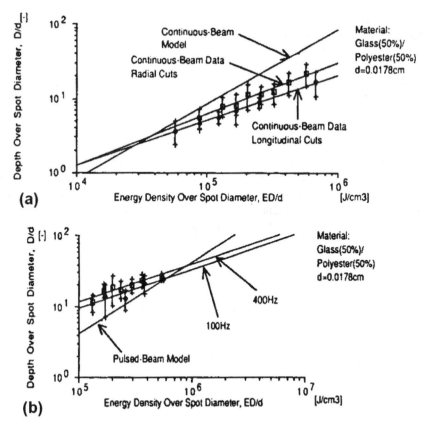

Fig. 6.31 Comparison of calculated and experimental depths of groove during laser blind cutting of glass/polyester composites using (**a**) a continuous, and (**b**) a pulsed laser beam. (Reprinted from Chryssolouris et al. 1990. With permission. Copyright American Society of Mechanical Engineers.)

groove depth shows the linear relationship with the laser energy density and that the analytical predictions are in close agreement with the experimental results. Similar analysis for the glass/polyester composites indicated that the model predictions consistently overestimated the groove depth values for continuous beam results and underestimated the groove depth values for pulsed beam results (Fig. 6.31). The deviations in the model predictions and the experimental values may be due to beam divergence effects and the high temperature chemical inter-actions. During initiation of grooving, the beam is generally focused on the surface of the workpiece. Hence, as the depth of the groove increases, the laser beam tends to diverge due to increasing distance between the nozzle and the groove surface. The divergence of the beam during grooving is associated with the decrease in the laser energy density incident on the surface. In addition, a portion of the incident laser energy may be consumed in associated endothermic reactions such as pyrolysis of polyester into gaseous products and carbon, and chemical reactions between silica in the fiber materials and carbon in the matrix char (Chrysolouris et al. 1990).

One of the important parameters during machining of ceramic materials is the surface quality of the machined parts described in terms of dimensional accuracy and surface finish. In the context of laser machining using intersecting beams where each beam produces a blind cut, the surface quality is primarily described by wall straightness and kerf width of the blind cuts. Various factors affecting the surface quality of the blind cuts include traverse speed of the workpiece, spatial profile of the beam, beam dimension, beam polarization effects, focusing parameters (focus position relative to workpiece surface, adjustments in the focal position during machining to compensate for the increasing depth of cut), and gas assist, etc. Figure 6.32 presents the influence of the supersonic nozzle

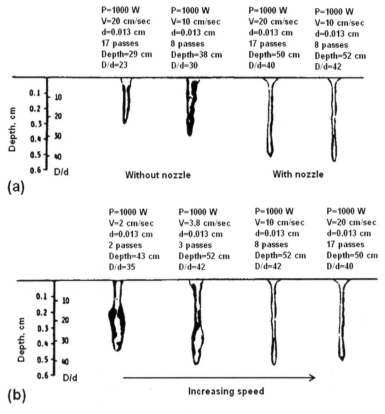

Fig. 6.32 Influence of **(a)** supersonic nozzle and **(b)** workpiece traverse speed on the profiles of laser machined blind cuts in alumina ceramic. (Reprinted from Chryssolouris et al. 1988. With permission. Copyright American Society of Mechanical Engineers.)

and increasing workpiece traverse speed on the profiles of blind cuts in Al_2O_3 ceramics. The use of supersonic nozzle and increased workpiece traverse speed enhances the materials removal and significantly improves the quality of cuts indicated by longer depth of cut, better wall straightness, and narrower kerf widths.

6.4 Applications of Three-Dimensional Laser Machining

Three-dimensional machining approaches are attractive for shaping of difficult-to-machine materials such as ceramics and composite materials. These approaches have demonstrated the utilization in the machining of complex geometries such as threads, grooves, slots, and complex patterns in ceramic workpieces. Figure 6.33a shows the square rod (10 × 10 mm²) of Si_3N_4 ceramic turned into cylindrical shape using Nd:YAG laser. The laser turning was reported to be associated with the material removal rate of $4 \times 10^{-2}\,mm^3/s$ and workpiece roundness error of 13 μm. The other potential application of three-dimensional machining which has attracted significant interests is the turning of threads on cylindrical ceramic workpieces. Figure 6.33b shows such laser turned threads in Si_3N_4 ceramic (Liu et al. 1999). In addition to the turning of screw threads, laser machining has been demonstrated for producing "O" ring grooves in alumina ceramic (Copley et al. 1983, Copley 1987). Figure 6.34 presents a pattern machined in silicon carbide ceramic using pulsed Nd:YAG laser. The pattern was produced by overlapping laser scans on the surface using computer numerical control (CNC) positioning of workpiece table. Laser machining is also attractive for the dressing or truing of the abrasive grinding wheels (such as alumina and cubic boron nitride (CBN) wheels) used in industrial machining practices (Harimkar and Dahotre 2006). In order to exploit

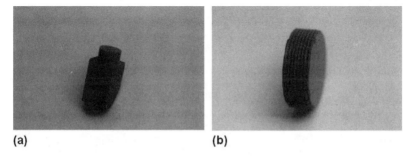

(a)   **(b)**

Fig. 6.33 Three-dimensional laser machining applications: **(a)** turning of cylindrical shape from a square Si_3N_4 rod, **(b)** turning of threads in Si_3N_4. (Reprinted from Liu et al. 1999. With permission. Copyright Elsevier.)

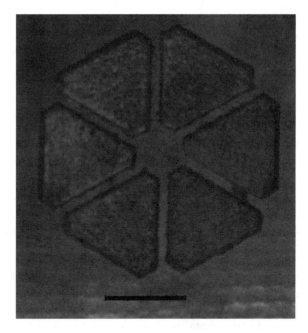

Fig. 6.34 Laser-machined pattern in silicon carbide using overlapped beam scanning. The length of magnification mark corresponds to 1 cm (Dahotre et al. 2006, unpublished data)

the full potential of three-dimensional machining, significant research efforts needs to be directed towards addressing the important challenges associated with the process so that improvements in the dimensional accuracy, surface finish, and materials removal rates can be achieved.

References

Anderson M, Patwa R, Shin YC (2006) Laser-assisted machining of inconel 718 with an economic analysis. International Journal of Machine Tools and Manufacture 46:1879–1891.

Barnes S, Pashby IR, Mok DK (1996) The effect of workpiece temperature on the machinability of an aluminum/SiC MMC. Journal of Mechanical Design 118:422–427.

Chrysolouris G, Bredt J, Kordas S, Wilson E (1988) Theoretical aspects of a laser machine tool. Journal of Engineering for Industry 110:65–70.

Chrysolouris G, Sheng P, Choi WC (1990) Three-dimensional laser machining of composite materials. Journal of Engineering Materials and Technology 112:387–392.

Chrysolouris G (1991) Laser Machining: Theory and Practice. Springer, New York.

Chrysolouris G, Anifantis N, Karagiannis S (1997) Laser assisted machining: an overview. Journal of Manufacturing Science and Engineering, Transactions of the ASME 119:766–769.

Cohen PH (1989) Forces, power, and stresses in machining. In: Davis JR (ed) ASM Handbook: Machining. ASM International, Ohio, vol. 16, pp. 13–18.

Copley S, Bass M, Jau B, Wallace R (1983) Shaping materials with lasers. In: Bass M (ed) Laser Materials Processing. North-Holland, Amsterdam, The Netherlands, pp. 297–336.

Copley SM (1987) Laser applications. In: King RI (ed) Handbook of High-speed Machining Technology. Chapman & Hall, New York, pp. 387–416.

Field M, Kahles JF, Koster WP (1989) Surface finish and surface integrity. In: Davis JR (ed) ASM Handbook: Machining. ASM International, Ohio, vol. 16, pp. 19–36.

Harimkar SP, Dahotre NB (2006) Prediction of solidification microstructures during laser dressing of alumina-based grinding wheel. Journal of Physics D: Applied Physics 39:1642–1649.

Hsu RK, Copley SM (1990) Producing three-dimensional shapes by laser milling. Journal of Engineering for Industry 112:375–379.

Lei S, Shin YC, Incropera FP (2000) Deformation mechanisms and constitutive modeling for silicon nitride undergoing laser-assisted machining. International Journal of Machine Tools and Manufacture 40:2213–2233.

Lei S, Shin YC, Incropera FP (2001) Experimental investigation of thermo-mechanical characteristics in laser assisted machining of silicon nitride ceramics. Journal of Manufacturing Science and Engineering, Transactions of the ASME 123:639–646.

Liu JS, Li LJ, Jin XZ (1999) Accuracy control of three-dimensional Nd:YAG laser shaping by ablation. Optics and Laser Technology 31:419–423.

Pfefferkorn FE, Rozzi JC, Incropera FP, Shin YC (1997) Surface temperature measurement in laser-assisted machining process. Experimental Heat Transfer 10:291–313.

Pfefferkorn FE, Shin YC, Tian Y, Incropera FP (2004) Laser-assisted machining of magnesia-partially-stabilized zirconia. Journal of Manufacturing Science and Engineering, Transactions of the ASME 126:42–51.

Rebro PA, Shin YC, Incropera FP (2002) Laser-assisted machining of reaction sintered mullite ceramics. Journal of Manufacturing Science and Engineering 124:875–885.

Rebro PA, Shin YC, Incropera FP (2004) Design of operating conditions for crackfree laser-assisted machining of mullite. International Journal of Machine Tools and Manufacture 44:677–694.

Rosenthal D (1946) The theory of moving sources of heat and its application to metal treatments. Transactions of the ASME 68:849–866.

Rozzi JC, Pfefferkorn FE, Incropera FP, Shin YC (1998a) Transient thermal response of a rotating cylindrical silicon nitride workpiece subjected to a translating laser heat source, Part I: Comparison of surface temperature measurements with theoretical results. Journal of heat transfer 120:899–906.

Rozzi JC, Incropera FP, Shin YC (1998b) Transient thermal response of a rotating cylindrical silicon nitride workpiece subjected to a translating laser heat source, Part II: Parametric effects and assessment of a simplified model. Journal of Heat Transfer 120:907–915.

Rozzi JC, Pfefferkorn FE, Incropera FP, Shin YC (2000a) Transient, three-dimensional heat transfer model for the laser assisted machining of silicon nitride: I. Comparison of predictions with measured surface temperature histories. International Journal of Heat and Mass Transfer 43:1409–1424.

Rozzi JC, Pfefferkorn FE, Incropera FP, Shin YC (2000b) Transient, three-dimensional heat transfer model for the laser assisted machining of silicon nitride: II. Assessment of parametric effects. International Journal of Heat and Mass Transfer 43:1425–1437.

Rozzi JC, Pfefferkorn FE, Shin YC, Incropera FP (2000c) Experimental evaluation of the laser assisted machining of silicon nitride ceramics. Journal of Manufacturing Science and Engineering, Transactions of the ASME 122:666–670.

Tian Y, Shin YC (2006) Thermal modeling for laser-assisted machining of silicon nitride ceramics with complex features. Journal of Manufacturing Science and Engineering 128:425–434.

Wallace RJ, Bass M, Copley SM (1986) Curvature of laser-machined grooves in Si_3N_4. Journal of Applied Physics 59:3555–3560.

Wallace RJ, Copley SM (1986) Laser machining of silicon nitride: energetics. Advanced Ceramic Materials 1:277–283.

Wallace RJ, Copley SM (1989) Shaping silicon nitride with a carbon dioxide laser by overlapping multiple grooves. Journal of Engineering for Industry 111:315–321.

Wang Y, Yang LJ, Wang NJ (2002) An investigation of laser-assisted machining of Al_2O_3 particle reinforced aluminum matrix composite. Journal of Materials Processing Technology 129:268–272.

Chapter 7
Laser Micromachining

7.1 Introduction

Miniaturization is an important trend in many modern technologies. The requirement for material processing with micron or submicron resolution at high speed and low unit cost is an underpinning technology in nearly all industries manufacturing high-tech microproducts for biotechnological, microelectronics, telecommunication, MEMS, and medical applications (Gruenewald et al. 2001). In view of this increasing trend toward miniaturization, micromachining becomes an important activity in the fabrication of microparts. Various technologies such as mechanical micromachining (microdrilling and micromilling), focused ion beam micromachining, laser micromachining are being used in microfabrication. Laser micromachining is a relatively recent process and offers better flexibility in dimensional design of microproducts. Advances in the laser technology combined with the better understanding of laser–matter interaction make laser micromachining a viable, attractive, cost-effective, and enabling technology to support these applications (Subrahmanyan 2003). At present, laser micromachining is extensively used to produce shapes with greater complexity and lesser material damage than competing micromachining approaches. This chapter provides a brief overview of the various approaches and applications of laser-based micromachining processes.

7.2 Laser Micromachining Mechanisms

The two important mechanisms of material removal with micron and submicron precision during laser micromachining are ablation and laser-assisted chemical etching. This section briefly explains the important aspects of these material removal mechanisms.

7.2.1   Laser Ablation

7.2.1.1   Thermal and Nonthermal Ablation

As explained in Chapter 2, the ablation during laser processing refers to the material removal due to thermal and/or photochemical (nonthermal) interactions. The laser–material interactions during ablation are complex and may involve the interplay between thermal and photochemical processes often referred to as photophysical processes. Figure 7.1 presents the various mechanisms of laser ablation (Bäuerle 2000).

In nonthermal ablation, the energy of the incident photon causes the direct bond breaking of the molecular chains in the organic materials resulting in material removal by molecular fragmentation without significant thermal damage. This suggests that for the ablation process, the photon energy ($h\nu$) must be greater than the bond energy. However, it has been observed that ablation also takes place when the photon energy is less than the dissociation energy of the molecular bond. This is the case for far ultraviolet radiation with longer wavelengths (and hence correspondingly smaller photon energies). Such an observation is due to multiphoton

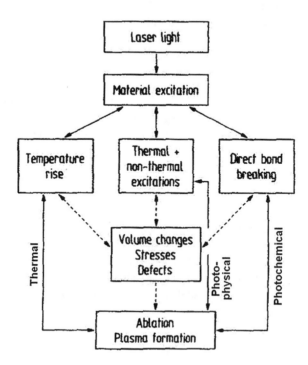

Fig. 7.1 Various mechanisms of laser ablation. (Reprinted from Bäuerle 2000. With permission. Copyright Springer.)

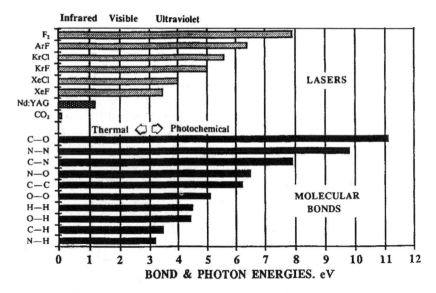

Fig. 7.2 Dissociation energies of some common molecular bonds and photon energies of some common lasers. (Reprinted from Gower 1993. With permission. Copyright Springer.)

mechanism for laser absorption. In multiphoton mechanism, even though the energy associated with each photon is less than the dissociation energy of bond, the bond breaking is achieved by simultaneous absorption of two or more photons. Figure 7.2 presents the photon energies associated with various laser radiations and the dissociation energies of various molecular bonds. As indicated in the figure, photon energies of various excimer lasers exceed the dissociation energies of many common molecular bonds resulting in effective ablation of various materials by direct bond breaking. For long wavelength lasers such as Nd:YAG ($\lambda = 1.06\,\mu$m, $hv \sim 1.2$ eV) and CO_2 ($\lambda = 10.6\,\mu$m, $hv \sim 0.12$ eV) lasers, the photon energies are significantly less than the dissociation energies of common molecular bonds. In such cases, ablation may result due to multiphoton absorption. However, it is often associated with undesirable effects such as combustion, charring, melting, and boiling of surrounding unirradiated material (Gower 1993).

During thermal ablation, the excitation energy is rapidly converted into heat, resulting in temperature rise. This temperature rise can cause the ablation of material by surface vaporization or spallation (due to thermal stresses). Thermal ablation mechanisms dominate the material removal during micromachining of metal and ceramics. Even though photochemical ablation is the dominant material removal during micromachining of polymers, some of the polymers can also be effectively ablated by thermal ablation mechanism. One of the important considerations during the laser–material interaction during ablation is the thermal relaxation time (τ), which is related with the dissipation of heat during laser pulse irradiation and is expressed as (Vogel et al. 1999):

$$\tau = \frac{d^2}{4\kappa},\qquad(7.1)$$

where d is absorption depth and κ is the thermal diffusivity. Thus, the two important parameters that determine the ease with which the ablation can be initiated are absorption coefficient (α) and thermal diffusivity (κ). The large value of absorption coefficient and small value of thermal diffusivity generally provide the high ablation efficiency of a material. The ablation of material by confinement of laser energy in thin layer can also be facilitated by using short pulses (pulse time shorter than thermal relaxation time). For longer pulses (pulse time longer than thermal relaxation time), the absorbed energy will be dissipated in the surrounding material by thermal processes. Thus, efficient ablation of the material during laser–material interactions necessitates the lasers operating with short pulses. Recently, significant interests have been attracted toward the use of ultrashort pulse lasers for micromachining of micrometer and submicrometer features. Ultrashort pulses produce a very high peak intensity ($> 10^{15}$ W/cm^2) and deliver energy before thermal diffusion occurs, thus giving high efficiency and precision to the process without significant thermal degradation (melting, spatter, recrystallization, etc.) to the surrounding region (Liu et al. 1997).

7.2.1.2 Ablation Threshold

The ablation process is characterized by the ablation threshold, which corresponds to the laser fluence at which ablation starts. Different materials have different ablation thresholds primarily due to differences in the optical and thermal properties. Ablation threshold is generally determined from the relationship between ablation rate and irradiated laser fluence. Ablation rate is expressed as ablation depth per pulse and is derived from the depth of ablated hole (h) and the number of laser pulses (n) used. Ablation rate is then given by h/n. A typical relationship between the ablation rate and the laser fluence for silver is presented in Fig. 7.3 using a femtosecond Ti:sapphire laser. The ablation of silver is characterized by two distinct fluence regimes: low and high. The ablation rates (L) in these two regimes can be described by the logarithmic laws (Furusawa et al. 1999):

$$L = d \ln\left[\frac{F}{F_{th}^{skin}}\right], \quad \text{for lower laser fluence}\qquad(7.2)$$

$$L = l \ln\left[\frac{F}{F_{th}^{thermal}}\right], \quad \text{for higher laser fluence}\qquad(7.3)$$

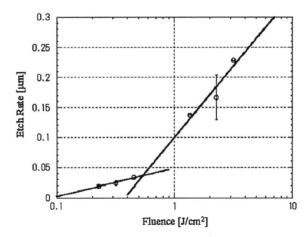

Fig. 7.3 Laser ablation characteristics of silver showing two distinct regimes in the variation of ablation rate with laser fluence. (Reprinted from Ref. Furusawa et al. 1999. With permission. Copyright Springer.)

where d and l are absorption depth (skin depth) and thermal diffusion length, respectively; and, F_{th}^{skin} and $F_{th}^{thermal}$ are the threshold laser fluences of two ablation regimes (lower and higher laser fluence regimes). The ablation thresholds can be deduced by fitting the data points in Fig. 7.3 using the above equations. The ablation thresholds for silver in two regimes are 93 and 397 mJ/cm^2, respectively. Similar relationships are reported for other metallic materials such as copper, aluminum, and gold (Furusawa et al. 1999; Harzik et al. 2005). At lower laser fluence, the energy transfer during laser–material interaction occurs only within the area characterized by the absorption depth ($d = 1/\alpha$), whereas at higher laser fluence, the ablation rate is dominated by the thermal diffusion length. The existence of two regimes of ablation dominated by either absorption length or thermal diffusion length is shown by the two-temperature model. The important observation during the ablation of metals is the reduction in the ablation threshold with decreasing pulse width of the incident pulse (Fig. 7.4). This observation can be the result of a decrease in heat penetration depth and/or an increase in the absorption coefficient due to multiphoton excitation (Bäuerle 2000). The reduction in the threshold fluence with decreasing pulse width suggests the effectiveness of the shorter pulses with relatively low energies for the micromachining applications.

7.2.1.3 Factors Influencing Ablation Rate

The ablation rates during laser micromachining depend primarily on the laser parameters and the materials properties. Various laser processing parameters include laser wavelength, laser fluence, and number of pulses. Figure 7.5

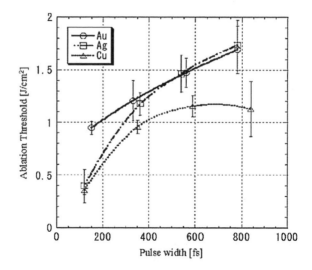

Fig. 7.4 Variation of ablation threshold with incident pulse width during laser ablation of various metals. Ablation threshold was calculated for laser fluences greater than 1 J/cm². (Reprinted from Furusawa et al. 1999. With permission. Copyright Springer.)

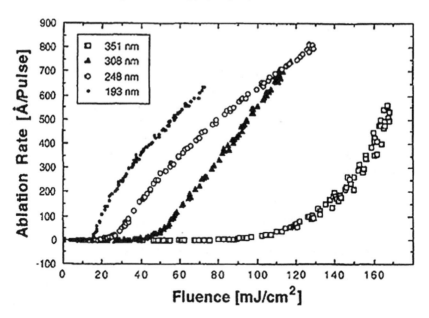

Fig. 7.5 Variation of ablation rate of polyimide with laser fluence for various excimer laser wavelengths. (Reprinted from Küper et al. 1993. With permission. Copyright Springer.)

presents the variation of ablation rate (ablated depth per pulse) of polyimide with laser fluence for various excimer laser wavelengths. For each wavelength, the ablation characteristics show the presence of threshold fluence at which ablation starts. The ablation threshold is particularly sharp at shorter wavelength

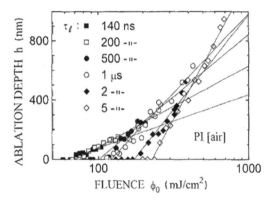

Fig. 7.6 Variation of ablation depth in polyimide with laser fluence for various pulse durations of Ar $^+$-laser. (Reprinted from Himmelbauer et al. 1996. With permission. Copyright Springer.)

(193 nm) compared with that at longer wavelengths (248, 308, 351 nm). The most important observation from the ablation characteristic is related with the influence of laser wavelength on the ablation threshold. As the laser wavelength increases, the threshold fluence at which ablation starts increases (Küper et al. 1993).

The influence of Ar $^+$ -laser pulse durations on the ablation characteristics of polyimide is presented in Fig. 7.6. The figure indicates that within the range of pulse durations (140 ns–5 μs), the threshold fluence decreases with decreasing pulse duration (Himmelbauer et al. 1996). However, at higher laser fluences, significant ablation can be obtained with longer pulses compared with that obtained with shorter pulses. This may be due to efficient absorption of laser energy if the optical penetration depth is longer (Bäuerle 2000).

In addition, the ablation characteristics are influenced by the number of laser pulses. Figures 7.7 and 7.8 present the typical variation of ablation depth with pulse number for various laser fluences and pulse repetition rates, respectively during micromachining of glass using ArF laser. As indicated in the figure, the ablation depth increases linearly with the number of pulses. The slope of the linear relationship depends on the laser fluence and pulse repetition rate (Tseng et al. 2004). However, many studies have observed that the relationship between the ablation depth and number of pulses generally deviate from linearity at a higher number of pulses. This is shown in Fig. 7.9 for the case of PZT micromachined using pulsed excimer laser. As indicated in the figure, the variation of groove depth with the number of pulses is linear up to around 1,000 pulses. Beyond 1,000 pulses, the variation deviates from linearity to approximately logarithmic, and the ablation rate decreases. Such decrease in the ablation rate at higher number of pulses may be due to absorption and scattering of the incident laser radiation by the ablating species resulting in overall decrease in the effective fluence on the sample surface and/or due to the inefficient transport of the ablating species resulting in material

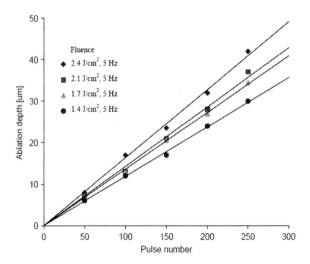

Fig. 7.7 Variation of ablation depth with number of laser pulses for various laser fluences during micromachining of glass using ArF laser. (Reprinted from Tseng et al. 2004. With permission. Copyright Elsevier.)

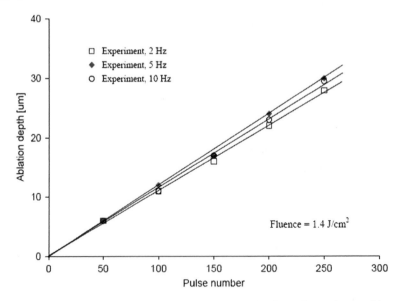

Fig. 7.8 Variation of ablation depth with number of laser pulses for various pulse repetition rates during micromachining of glass using ArF laser. (Reprinted from Tseng et al. 2004. With permission. Copyright Elsevier.)

recondensation. These effects become pronounced as the depth of the ablated grooves increases (Eyett et al. 1987).

One of the important observations during the ablation of weak absorbing materials such as polymethyl methacrylate (PMMA) is the "incubation pulses."

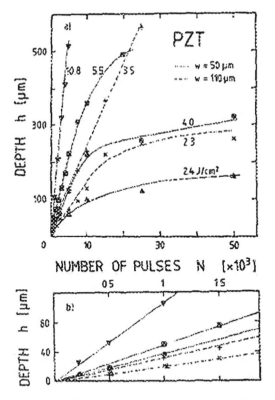

Fig. 7.9 (a) Variation of groove depth with number of laser pulse for different laser fluences (Φ) and widths of focus (w), and (b) enlarged scale in the region of low number of pulses. Laser parameters – Focus $w = 5\,\mu m$: $\Delta\ \Phi = 2.4\,J/cm^2$; $\bullet\ \Phi = 4.0\,J/cm^2$; $\blacksquare\ \Phi = 5.5\,J/cm^2$; $\nabla\ \Phi = 10.8\,J/cm^2$; Focus $w = 110\,\mu m$: $\times\ \Phi = 2.3\,J/cm^2$; $+\ \Phi = 3.5\,J/cm^2$. (Reprinted from Eyett et al. 1987. With permission. Copyright American Institute of Physics.)

Incubation pulses refer to the initial pulses of laser radiation which result in zero or significantly less etch depth per pulse than the subsequent laser pulses. Figure 7.10 presents the variation of etch depth with the number of pulses during UV laser ablation of PMMA. The figure also presents the predictions based on a dynamic model. The figure indicates that the ablation process at each laser fluence starts only after a certain number of incubation pulses (more that 30 pulses for laser fluence of $400\,mJ/cm^2$). This incubation period is followed by the transient and steady-state regimes in the ablation behavior (Sutcliffe and Srinivasan 1986). Laser irradiation during incubation pulses causes the photodecomposition of polymer to oligomers and volatile products to much greater depth than the unirradiated material (observed by increased solubility of the irradiated material). It seems that incubation pulses break up the polymer but do not create the same fraction of gaseous products as pulses following the incubation (Srinivasan et al. 1990). The number of incubation pulses strongly depends on the laser fluence.

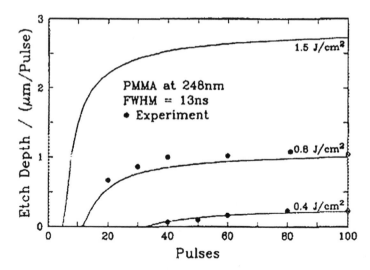

Fig. 7.10 Dependence of etch depth on the number of pulses during laser ablation of PMMA indicating the incubation region before starting of significant ablation. (Reprinted from Sutcliffe and Srinivasan 1986. With permission. Copyright American Institute of Physics.)

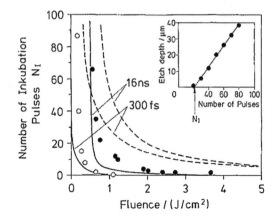

Fig. 7.11 Dependence of incubation pulses (N_I) on the laser fluence for 16 ns and 300 fs pulses during excimer laser ablation of PMMA. Broken lines are calculated according to dynamic model (Sutcliffe and Srinivasan 1986) and solid lines are calculated according to ablation model (Küper and Stuke 1987). Inset shows the variation of etch depth with number of pulses for 16 ns ablation at 750 mJ/cm². Solid line in the insert corresponds to linear regression line. (Reprinted from Küper and Stuke 1987. With permission. Copyright Springer.)

Figure 7.11 presents the dependence of a number of incubation pulses on the laser fluence during excimer laser micromachining of PMMA. The number of incubation pulses decreases with increasing laser fluence and the curve shifts to the lower fluences for shorter pulses. Figure 7.12 presents the schematic of the incubation and ablation process for PMMA (Küper and Stuke 1987).

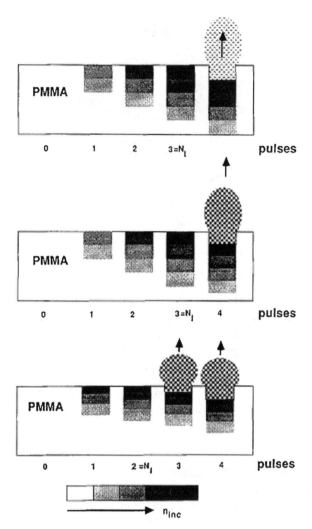

Fig. 7.12 Schematic influence of the number of laser pulses on the ablation behavior of PMMA: (a) ablation with constant transmission and formation of plume, (b) increased attenuation by the plume without affecting incubation pulse number, N_I (c) ablation with nonconstant absorption due to incubation and multiphoton effects resulting in changes in incubation pulse number and etch rate per pulse. (Reprinted from Küper and Stuke 1987. With permission. Copyright Springer.)

7.2.1.4 Ablation Damage

The ablation of materials may be associated with a number of phenomena which may limit the precision during micromachining. The material damage during laser ablation has been extensively studied for various materials, from biological materials to dielectrics and metals.

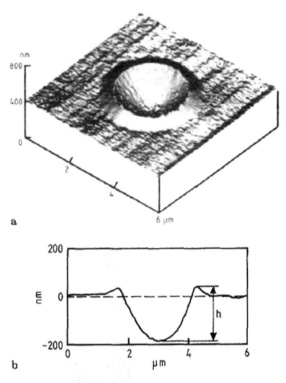

Fig. 7.13 AFM pictures of a typical hole ablated in polyimide using single-shot Ar $^+$ laser with pulse length of 1 μs and laser fluence of 185 mJ/cm^2. (Reprinted from Himmelbauer et al. 1996. With permission. Copyright, Springer.)

Typical morphology of the single-shot ablated hole in polyimide is presented in Fig. 7.13. As indicated in the figure, the ablated hole is surrounded by a rim or hump which may be due to amorphization of the polymer or swelling due to formation of product species (Himmelbauer 1996).

Important processes such as modification (amorphization/oxidation), annealing, and ablation, and corresponding regions during irradiation of silicon with single-pulse Ti:sapphire laser are presented in Fig. 7.14 (Bonse et al. 2002). In contrast to single-pulse irradiation, the multiple-pulse irradiation is characterized by the formation of periodic surface structures (ripples) in the central region. The ripples are generally oriented perpendicular to the electric field vector of the incident radiation and its period is of the laser wavelength suggesting its origin to the interference phenomena. For larger number of pulses, the surface morphology is characterized by the formation of pillars or columns. Figure 7.15 presents the evolution of the surface morphology in silicon with laser fluence irradiated with 100 pulses of Ti: sapphire laser. At lower laser fluences (above ablation threshold), the surface morphology consisted of uniform ablated craters with periodic ripple structure. As the laser fluence increases, conical structures appear at the bottom of the craters

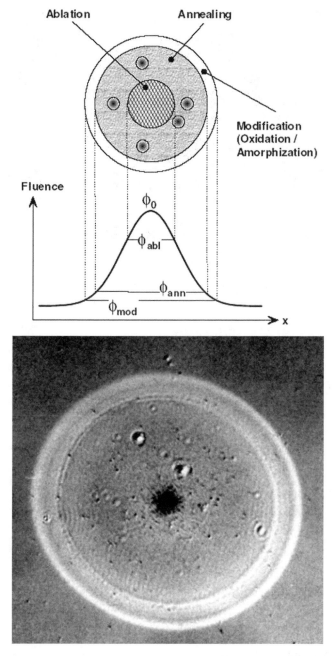

Fig. 7.14 Various processes and corresponding regions during single-pulse femtosecond laser ablation of silicon (wavelength 800 nm; pulse length 130 fs; fluence 1.5 J/cm²; the outermost ring in bottom figure is 45 μm in diameter). (Reprinted from Bonse et al. 2002. With permission. Copyright Springer.)

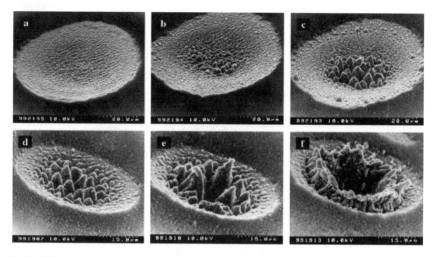

Fig. 7.15 SEM micrographs of ablation damage in silicon generated with Ti:sapphire laser pulses in air: (**a**) $1.0 J/cm^2$, (**b**) $1.3 J/cm^2$, (**c**) $1.8 J/cm^2$ (wavelength = 780 nm, pulse length = 100 fs, number of pulses = 100); (**d**) $2.0 J/cm^2$, (**e**) $2.8 J/cm^2$, (**f**) $4.1 J/cm^2$ (wavelength = 800 nm, pulse length = 130 fs, number of pulses = 100). (Reprinted from Bonse et al. 2002. With permission. Copyright Springer.)

leading to the formation of microcolumns. The height and spacing of the columns is determined by the local laser fluence. Furthermore, at a given fluence, the morphology of the columns depends on the location within the ablated crater: columns are wider, taller, and sparser at the center compared to the border region. Above certain laser fluence, the columns protrude above the original surface suggesting the origin of formation of microcolumns as redeposition or recrystallization. At very high laser fluence, a volcano-like structure is observed within the ablated region which may be due to incomplete redeposition of the ejected material on the crater walls (Bonse et al. 2002). Similar structures can also be observed in the ablation region of metallic materials. Figure 7.16 presents a micrograph of a groove machined in aluminum using 20 scans of femtosecond Ti:sapphire laser. The figure indicates the formation of periodic spike-like solidified structures at the bottom and significant burr at the edge (Harzik et al. 2005).

7.2.2 Laser-Assisted Chemical Etching

7.2.2.1 Dry and Wet Etching

The material removal during micromachining can also be carried out by using suitable etchant (precursors) in combination with selective laser irradiation. When gaseous precursors such as Cl_2 and Br_2 are used, the etching process is referred to

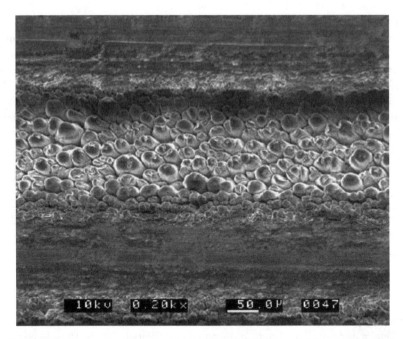

Fig. 7.16 SEM micrograph of a groove machined in aluminum using 20 scans of femtosecond Ti:sapphire laser (scan speed = 2 mm/s, pulse duration = 120 fs, energy density = 9 J/cm²). (Reprinted from Harzic et al. 2005. With permission. Copyright Springer.)

as dry etching, whereas when liquid precursors such as HCl, HNO_3, H_2SO_4, NaCl, and K_2SO_4 are used, the etching process is referred to as wet etching. The etching process may be initiated by the absorption of laser radiation by solid material in contact with the etchant, or the etchant in contact with the solid material, or by both solid material and the etchant simultaneously. The laser radiation influences the reaction between the material and etchant by excitation of the etchant molecules, excitation of the material surface, or adsorbate–adsorbent complexes (Bäuerle 2000). The etching may be thermal and photochemical process. In photothermal etching, the chemical reactions are enhanced by heating, whereas in photochemical etching, the laser radiation causes the photogeneration of radicals followed by interactions between the radicals and the material surface. The etching process generally involves the complex interactions between the photothermal and photochemical processes.

7.2.2.2 Etching Process

Halides and halogen compounds are the most commonly used precursors for the laser etching. Excimer lasers are extensively used for micromachining by selective etching. Various materials such as metal, insulators, and semiconductors have been etched with the excimer lasers in combination with suitable etchants. A representative reaction of the etching of crystalline silicon in chlorine atmosphere is discussed here (Bäuerle 2000). The first step for the etching process is the photodissociation of chlorine to generate halogen radicals. Chlorine show strong absorption band centered at 330 nm resulting primarily from active dissociative transition $^1\Pi_u \leftarrow {}^1\Sigma_g^+$. At wavelengths longer than 480 nm, chlorine has weak vibrational structure superimposed on this band (Affrossman et al. 1989). Thus, efficient photodissociation reaction requires the wavelength of laser radiation to be shorter than around 500 nm. The photodissociation reaction of chlorine can be represented as:

$$Cl_2 + h\nu \rightarrow 2Cl.$$

The halogen radicals produced by the photodissociation of chlorine strongly chemisorb on the silicon surface where chemical reaction takes place. The etching of silicon on the surface can be written as:

$$Si + xCl \rightarrow SiCl_x \uparrow,$$

where $x \leq 4$. Subsequently, the products of the above reaction desorb from the surface resulting in micromachined area on silicon surface (Bäuerle 2000). Thus, three important steps in the micromachining using laser-assisted chemical etching are formation of chemical reactive species, reaction of these reactive species with the surface to be micromachined, and the removal of reaction products from the surface of the material (Brannon 1997). The silicon can also be etched in aqueous solutions of HF, NaOH, and KOH using Ar^+, Nd:YAG, and CO_2 laser radiation (wet etching).

7.2.2.3 Factors Influencing Etch Rate

The extent of etching is determined by a number of important parameters such as exposure time, laser power, focusing conditions, laser wavelength, and crystal orientation. The important measure during the micromachining is the etch rate, which is expressed as the volume of material removed per unit time. In case of symmetric cross sections such as circular holes, the etch rate may be expressed as the depth of material removed per unit time provided the diameter of the hole remains approximately constant during the tests. The micromachining by etching have been investigated extensively for various materials such as metals, insulators, and semiconductors. This section primarily discusses the factor influencing the laser-assisted chemical etching of silicon in chlorine atmosphere.

Figure 7.17 presents the variation of the depth of the hole with exposure time for various focal conditions during Ar^+ laser-assisted chemical etching of single crystal silicon in chlorine atmosphere. The depth of etching is generally proportional to the exposure time for various focusing conditions. The displacement of the sample from the focal position tends to reduce etch rate (etch depth per unit time). Similar linear

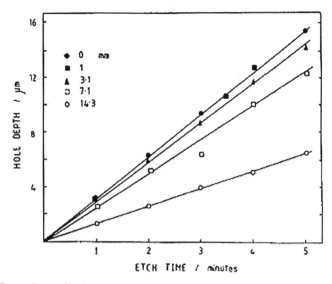

Fig. 7.17 Dependence of hole depth on etch time for various positions of sample relative to focal plane during laser-assisted chemical etching of silicon in chlorine (Si (100), 3 W multiline Ar+ ion laser, 200 Torr Cl_2). (Reprinted from Affrossman et al. 1989. With permission. Copyright Springer.)

relationships between the hole depths and exposure times were observed for a wide range of laser powers and chlorine pressures (Affrossman et al. 1989).

One of the important parameters during the laser-induced chemical etching of silicon is the laser fluence. The influence of laser fluence on the etch rate during excimer laser etching of silicon for various chlorine pressures is presented in Fig. 7.18. The figure indicates the three regimes of variation of etch rate with laser fluence (Kullmer and Bäuerle 1987).

Regime 1: Low energy density regime ($\phi < 150\,mJ/cm^2$)
In this regime, the etch rate increases linearly with the energy density (slope ~ unity). A very small temperature rise at such low energy densities and linear increase in the etch rate suggest the nonthermal activation of the etching process (electron–hole pair generation) and the generation of chlorine radicals as the dominating processes in this fluence range. The etching mechanism seems to involve the transfer of photoelectrons to the impinging chlorine radicals on the silicon surface, resulting in generation of Cl^- ions. The Cl^- ions diffuse into the Si lattice and break the Si–Si bonds. The $SiCl_4$ formed is subsequently desorbed from the surface. The linear increase in the etch rate in this fluence range seems to be due to linear increase in the chlorine atom flux to the surface with fluence (Kullmer and Bäuerle 1987).

Regime 2: Medium energy density regime ($150\,mJ/cm^2 < \phi < 440\,mJ/cm^2$)
In this regime, the etch rate increases nonlinearly with laser fluence. Due to higher laser fluence, the temperature rise is significantly higher. This suggests the importance

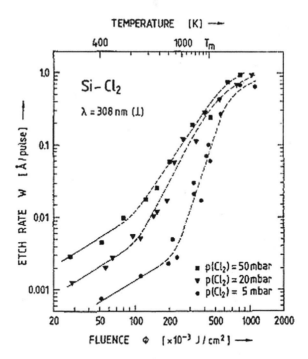

Fig. 7.18 Dependence of etch rate on 308 nm XeCl excimer laser fluence for various chlorine pressures during laser-assisted chemical etching of silicon. The calculated temperature is indicated on the upper scale. (Reprinted from Kullmer and Bäuerle 1987. With permission. Copyright Springer.)

of photothermal activation of the etching process, in addition to the photodissociation of chlorine, in producing nonlinear increase in etch rates. The higher temperature rise at the surface seems to facilitate the diffusion of Cl⁻ ions within Si surface, Si–Si bond breaking by chlorine, and desorption of reaction products (Kullmer and Bäuerle 1987).

Regime 3: High energy density regime ($\phi > 440\,\text{mJ/cm}^2$)
In this regime, the etch rates almost saturate and become only slightly dependent on the chlorine pressure. The high fluence results in the surface melting. It seems that the etching process in this regime is dominated primarily by thermally activated reactions involving adsorbed chlorine and not by the photodissociation of chlorine. The influence of chlorine pressure on the etch rate for three laser fluences (corresponding to three regimes discussed) is presented in Fig. 7.19. The figure indicates that for the fluences below the melting threshold (Φ_m), the etch rate is proportional to the chlorine pressure. Above the melting threshold, the etch rate increases slightly with the chlorine pressure indicating the presence of chlorine atoms or molecule on the surface before the beginning of laser pulse (Kullmer and Bäuerle 1987).

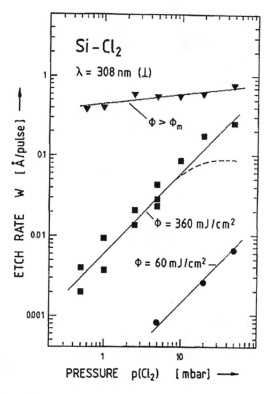

Fig. 7.19 Dependence of etch rate on chlorine pressure for three different laser fluences during laser-assisted chemical etching of silicon. (Reprinted from Kullmer and Bäuerle 1987. With permission. Copyright Springer.)

7.3 Laser Micromachining Techniques

As explained in Section 7.2, the primary mechanisms of material removal during precision micromachining of materials are ablation and etching. The material removal by these mechanisms can be performed in various ways. The three important techniques of micromachining are direct writing, mask projection, and interference. These techniques along with corresponding recent advances are explained in the following sections.

7.3.1 Direct Writing Technique

In direct writing technique, the laser beam is focused on substrate surface. The micromachining of desired pattern is carried out either by translating the substrate with respect to the fixed laser beam or by scanning the laser beam with

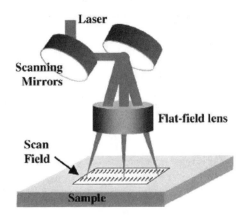

Fig. 7.20 Schematic of laser direct writing technique utilizing scanning mirrors. (Reprinted from Rizvi and Apte 2002. With permission. Copyright Elsevier.)

respect to stationary substrate. In the former case, programmable positioning stage can be used for mounting the sample, whereas in the latter case, additional galvanometer-controlled scanning mirrors can be used to continuously deflect the beam in the desired pattern. With the laser beam scanning system using mirrors, it is important to maintain the focal conditions at various angles of beam deflection. This is generally achieved using flat-field lens (Rizvi and Apte 2002). A schematic of the direct writing technique using scanning mirrors is presented in Fig. 7.20.

The important parameters during direct writing technique are size of the focus, the working distance, and the depth of focus. The minimum spot size is limited by the diffraction phenomenon. One of the major difficulties encountered during direct writing technique for large area processing is that the desired pattern may not be machined in a single scanning step. It is often required to make multiple adjacent and/or overlapping scans necessitating the repositioning of the beam after each scan. Thus, the direct writing process is often termed as step-and-scan process involving switching off of laser and repositioning of sample between two laser scans. This discontinuous laser scanning may result in significant increase in the total machining time. Recently, a technique called sync scan has been developed in which the sample is continuously moved while the scanning mirrors scan the beam on the substrate. As the sample is moved, the scan field is continuously updated allowing the continuous writing of the pattern (Rizvi and Apte 2002).

Laser direct writing is a simple technique and offer significant advantages such as simple and inexpensive optics, and ease of interfacing with CAD files. This technique is extensively used for micromachining application such as hole drilling, cutting, scribing, and marking.

7.3.2 *Mask Projection Technique*

In mask projection technique, a mask consisting of the shape of pattern to be produced on the substrate is illuminated with the laser light. The masks are generally made of quartz substrate covered with a thin layer of chrome metal. The laser beam is shaped and homogenized to produce uniform intensity distribution across the exposed area of mask. The pattern on the mask is then projected and shrunk on the substrate using a projection lens (Fig. 7.21). The projection patterning method allows the generation of a large pattern on the surface with single or multiple laser shots. The number of shots determines the depth of structures machined on the substrate surface. The resolution of the features in the micromachined structures are determined by the mask and projection systems. Excimer lasers are extensively used for micromachining using the mask projection technique. In addition to the high resolution, better reproducibility, and fine depth control, the mask projection techniques using excimer lasers allows the micromachining of large substrate areas (Rizvi and Apte 2002).

The simplest mask projection technique explained above consisted of a stationary mask. The method can be used for producing features such as micro holes and channels. A hole can be produced by firing the laser shots while the mask and the substrate are stationary, whereas a channel can be produced by firing the laser shots while the mask is stationary and the substrate is continuously moving. By synchronizing the substrate positioning and laser firing sequence, periodic discrete features such as arrays of suitably spaced holes or channels can be produced. In this case, the substrate is translated between each processing step. Furthermore, by changing

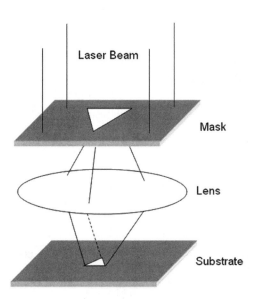

Fig. 7.21 Schematic of mask projection technique

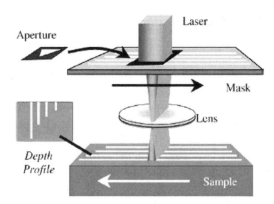

Fig. 7.22 Schematic of synchronized overlay scanning. (Reprinted from Rizvi and Apte 2002. With permission. Copyright, Elsevier.)

the mask between the steps, combinations of various features can be produced (Rizvi 1999). One of the important characteristics of the features produced by simple mask projection technique is the uniform feature depth. For many applications such as microfluidic systems and optical devices contoured depth profiles of the micromachined features are desired. Such complex geometric features necessitate the modifications in the standard mask projection technique. One such development is the synchronized overlay scanning technique. The basic arrangement for the synchronized overlay scanning is presented in Fig. 7.22. In this technique, the motion of the mask and the substrate is synchronized. Most importantly, the depth profiles are obtained on the substrate by placing an aperture of suitable shape above the mask. Various combinations of masks and the apertures can be used to obtain desired features. The shape of the mask determines the features type, whereas the shape of the aperture determines the depth profile of the features. As indicated in Fig. 7.22 the mask consists of open slots which produce channels on the substrate. The depth profile of the channels is determined by the shape of the apertures. The triangular aperture in Fig. 7.22 gives the triangular variation of depth profiles with highest and smallest depths of the channels at two extremes (Rizvi and Apte 2002).

7.3.3 Interference Technique

Laser interference technique involves splitting of a laser beam using a beam splitter followed by superposition of the beams to generate interference patterns. The interference pattern thus produced shows unique intensity variation which can be used for periodic micromachining of the substrates. Typical experimental arrangement for materials processing using two-beam interference technique is shown in

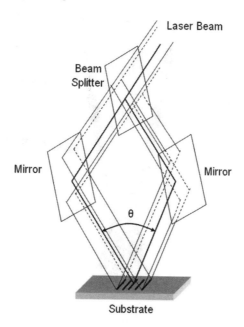

Laser Beam

Beam
Splitter

Mirror

Mirror

θ

Substrate

Fig. 7.23 Schematic of laser interference technique (Bäuerle 2000)

Fig. 7.23. The various elements of this arrangement are the laser source, interferometer (mirrors, beam splitter, etc.) and the imaging surface.

The geometry of the interference patterns formed by the superposition of two or more coherent and linearly polarized beams depends on the wavelength and the angle between the beams. The intensity distribution resulting from the superposition of two linearly polarized beams with their E vectors in the x-direction can be expressed as (Daniel et al. 2003):

$$I(x) = 2I_0 \left[\cos\left(\frac{2\pi x}{1} \right) + 1 \right],$$
(7.4)

where I_0 is the intensity of a laser beam, λ is the wavelength, θ is the angle between the beams, and l is the period. The superposition of two beams produces an interference pattern with spatially modulated light field with intensity distribution oscillating between 0 and $4I_0$. The interference of two beams generates a one-dimensional periodic pattern. Two- and three-dimensional periodic patterns can be obtained by increasing the number of beams. The periodicity of the two beam interference pattern is expressed as:

$$l = \frac{\lambda}{2 \sin(\theta/2)}.$$
(7.5)

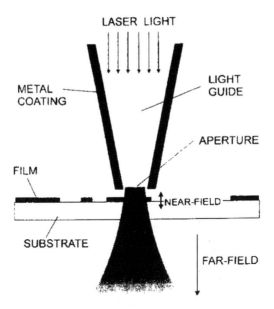

Fig. 7.24 SNOM (scanning near-field optical microscopy) arrangement for nanopatterning. (Reprinted from Bäuerle 2000. With permission. Copyright Springer.)

7.3.4 Combined Techniques

Recently, a combined technique based on scanning near-field optical microscopy (SNOM) and atomic force microscopy (AFM) has demonstrated applications in micromachining (by ablation and etching). The setup involves the coupling of the laser light into the tip of solid or hollow fiber (Fig. 7.24). It is possible to create the nanostructures on the substrate by placing it within the near field on the fiber tip and irradiating with laser radiation. Most important characteristic of this combined technique is that widths of the patterns are not limited by optical diffraction effects (Bäuerle 2000).

Significant interests have recently been directed toward laser-induced nano-patterning by means of interference subpatterns generated by microspheres (Fig. 7.25). The technique is based on the formation of a regular two-dimensional (2D) lattice of microspheres by self assembly processes (e.g., from colloidal suspensions). The lattice is used as an array of microlenses on a transparent support and focus the incident radiation on the substrate (with significant spherical aberration). The distance between the center of the microspheres and substrate surface (f) is tuned by a spacer such that:

$$f \approx \frac{r}{2}\frac{n}{n-1},$$

(7.6)

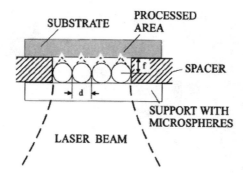

Fig. 7.25 Schematic of the setup for laser-induced surface patterning by means of microspheres. (Reprinted from Denk et al. 2003. With permission. Copyright Springer.)

where n and r are the refractive index and radius of the microspheres, respectively. The technique allows the single-step maskless surface patterning of the substrate to produce submicron holes by ablation mechanism (Denk et al. 2003).

7.4 Laser Micromachining Applications

Laser micromachining is extensively used for manufacturing of microparts with micron and submicron resolution in a wide range of applications such as biotechnological, microelectronics, telecommunication, MEMS, and medical applications (Gower 2001). A wide range of materials such as metals, ceramics, and polymers have been successfully micromachined. This section briefly explains some of the important applications of lasers in micromachining.

7.4.1 Microvia Drilling

Laser microdrilling is an important technology capable of drilling holes with micron or submicron resolution for a variety of applications. The process is attractive from the aspects of flexibility, processing speed, feature resolution, and accuracy. Laser drilling is becoming increasingly acceptable method for forming microvias in high density electronic interconnect and chip packaging devices. Multilayer printed circuit boards (PCBs) technology requires reliable and economical production of through and blind holes to establish electrical connection between the various layers. CO_2 lasers, with wavelength in the far infrared region, are widely used in PCB industry for drilling 100 μm microvias through dielectric layers, whereas UV lasers are used when the microvia design requires less than 100 μm

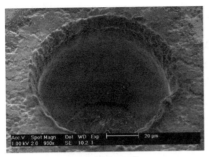
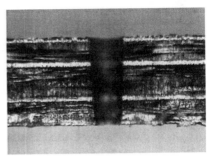

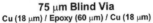

75 μm Blind Via   **100 μm Through Via**
Cu (18 μm) / Epoxy (60 μm) / Cu (18 μm)   Glass-fiber-reinforced multilayer

Fig. 7.26 Typical blind and through vias formed by trepanning. (Reprinted from Dunsky 2002. With permission. Copyright Institute of Electrical and Electronics Engineers.)

diameter (Raman 2001). Lasers are now capable of drilling vias down to 25–30 μm in the production with throughputs as high as 250 vias/s. These developments are primarily driven by improvements in the laser power, beam positioners, and optical designs. Various methods such as percussion, trepanning, and spiraling are used for microvia drilling. Typical blind and thorough via produced by trepanning are presented in Fig. 7.26. The trepanning and spiraling methods require precise control of the movement of focused laser beam which may limit the drilling speed. Hence, percussion drilling is favored over trepanning and spiraling. Significant developments have been done to minimize the laser to laser variation in beam size, roundness, astigmatism, etc. by implementing image projection techniques. In one of the image projection technique, an aperture is used which masks the low energy "wings" of the Gaussian beam and allows the central, high energy to pass onto the work surface. This clipped Gaussian configuration results in better control of the vias but at the expense of rejecting significant fraction of energy at the aperture mask. In other configuration, the Gaussian beam can be shaped into near uniformed "tophat" profile (Dunsky 2002).

7.4.2 Drilling of Inkjet Nozzle Holes

Recently, excimer lasers are also finding increased utilization in the fabrication of high-quality inkjet nozzle plates that meet requirements for inkjet and fluid-dispensing applications. Inkjet nozzle plate which is an important component of the printer head comprise of a row of small tapered holes through which ink droplets are squirted onto paper. Adjacent to each nozzle, a tiny resistor rapidly heats and boils ink ejecting the ink droplet through specific nozzle. The printer quality can be increased by simultaneously reducing the nozzle diameter,

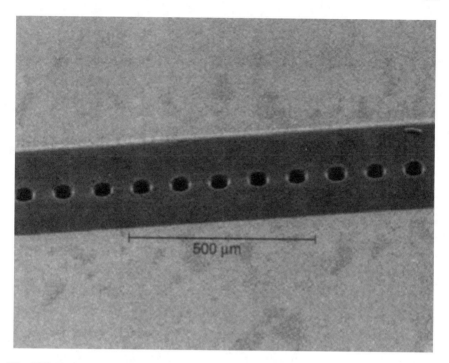

Fig. 7.27 Excimer laser-drilled nozzles along the central ink channel for an inkjet printer head (Lizotte et al. 2002)

decreasing the hole pitch, and lengthening the head. The most common material for the nozzle plate is polyimide. Excimer lasers can be used to generate a variety of nozzle shapes such as round, square, or elliptical. The laser microdrilling of nozzles in the polyimide plate is based on the photoablation mechanism which involves the chemical bond breaking and ejection of fragments upon laser irradiation. Modern printers have around 28 µm diameter nozzles giving a resolution of 600 dots per inch (dpi). For a resolution of 1,200 dpi, the exit-nozzle diameter needs to be well below 20 µm. One of the important considerations during excimer laser drilling of nozzle plates is the cleaning of the carbon debris formed around the nozzles after drilling (Lizotte et al. 2002; Gower 2000).

7.4.3 Resistor Trimming

Resistors form the integral part of almost all electronic circuits and often needs to be adjusted for resistance values to ensure the proper functioning of the

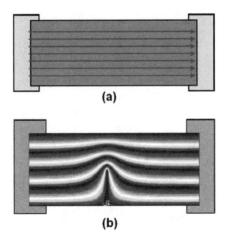

Fig. 7.28 Schematic of **(a)** untrimmed and **(b)** trimmed resistor showing current flow lines. (Reprinted from Deluca 2002. With permission. Copyright Institute of Electrical and Electronics Engineers.)

circuits. The resistors come in both thin film and thick film type. The thin film resistors, with thickness usually less than 1 μm, are made from metal-based alloys deposited on the board surface, whereas the thick film resistors are printed or screened on the substrate using a liquid ink, dried, and fired (Schaeffer 2002). In a manufactured form, these components may not have exactly the required resistance. Laser is used to trim the resistor until its resistance increases to specific value. Since the laser can only increase the resistance, the resistors are fabricated with intentionally low values of resistance and then trimmed by removing the material. While processing, the resistor is connected to a high-speed measurement system via a suitable probe system, and a laser is directed to machine a cut through the resistor thickness in a direction orthogonal to the current flow. As the laser cut forms, the measurement system detects a decreased current flow relative to the applied potential, and interrupts the laser radiation when the desired resistor value given by Ohm's law has been reached (Fjeldsted and Chase 2002). Figure 7.28 presents the schematic of trimmed and untrimmed resistors showing current flow lines. Various trim shapes (Fig. 7.29) have been used according to specific requirements. A commonly used geometry is L-shaped cut (Fig. 7.30) which provide the most accurate trim with least significant change to resistor shape or morphology (Deluca 2002). Laser trimming of both thick and thin film embedded resistors can be achieved with better than 1% accuracy. Q-switched Nd:YAG lasers emitting a train of pulses with peak power in the kilowatt range, pulse duration in nanosecond range, and pulse repetition rates in kilohertz range are often used for laser trimming of resistors.

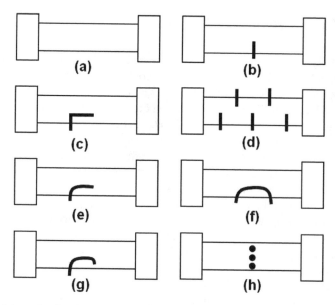

Fig. 7.29 Schematic of various cut geometries during laser trimming of resistors: (**a**) untrimmed, (**b**) single-slot, (**c**) L-cut, (**d**) serpentine, (**e**) curved L-cut, (**f**) curved U-cut, (**g**) curved J-cut, and (**h**) holes

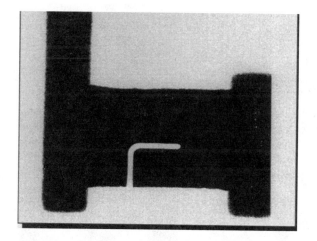

Fig. 7.30 Typical L-cut geometry in laser trimmed resistor. (Reprinted from Deluca 2002. With permission. Copyright Institute of Electrical and Electronics Engineers.)

7.4.4   Laser Scribing and Dicing

Laser scribing is an alternative process to the traditional saw dicing for die separation of thin silicon wafers, delicate III-V materials, and hard materials such as sapphire, glass, and ceramics. The traditional saw dicing process is associated with various disadvantages due to abrasive action of the saw blade and the introduction of high-pressure cooling water such as damage to the circuits, pollution of sensitive devices, and the residual stresses which may lead to subsequent cracking. Laser scribing is based on the creation of either a continuous groove at a certain depth or a series of equally spaced blind holes through the wafer. The scribe line creates the stress concentration facilitating easy breakage of the wafer along it after application of force. The laser scribing readily achieve the kerf widths of 30 μm and as small as 10 μm as compared to the more routinely produced kerf widths of 250 μm with diamond saw dicing. Figure 7.31 illustrates the good quality, with no evidence of cracking and chipping of the component edge, during laser scribing of the LED components on sapphire substrate using 255 nm UV laser. Figure 7.32 presents the I–V characteristics of micro-LED devices before and after UV laser dicing. The figure indicates reduction in the maximum current after laser scribing for both UV wavelengths (Gu et al. 2004; Illy et al. 2005)

Fig. 7.31   SEM image illustrating the quality of ~25 μm deep scribe of LED components on sapphire substrate using 255 nm UV laser. (Reprinted from Illy et al. 2005. With permission. Copyright, Elsevier.)

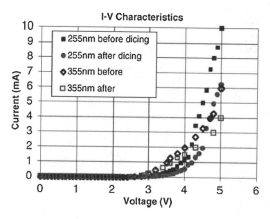

Fig. 7.32 I-V characteristics of LED devices before and after UV laser dicing (scribing velocity 5 mm/s). (Reprinted from Illy et al. 2005. With permission. Copyright Elsevier.)

Laser scribing is also attractive for the thin film photovoltaic (PV) materials due to narrow scribe widths and superior profile. Excellent scribe profiles were demonstrated on materials such as semiconductors (cadmium telluride, copper indium (gallium) diselenide, and silicon), transparent conducting oxides (fluorine-doped tin oxide and aluminum-doped zinc oxide), and metals (molybdenum and gold) using excimer lasers (Compaan et al. 2000). Recently, water jet-guided laser technology has been devised for the improved scribing characteristics of silicon PV-cells. The total internal reflection of the laser beam guided through the hair-thin water jet offers several advantages in terms of working distance, heat control, and material removal and cleanliness. Scribing of silicon PV-cells at speeds up to 300 mm/s with a groove depth of 30 μm were achieved with the water jet-guided laser scribing (Perrottet et al. 2005).

7.4.5   Laser Marking and Engraving

Various conventional methods such as ink marking, electrochemical marking, stamping, and mechanical engraving are used for the marking of manufactured products with logos and product information for the purpose of identification and theft prevention. Laser marking is increasingly used due to various advantages over the conventional marking methods such as no wear of tool, high degree of automation, free programming, and choice of characters (Qi et al. 2003). The process is particularly suitable for the fast and precision marking in automated manufacturing environment. Various applications include but not limited to precision marking of calipers; ID marks such as bar codes, date codes, part numbers, 2D matrix codes, and graphics; company logos; and aesthetic laser marking. Almost all the classes of materials can be laser marked by optimizing the process parameters such as pulse frequency, electric current, and scanning speed to ensure the acceptable mark quality.

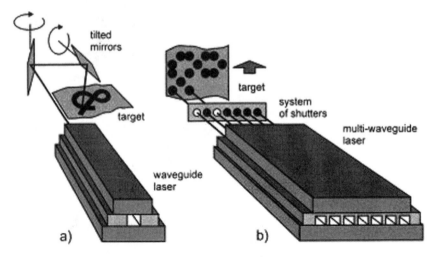

Fig. 7.33 Two approaches to laser marking based on (**a**) single waveguide CO_2 laser and (**b**) multi-waveguide RF excited CO_2 laser. (Reprinted from Plinski et al. 2000. With permission. Copyright Elsevier.)

Two main approaches to laser marking are shown in Fig. 7.33. First is based on single wave guide CO_2 laser in which the output laser beam is controlled using a system of movable mirrors to obtain a continuous marking on a stationary target (Fig. 7.33a). Second is based on the multiwave guide RF excited CO_2 laser in which required shapes are formed by series of dots (Fig. 7.33b) (Plinski et al. 2000). Recently, a new method of marking based on spectral properties of CO_2 laser was presented which avoids the complicated and fallible mechanical elements in conventional laser marking systems. The only executive elements of this new marker are diffraction grating and piezoelectric transducer (Fig. 7.34). The laser beam is deflected on the diffraction grating and then directed on the movable target producing the predictable distributed spots. It had been shown that the slab-waveguide configuration of RF excited CO_2 laser equipped with unstable resonator applied to diffraction marker produces well-ordered signature due to high spectral purity thus making the programming of the PZT system easy (Plinski et al. 2000).

One of the most important applications of laser marking, which is increasingly used in micromachining, is the marking of identification numbers or bar codes on the backside of the wafer at the die level (Fresonke 1994). Laser marking of identification numbers on the silicon wafers facilitates the traceability of the manufacturing process for fault analysis. The laser marks must be machine readable, miniaturized, and have no adverse effect on the further processing steps, and still permit the clear identification at the end on the manufacturing chain. With the increasing trend toward very thin wafers (150 µm) in consumer electronics, laser marking requirements are becoming more and more stringent. Q-switched Nd:YAG lasers are conventionally used for laser marking of silicon wafers to produce about 2.5 µm deep marks by melting. Permanent, debris-free, and high-contrast shallow marks (less than 1 µm) on the

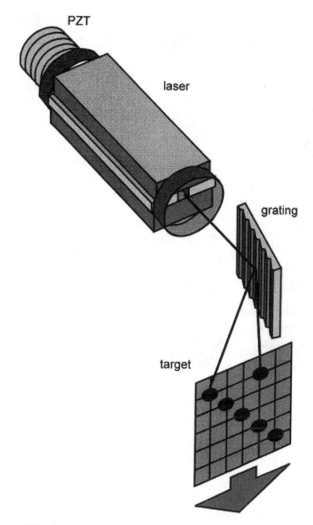

Fig. 7.34 Laser diffraction marker using piezoelectric transducer and diffraction grating. (Reprinted from Plinski et al. 2000. With permission. Copyright Elsevier.)

backside of the wafers can be achieved. This new micromarking technology allows the marking font size to be much less than 0.3 mm (Gu 2005).

Among the various laser parameters, one of the important parameters is the laser pulse frequency. Qi et al. (2003) studied the effect of laser pulse frequency on the various marking characteristics (mark depth, width, and contrast) during laser marking of stainless steel using Q-switched Nd:YAG laser. Pulse frequency of the laser had the most significant impact on the mark depth and contrast. Maximum mark depth was found to be at pulse frequency of 3 kHz, whereas the maximum contrast was obtained at 8 kHz. Figure 7.35 shows the SEM features of the marked surface at various pulse frequencies (Qi et al. 2003).

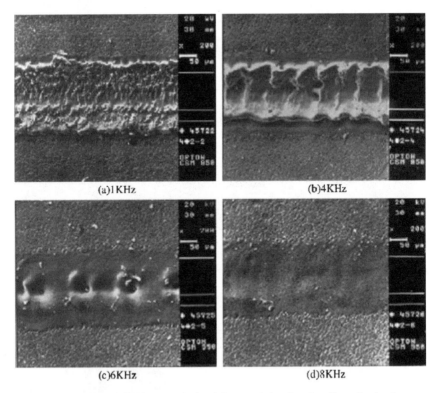

Fig. 7.35 Surface features of laser marked stainless steel showing the effect of pulse frequency of laser on mark quality. (Reprinted from Qi et al. 2003. With permission. Copyright Elsevier.)

7.4.6 Biomedical Applications

Lasers are now increasingly used in the fabrication of medical devices such as stents, catheters, and microfluidic devices. The important characteristic of lasers is their ability to be used for a variety of medical device materials provided the laser parameters are carefully selected based on the properties of the materials.

Stents are medical implants used to physically expand the blood vessels narrowed by plaque accumulation. Coronary stents are the most common type of stents used in the treatment of coronary artery diseases. Stents are generally fabricated by laser cutting of stainless-steel tubes (usually 316L steel). Conventionally, Nd: YAG lasers with pulse durations in the nanosecond and millisecond range have been used for the fabrication of stents. However, conventional laser processes are associated with significant quality issues related to a large heat-affected zone and melting. The melting at the cutting edge leads to the formation of burr and resolidified droplets. This may necessitate the additional processing steps. Conventional laser cutting also have the limitation of the minimum stent size that can be cut with acceptable quality (Momma et al. 1999). Recently, ultrashort pulse laser processes have demonstrated the abilities to fabricate the burr-free and damage-free stents (Rizvi 2003). In addition to stainless steel, the ultrashort laser systems are capable of processing various other alloys and organic materials. Figure 7.36 present the prototype stent (used in cardiovascular surgery) fabricated by ultrashort pulse laser indicating the damage-free and burr-free surfaces (Nolte et al. 2000).

Laser micromachining has also been described for the fabrication of liquid handling systems on polymer substrate chips. The microchannels fabricated by UV laser ablation with their ability to generate electro-osmotic flow are expected to find applications in miniaturized capillary electrophoresis instrumentation as

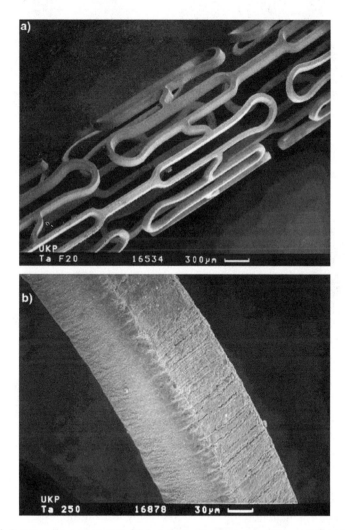

Fig. 7.36 Prototype coronary stent fabricated by ultrashort pulse laser cutting of tantalum. (Reprinted from Nolte et al. 2000. With permission. Copyright Wiley.)

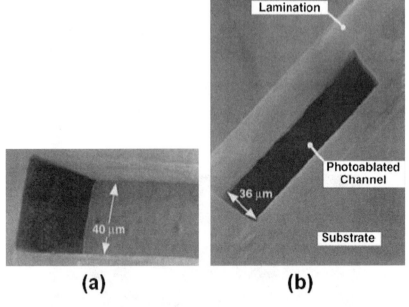

Fig. 7.37 (a) Photoablated microchannel in polymer substrate, and (b) microchannel entrance between polycarbonate substrate and PET/PE laminate. (Reprinted from Roberts et al. 1997. With permission. Copyright American Chemical Society.)

liquid handling platforms. These diagnostic and analysis systems are often termed as microscale total analytical systems (μ-TAS) (Roberts et al. 1997; Dario et al. 2000). Fabrication of microchannels with straight walls and high aspect ratios is demonstrated in a variety of substrates including polystyrene, polycarbonate, cellulose acetate, and poly (ethylene terephtalate) (PET) using mask projection technique with ArF excimer laser radiation. The resultant photoablated channels were sealed by two-layer PET/polyethylene film lamination technique. Figure 7.37 presents the micrographs of typical microchannels produced by photoablation in polymer.

7.4.7 Thin Film Applications

Thin films are extensively used in microelectronics, tribology, solar cells, and decorative coating applications. In many of these applications, it is necessary to pattern the thin films. Laser micromachining using ablation mechanism presents the tremendous potential for precision patterning of the thin films. For example, selective patterning of TiN films has been demonstrated using excimer lasers. Direct exciemr laser ablation of thin films offer significant advantages such as fewer number of processing steps compared to conventional etching using solutions of reactive gases. Processing steps for the micromachining of TiN films from the sacrificial layers on silicon substrate are presented in Fig. 7.38. The results indicated that TiN films can be selectively removed from the 400 to 1000 nm thick chromium sacrificial layers (Dowling et al. 2002). Laser micromachining by ablation has also been demonstrated for the patterning of $Tb_{0.4}Fe_{0.6}/Fe_{0.5}Co_{0.5}$, $Fe_{0.6}Co_{0.4}/SiO_2$ multilayers and SiNy films. Figure 7.39 presents the micrographs of the trenches micromachined in $Tb_{0.4}Fe_{0.6}/Fe_{0.5}Co_{0.5}$ multilayers on SiO_2 substrate (Pfleging et al. 2000).

7.4.8   Fuel Injector Drilling

Drilling of fuel injector nozzles is traditionally done by electrical discharge machining (EDM) and the hole diameters in the range of 150–200 µm are achieved. Increasingly, stringent environmental issues to reduce emissions necessitate the employment of smaller injection holes and/or specially shaped

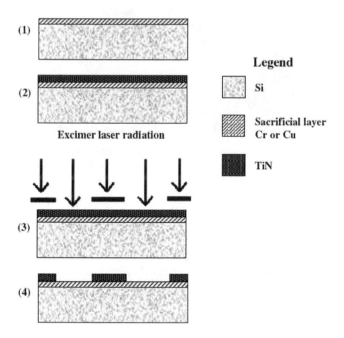

Fig. 7.38 Processing steps in the micromachining of TiN films from sacrificial layers using KrF laser (248 nm): (1) deposition of sacrificial layer, (2) deposition of TiN film, (3) laser patterning, and (4) micromachined TiN film on sacrificial layer. (Reprinted from Dowling et al. 2002. With permission. Copyright Institute of Physics.)

trenches:

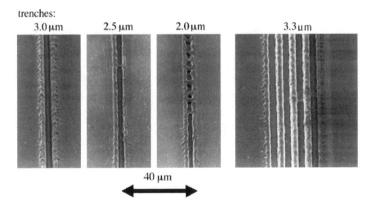

Fig. 7.39 Excimer laser micromachining of trenches in $Tb_{0.4}Fe_{0.6}/Fe_{0.5}Co_{0.5}$ multilayers on SiO_2 substrate (film thickness: 1 µm). (Reprinted from Pfleging et al. 2000. With permission. Copyright Elsevier.)

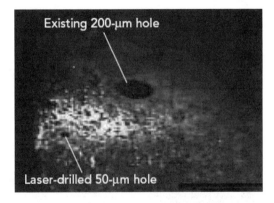

Fig. 7.40 Comparison of the holes drilled by EDM and laser (Reprinted from Herbst et al. 2004. With permission. Copyright Pennwell)

holes which are beyond the capabilities of the EDM. Laser is capable of drilling smaller holes (as small as 20 µm in diameter) in specially engineered configurations such as shape and taper. Figure 7.40 compares the 200 µm hole produced by EDM with a 50 µm hole created by helical laser drilling (Herbst et al. 2004). In addition, Fig. 7.41 shows the results of copper vapor lasers (CVLs) for drilling fuel injector nozzles. CVLs are inherently pulsed lasers with two laser transitions at 511 and 578 nm. Typical pulse lengths are in the range of 10–50 ns and pulse repetition frequencies of 5–30 kHz. The typical hole diameters achieved by CVL drilling are in the range of 40–500 µm. Laser micromachining is also demonstrated for the fabrication spinnerets for the production of synthetic fibers.

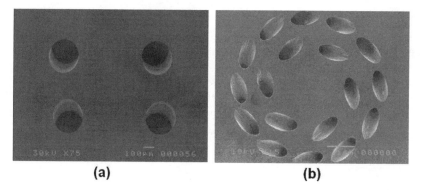

Fig. 7.41 Laser drilled holes in gasoline fuel injector (a) at small angle to the plate and (b) at high angle to the plate. (Courtesy of Andrew Kearsley, Oxford Laser Ltd.)

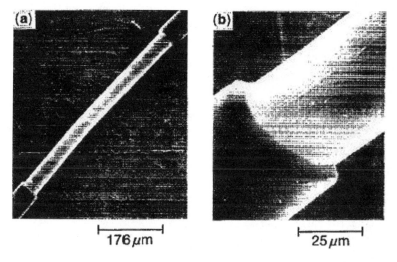

Fig. 7.42 Excimer laser stripping of wire insulation: (a) low and (b) high magnification views indicating the clean and precise stripping. (Reprinted from Brannon et al. 1991. With permission. Copyright American Institute of Physics.)

7.4.9 Stripping of Wire Insulation

The plastic insulation of the electrical wires needs to be removed during electrical connections to other wires or conductors. The wire stripping can be done with a variety of mechanical, thermal, electrical, or chemical methods. These methods are generally used where the precision and cleanliness of the stripped region are not critically important. However, for applications in microelectronics and computer industry, these quality parameters are of great importance. In such applications, laser stripping of insulations offer significant advantages such as precision and cleanliness. Various lasers such as CO_2 and KrF excimer lasers have been used for the stripping of wire insulation. Excimer laser radiation is strongly absorbed by most of the wire insulation material and offers advantages such as higher efficiency compared to CO_2 lasers. Figure 7.42 presents the micrographs of the excimer laser-stripped wire indicating the precise and clean stripping (Brannon et al. 1991).

References

Affrossman S, Bailey RT, Cramer CH, Cruickshank FR, McAllister JM, Alderman J (1989) Laser photochemical etching of silicon. Applied Physics A 49:533–542.

Bäuerle D (2000) Laser Processing and Chemistry. Springer, Berlin.

Bonse J, Baudach S, Krüger J, Kautek W, Lenzner M (2002) Femtosecond laser ablation of silicon-modification thresholds and morphology. Applied Physics A 74:19–25.

Brannon JH, Tam AC, Kurth RH (1991) Pulsed laser stripping of polyurethane-coated wires: a comparison of KrF and CO_2 lasers. Journal of Applied Physics 70:3881–3886.

Brannon J (1997) Excimer laser ablation and etching. IEEE Circuits and Devices 13:11–18.

Compaan AD, Matulionis I, Nakade S (2000) Laser scribing of polycrystalline thin films. Optics and Lasers in Engineering 34:15–45.

Daniel C, Mucklich F, Liu Z (2003) Periodic micro-nano-structuring of metallic surfaces by interfering laser beams. Applied Surface Science 208–209:317–321.

Dario P, Carrozza MC, Benvenuto A, Menciassi A (2000) Micro-systems in biomedical Applications. Journal of Micromechanics and Microengineering 10:235–244.

Deluca P (2002) A review of thirty-five years of laser trimming with a look to the future. Proceedings of the IEEE 90:1614–1619.

Denk R, Piglmayer K, Bäuerle D (2003) Laser-induced nano-patterning by means of interference subpatterns generated by microspheres. Applied Physics A 76:1–3.

Dowling AJ, Ghantasala MK, Hayes JP, Harvey EC, Doyle ED (2002) Excimer laser micromachining of TiN films from chromium and copper sacrificial layers. Smart Materials and Structures 11:715–721.

Dunsky C (2002) High-speed microvia formation with UV solid-state lasers. Proceedings of IEEE 90:1670–1680.

Eyett M, Bäuerle D, Wersing W, Thomann H (1987) Excimer-laser-induced etching of ceramic $PbTi_{1-x}Zr_xO_3$. Journal of Applied Physics 62:1511–1514.

Fjeldsted K, Chase SL (2002) Embedded Passives: Laser Trimmed Resistors. CircuiTree, March 2002.

Fresonke D (1994) In-Fab identification of silicon wafers with clean, laser marked barcodes, Advanced Semiconductor Manufacturing Conference, IEEE/SEMI, pp. 157–160.

Furusawa K, Takahashi K, Kumagai H, Midorikawa K, Obara M (1999) Ablation characteristics of Au, Ag, and Cu metals using a femtosecond Ti:sapphire laser. Applied Physics A 69: S359-S366.

Gower MC (1993) Excimer lasers: current and future applications in industry and medicine. In: Crafer RC, Oakley PJ (eds) Laser Processing in Manufacturing. Chapman & Hall, London, pp. 189–272.

Gower MC (2000) Industrial applications of laser micromachining. Optics Express 7:56–67.

Gower MC (2001) Laser micromachining for manufacturing MEMS devices. Proceedings of the SPIE 4559:53–59.

Gruenewald P, Cashmore J, Fieret J, Gower M (2001) High-resolution 157nm laser micromachining of polymers. Proceedings of the SPIE 4274:158–167.

Gu E, Jeon CW, Choi HW, Rice G, Dawson MD, Illy EK, Knowles MR (2004) Micromachining and dicing of sapphire, gallium nitride and micro LED devices with UV copper vapor laser. Thin Solid Films 453–454:462–466,

Gu B (2005) New laser marking technology using ultrafast lasers. Proceedings of SPIE 5713:132–136.

Harzik RL, Breitling D, Weikert M, Sommer S, Fohl C, Dausinger F, Valette S, Donnet C, Audouard E (2005) Ablation comparison with low and high energy densities for Cu and Al with ultra-short laser pulses. Applied Physics A 80:1589–1593.

Herbst L, Lindner H, Heglin M, Hoult T (2004) Targeting diesel engine efficiency. Industrial Laser Solutions, 19(10).

Himmelbauer M, Arenholz E, Bäuerle D (1996) Single-shot UV-laser ablation of polyimide with variable pulse lengths. Applied Physics A 63:87–90.

Illy EK, Knowles M, Gu E, Dawson MD (2005) Impact of laser scribing for efficient device separation of LED components. Applied Surface Science 249:354–361.

Kullmer R, Bäuerle D (1987) Laser-induced chemical etching of silicon in chlorine atmosphere. Applied Physics A 43:227–232.

Küper S, Brannon J, Brannon K (1993) Threshold behavior in polyimide photoablation: single-shot rate measurements and surface temperature modeling. Applied Physics A 56:43–50.

Küper S, Stuke M (1987) Femtosecond UV excimer laser ablation. Applied Physics B 44:199–204.

Lizotte T, Ohar O, Waters SC (2002) Excimer lasers drill inkjet nozzles. Laser Focus World 5:165–168.

Liu X, Du D, Mourou G (1997) Laser ablation and micromachining with ultrashort laser pulses. IEEE Journal of Quantum Electronics 33:1706–1716.

Momma C, Knop U, Nolte S (1999) Laser Cutting of Slotted Tube Coronary Stents-State-of-the-Art and Future Developments. Progress in Biomedical Research 4:39–44.

Nolte S, Kamlage G, Korte F, Bauer T, Wagner T, Ostendorf A, Fallnich C, Welling H (2000) Microstructuring with femtosecond lasers. Advanced Engineering Materials 2:23–27.

Perrottet D, Boillat C, Amorosi S, Richerzhagen B (2005) PV processing: Improved PV-cell scribing using water jet guided laser. Refocus 6:36–37.

Pfleging W, Ludwig A, Seemann K, Preu R, Mackel H, Glunz SW (2000) Laser micromachining for applications in thin film technology. Applied Surface Science 154–155:633–639.

Plinski EF, Witkowski JS, Abramski KM (2000) Diffraction scanning mechanism for laser marker. Optics and Laser Technology 32:33–37.

Qi J, Wang KL, Zhu YM (2003) A study on the laser marking process of stainless steel. Journal of Materials Processing Technology 139:273–276.

Raman S (2001) Laser microvia productivity: dual head laser drilling system. CircuiTree 7:30–36.

Rizvi NH, Apte P (2002) Developments in laser micro-machining techniques. Journal of Materials Processing Technology 127:206–210.

Rizvi NH (1999) Production of novel 3D microstructures using excimer laser mask projection techniques. Proceedings of the SPIE 3680:546–552.

Rizvi NH (2003) Femtosecond laser micromachining: Current status and applications. Riken Review 50:107–112.

Roberts MA, Rossier JS, Bercier P, Girault H (1997) UV laser machined polymer substrates for the development of microdiagnostic systems. Analytical Chemistry 69:2035–2042.

Schaeffer RD (2002) Seeing the light: embedded resistor trimming. CircuiTree, June 2002.

Srinivasan R, Braren B, Casey KG (1990) Nature of "incubation pulses" in the ultraviolet laser ablation of polymethyl methacrylate. Journal of Applied Physics 68:1842–1847.

Subrahmanyan PK (2003) Laser micromachining in the microelectronics industry: emerging applications. Proceedings of the SPIE 4977:188–197.

Sutcliffe E, Srinivasan R (1986) Dynamics of UV laser ablation of organic polymer surfaces. Journal of Applied Physics 60:3315–3322.

Tseng AA, Chen YT, Ma KJ (2004) Fabrication of high-aspect-ratio microstructures using excimer laser. Optics and Lasers in Engineering 41:827–847.

Vogel A, Noack J, Nahen K, Theisen D, Busch S, Parlitz U, Hammer DX, Noojin GD, Rockwell BA, Birngruber R (1999) Energy balance of optical breakdown in water at nanosecond to femtosecond time scales. Applied Physics B 68:271–280.

Part III
Laser Fabrication

Chapter 8
Laser Forming

8.1 Introduction

Forming techniques, in a broader sense, comprise a variety of metal working processes in which the material is shaped in solid state by plastic deformation. The forming techniques are generally classified as bulk forming and sheet forming processes. Common bulk forming processes are rolling, extrusion, drawing, forging, etc., whereas common sheet metal forming processes are bending, surface contouring, linear contouring, shallow recessing, etc. The end products of forming processes can either be final components or basic shapes such as rods, bars, tubes, sheets, plates, etc. (Semiatin 1988). In the present context, forming process is primarily referred to as sheet material forming such as bending. Conventional mechanical bending process for a sheet material involves a set of bending die and punch with a sheet material placed between them. During bending, the sheet material is plastically deformed into desired shapes by application of suitable forces such that the shape of the sheet material conforms to the contours of the die and punch (Fig. 8.1). The mechanical sheet material forming processes are primarily suitable for mass production due to high cost of dies and punches. For a rapid production of few parts such as those required for test prototypes and special shapes, the conventional sheet metal forming processes are often uneconomical due to high cost of dies, and longer time for the fabrication and error corrections of dies. Also, the mechanical sheet metal forming processes are often associated with the inherent effects such as spring-back effects where the actual bending angle is always less than the desired bending angle defined by the dies.

The idea of using lasers for forming of sheet material was first conceived by Kitamura in Japan. This is followed by successful bending of 22 mm thick steel plates using a 15 kW CO_2 laser (Kitamura 1983). Since then, significant interests have been attracted toward use of laser for forming of sheet materials particularly used for the applications in aerospace, automotive, and electronic industries. The key idea here is to establish the steep thermal gradients that set up the stresses for rapid expansion of material in a confined region. The laser forming is based on the earlier flame bending processes used for ship construction. In flame bending, a part of the material is heated by a flame (such as oxyacetylene torch) in a localized way

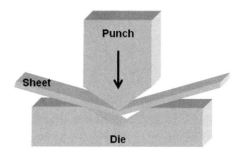

Fig. 8.1 Schematic of sheet metal forming process for typical V-bends

such that the expansion of the heated material and subsequent yielding against the
cooler region results in overall bending of the part (Moshaiov and Vorus 1987).
Even though flame bending has been extensively used in the shipbuilding industry,
it is often associated with the difficulties in bending thin sections of materials and
sections of high-conductivity materials. This is primarily due to the diffuse nature
of the heating source resulting in moderate temperature gradients (Magee et al.
1998). With the development of high-power lasers with an enhanced ability to
deliver intense beam on highly localized regions, steep temperature gradients can
efficiently be established in the materials for bending action. Laser forming of sheet
materials is generally free from spring-back effects and also associated with typical
advantages of laser materials processing such as rapid manufacturing, ease of automation, and flexible manufacturing.

8.2   Laser Forming Processes

During laser forming, the surface of the sheet material is scanned with a defocused
laser beam such that laser–material interaction causes localized heating of the surface without melting. The heating of the material causes the expansion of the material in a confined region. Due to continuity of the heated region with the surrounding
material, the free expansion of the hot region is resisted, resulting in bending of the
part. The overall deformation in the sheet material is determined by the complete
thermal cycle (heating and cooling) associated with laser processing. The thermal
effects during laser–material interaction are complex and depend on both the laser
parameters and the material properties. Depending on the nature of the thermal
effects during laser forming, the strains can be induced either within or out of the
sheet plane (Fig. 8.2). If the strains induced are uniform throughout the thickness
of the material, the material tends to shrink or shorten in length, whereas if the
strains are not uniform throughout the thickness, the material tends to deflect resulting in out-of-plane deformation or bending (Magee et al. 1998).

When the material is scanned with a high-power laser during laser forming,
temperature distribution within the material is generally nonuniform in three

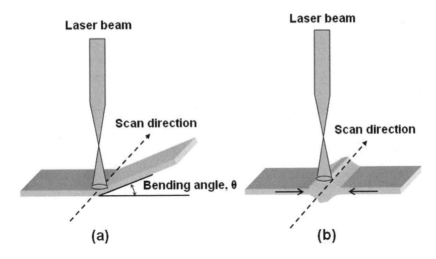

Fig. 8.2 (a) Out-of-plane and (b) in-plane deformation during laser forming

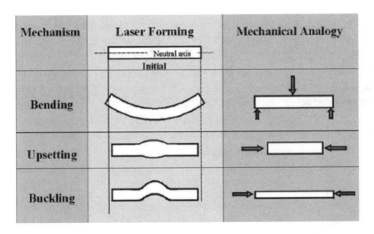

Fig. 8.3 Mechanical analogy of three laser forming mechanisms (Reprinted from Marya and Edwards 2001. With permission. Copyright Elsevier.)

dimensions. The formation of various shapes in laser forming can be better understood from the rectangular components of forces if the process is considered to be two-dimensional. Figure 8.3 presents the schematic of the shapes formed during various laser forming operations such as bending, buckling, and upsetting. The corresponding mechanical analogy illustrates the desired direction of force application relative to the workpiece neutral axis. The figure indicates that shape change during bending is associated with stresses perpendicular to the neutral axis. The shape change can also be induced if the stresses are parallel to the neutral axis. In this case, the magnitude of stresses determines whether the shape change is a result of buckling or upsetting. Thus, the three mechanisms of shape change during laser forming can be identified as bending, buckling, and upsetting mechanisms (Marya and Edwards 2001).

8.2.1 Bending or Temperature Gradient Mechanism

In the most typical laser bending arrangement, one side of the sheet material is clamped in a fixture and the laser beam is scanned linearly parallel to the free side of the sheet (Fig. 8.4). Such a simplified arrangement is useful for typical V-shape bends in the sheet. However, more complex, and nonlinear scanning patterns are possible for complex shapes of the bends. The final shape of the bend is determined by the actual laser processing strategy employed. This includes the consideration of a number of parameters such as laser power, scanning velocity, number of scanning passes, offset between the successive laser scanning passes, etc.

Among the various mechanisms of laser forming, TGM is most extensively studied and reported in the literature. When the sheet surface is irradiated with a laser beam, a fraction of the laser energy is absorbed in the material. The absorption coefficient of metals is of the order of 10^5 cm^{-1}. Hence the laser energy is deposited in a very thin layer of material (Ready 1997). The heat energy is simultaneously conducted into the material. Since enormous laser energy is absorbed by the surface and inefficient conduction of this energy during heating, steep temperature gradients are established in the material. In general, the yield strength of the material decreases with increasing temperature as indicated in Fig. 8.5 (Hertzberg 1976). The steep temperature gradients result in the differential thermal expansion of the material. The layers of the material close to the surface are at a higher temperature than those away from the surface and hence expand more. This differential thermal

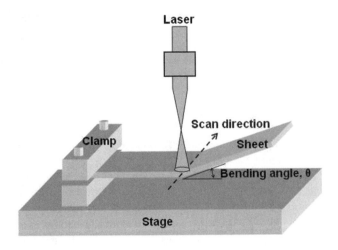

Fig. 8.4 Schematic of the laser bending setup

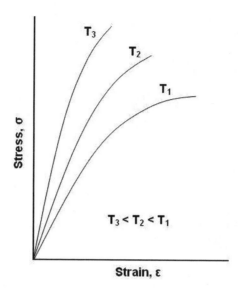

Fig. 8.5 Effect of temperature on yield strength of a material

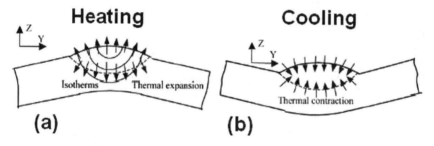

Fig. 8.6 Thermal gradient mechanism of laser bending: (**a**) heating process and (**b**) cooling process (Reprinted from Hu et al. 2002. With permission. Copyright Elsevier.)

expansion produces thermal stresses in the material leading to counter-bending of the sheet away from the laser beam. As the material is continuously heated, the bending moment opposes the counter-bending of sheet. When the thermal stress exceeds the temperature-dependent yield stress of the material, further expansion of the heated region is converted into plastic deformation. Since the free expansion of the heated region is resisted by the surrounding cooler material, the strains at the surface are compressive in nature. After the laser scan, the surface is rapidly cooled due to conduction of heat into the material, and convection and radiation of heat from the surface. The cooling of the surface layer causes the thermal contraction. Since the surface of the sheet had been compressed, the thermal contraction results in the local shortening of the surface layers. Thus local shortening of the surface layers causes the sheet to bend towards the laser beam (Fig. 8.6). During this cooling phase, the yield strength further increases and very little plastic restraining takes

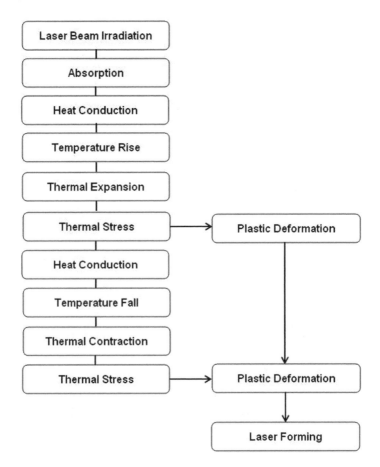

Fig. 8.7 Sequence of various effects taking place during temperature gradient mechanism of laser bending (Namba 1986)

place. The overall bending angle is greatly determined by the continuous heating and cooling events, and the associated thermal effects (Hu et al. 2002). Figure 8.7 presents the sequence of events taking place during the forming (bending) of sheet material based on TGM (Namba 1986). Furthermore, the bending angle can be increased by repeating the laser scanning of the surface.

8.2.2 Buckling Mechanism

The formation of characteristic shapes during laser forming is primarily influenced by the laser processing parameters and the geometric condition of the sheet material. Even though out-of-plane bending of the sheet by TGM is the most often observed deformation mode, additional deformation modes such as buckling of sheet can also be observed. The previously described TGM for bending of sheet is primarily operative in conditions where steep temperature gradients are developed in the material. Such conditions are dominant when the diameter of the laser beam is nearly equal to the thickness of the sheet. However, when the diameter of the laser beam is significantly larger than (nearly 10 times) the thickness of the sheet (i.e., for thin sheets) and when the ratio of thermal conductivity to the thickness of the material is large, negligibly small temperature gradient is developed in the thickness direction. The condition of negligibly small thermal gradient in thickness direction is also favored where the ratio of the thermal conductivity to the thickness of the material is large. Such conditions facilitate the deformation by BM. The various steps in laser forming by BM can be summarized as (Vollertsen et al. 1995; Hu et al. 2002):

1. Initial heating: When the thin sheet is irradiated with the laser beam, the absorbed energy causes the rapid heating of the material throughout the thickness of the sheet. The laser parameters and the geometry of the sheet result in negligibly small temperature gradient in the thickness direction such that nearly uniform thermal expansion takes place along the thickness of the sheet. The free expansion of the sheet is restricted by the surrounding material resulting in compressive stresses in the laser-heated region (Fig 8.8a).

2. Bulging: The sheet under compressive stresses tends to buckle when the stresses exceed a certain value. The tendency to buckle is enhanced when the sheet is thin and the temperature-dependent flow stress is not too low. If the temperature-dependent flow is too low, the sheet will compress plastically instead of buckling, whereas if the sheet is thick, the force of buckling will not be reached because the force of buckling increases with the square of sheet thickness (Fig. 8.8b).

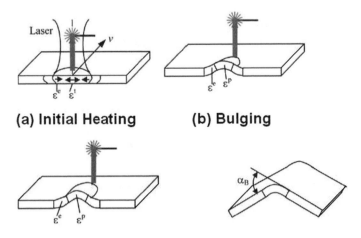

(a) Initial Heating   **(b) Bulging**

(c) Growth of buckle   (d) Development of bending angle

Fig. 8.8 Steps in the laser forming of sheet by buckling mechanism (BM). ε^e, ε^t, and ε^p refers to the elastic, thermal, and plastic strain respectively (Reprinted from Hu et al. 2002. With permission. Copyright Elsevier.)

3. Growth of buckle: Once buckling is initiated, the buckle is extended in length along the laser scanning path due to thermal expansion of the subsequent material. The buckling is associated with a combination of elastoplastic regions. Since the temperature-dependent flow stress is low at the top region of the buckle (top region being heated to higher temperature), the deformation is nearly pure plastic in this region, whereas in the surrounding regions of the buckle (which are not directly heated by laser beam), the deformation is nearly pure elastic (Fig. 8.8c).
4. Development of bending angle: When the laser beam leaves the scanning path after traversing the whole width of the sheet, the sheet develops the bending angle. The bending angle, α_b, is primarily determined by the plastic region of the bend which remains in the sheet. However, this is also facilitated by the straightening of the elastic bending of the sheet due to no resisting force (Fig. 8.8d).

One of the most interesting and important features associated with the BM is related with the direction of bending. The bending may take place toward (positive bending) or away (negative bending) from the laser beam, resulting in concave or convex shapes of the laser-formed sheet respectively. The formation of convex shapes cannot be explained by the TGM of laser bending. The bending behavior can be effectively controlled if the influence of the various factors is understood. Various important factors which may influence the bending direction during laser forming by BM are elastic or plastic prebending of sheet, preexisting residual stresses, and counter-bending due to temperature gradients. Schematic of these effects along with the relative probability toward positive and negative bending is presented in Fig. 8.9 (Vollertsen et al. 1995). Prebending of sheet almost always tends to increase the initial bending angle during subsequent laser forming. The forming of the prebent sheets may be positive or negative depending on the prebent

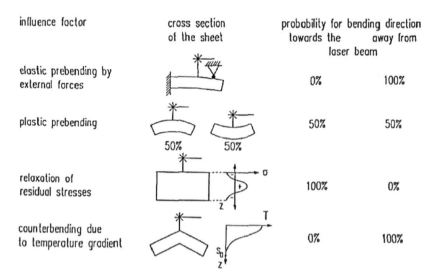

influence factor	cross section of the sheet	probability for bending direction towards the laser beam	away from laser beam
elastic prebending by external forces		0%	100%
plastic prebending	50% 50%	50%	50%
relaxation of residual stresses		100%	0%
counterbending due to temperature gradient		0%	100%

Fig. 8.9 Factors influencing the bending direction during laser forming by buckling mechanism (Reprinted from Vollertsen et al. 1995. With permission. Copyright Institute of Physics.)

angle. Generally, the positive bending of concave sheets and the negative bending of convex sheets are most easy cases. However, the positive bending of convex sheets and the negative bending of concave sheets may be possible by controlling the laser forming parameters such as laser power, scan speed, and beam size and shape. The other important factor influencing the bending direction during laser forming is the presence of preexisting residual stresses in the sheet. If the surface of the sheet has preexisting compressive stresses as in the case of rolled sheets, the subsequent laser forming facilitates the positive bending by relieving the surface compressive stresses during heating. The bending direction during buckling can also be influenced by the counter-bending during heating operation. Even though BM is not associated with temperature gradients developed in the material, a small number of temperature gradients are always present in the material during heating. The negligibly small counter-bending during the heating phase may facilitate the development of negative bending operation by BM (Vollertsen et al. 1995).

8.2.3   Upsetting Mechanism

Similar to the BM, the UM of laser forming is operative when the temperature gradient in the laser-heated region is negligibly small. However, the thickness of the sheet for UM is much larger than the heated region of the sheet. The localized heating of the sheet causes uniform thermal expansion of the material throughout the thickness of the laser-heated region. The resistance to expansion sets up the uniform compressive strains in the sheet. The geometry of the sheet prevents the buckling of the sheet. Instead, local shortening of the sheet in the laser-heated region takes place resulting in the local increase in thickness of the sheet. The schematic of the process is presented in Fig. 8.10.

The three mechanisms of laser forming explained above are the cases for a simplistic condition. In practice, however, the final shape of the sheet is governed by the complexity of interactions between the various mechanisms. Also, since the buckling and upsetting of sheet are results of buckling instabilities, there may be conditions where transition from conventional TGM to BMs and UMs take place. The influence of the interactions between the various BMs on the final bending angle is schematically illustrated in Fig. 8.11. As indicated in the figure, a combination of convex buckling and positive bending have a negative influence on the final bending compared with a combination of concave buckling and positive bending (Vollertsen et al. 1995).

8.3   Analysis of Laser Forming Processes

Extensive efforts have been directed toward modeling the laser forming process by both analytical and numerical approaches. Most of these efforts were primarily concentrated on the thermal gradient mechanism and BM of laser forming. This section explains the important models of laser forming by TGM and BM.

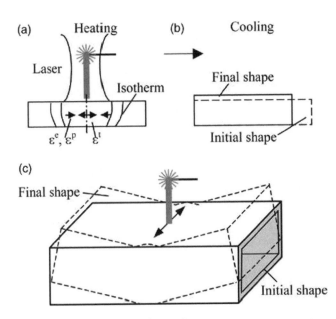

Fig. 8.10 Laser forming by upsetting mechanism (UM): (**a**) Initial heating, (**b**) cooling, and (**c**) final shape (Reprinted from Hu et al. 2002. With permission. Copyright Elsevier.)

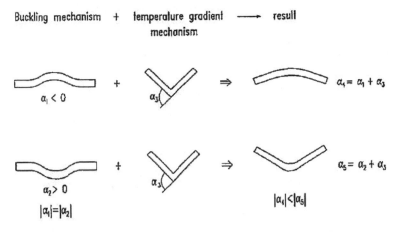

Fig. 8.11 Schematic of the influence of interaction between the various buckling mechanisms (TGM and BM) on final bending angle (Reprinted from Vollertsen et al. 1995. With permission. Copyright Institute of Physics.)

8.3.1 Temperature Gradient Mechanism

TGM is the most widely reported mechanism of laser forming. Early efforts for modeling the TGM were primarily conducted by Vollertsen and coworkers (Geiger and Vollertsen 1993; Vollertsen 1994a, b; Vollertsen and Rodle 1994). Vollertsen proposed an analytical model of laser forming based on the simple bending theory and an energy approach to temperature fields. Even though the model has been extensively referred in the literature, it is based on numerous simplified assumptions related with temperature field, temperature-dependent material properties, and geometry of the sheet. Recently, Cheng and coworkers addressed several assumptions of the Vollertsen model and proposed an analytical model of laser forming by TGM while taking into account three-dimensional temperature distribution in the sheet and temperature-dependent mechanical properties of material (Cheng and Lin 2000a; 2001). Most of this section is based on the model of laser forming proposed by Cheng. The important steps in the modeling of laser forming process are the determination of temperature distribution in the material and the determination of bending angle during heating and cooling phases of laser forming.

8.3.1.1   Determination of Temperature Distribution

Figure 8.12 presents the coordinate system for laser forming of sheet. During laser forming, a Gaussian laser beam is scanned along the positive x-axis with a constant traverse velocity, v (Cheng and Lin 2000a).

The governing equation for three-dimensional heat conduction is given by:

$$\rho c \frac{\partial T}{\partial t} = \frac{\partial}{\partial x}\left(k(T)\frac{\partial T}{\partial x} \right) + \frac{\partial}{\partial y}\left(k(T)\frac{\partial T}{\partial y} \right) + \frac{\partial}{\partial z}\left(k(T)\frac{\partial T}{\partial z} \right), \qquad (8.1)$$

where k is thermal conductivity, ρ is density, and c is specific heat of the material.

Linearizing the above equation using Kirchoff's transformation gives the transformed temperature, Θ, as:

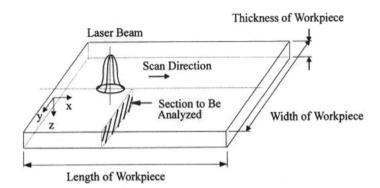

Fig. 8.12 Model geometry for laser forming by temperature gradient mechanism (Reprinted from Cheng and Lin 2001. With permission. Copyright Elsevier.)

$$\Theta = \int_{T_0}^{T} \frac{k(\tau)}{k_0} d\tau, \tag{8.2}$$

where k_0 is the thermal conductivity at temperature T_0 and $k(\tau)$ is the temperature-dependent conductivity.

The governing equation in terms of transformed temperature, Θ, can be expressed as:

$$\frac{1}{\alpha} \frac{\partial \Theta}{\partial t} = \frac{\partial^2 \Theta}{\partial x^2} + \frac{\partial^2 \Theta}{\partial y^2} + \frac{\partial^2 \Theta}{\partial z^2}, \tag{8.3}$$

where α is thermal diffusivity of material and given by $k/\rho c$.

In most of the laser forming experiments, a laser beam with Gaussian energy distribution is scanned along the predefined path. The Gaussian energy distribution can be expressed as:

$$Q(x,y) = \frac{q}{\pi \omega^2} \exp\left[-\frac{x^2 + y^2}{\omega^2}\right], \tag{8.4}$$

where q is total laser energy deposited and ω is laser beam radius.

For a moving Gaussian laser beam on a plate with finite thickness, the solution of the heat conduction equation can be expressed as (Cheng and Lin 2000a):

$$
\begin{aligned}
\Theta - \Theta_0 = \frac{\eta q}{\rho c \pi^{1.5} \alpha^{0.5}} \Bigg(&\int_{\tau=0}^{\tau=t} \frac{1}{\left[4\alpha(t-\tau)+\omega^2\right]\sqrt{t-\tau}} \\
&\times \exp 0\left[-\frac{(x-vt)^2+y^2}{4\alpha(t-\tau)+\omega^2} - \frac{z^2}{4\alpha(t-\tau)}\right] d\tau \\
+ \sum\Bigg(&\int_{\tau=0}^{\tau=t} -\frac{1}{\left[4\alpha(t-\tau)+\omega^2\right]\sqrt{t-\tau}} \\
&\times \exp\left[-\frac{(x-vt)^2+y^2}{4\alpha(t-\tau)+\omega 2} - \frac{z_n^2}{4\alpha(t-\tau)}\right] d\tau \\
+ &\int_{\tau=0}^{\tau=t} \frac{1}{\left[4\alpha(t-\tau)+\omega^2\right]\sqrt{t-\tau}} \\
&\times \exp\left[-\frac{(x-vt)^2+y^2}{4\alpha(t-\tau)+\omega^2} - \frac{z_n'^2}{4\alpha(t-\tau)}\right] d\tau \Bigg)\Bigg).
\end{aligned}
\tag{8.5}
$$

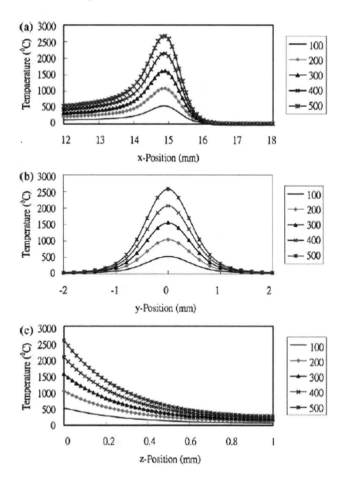

Fig. 8.13 Calculated temperature distributions in (**a**) *x* direction, (**b**) *y* direction, and (**c**) *z* direction as a function of laser power during laser forming of 304 stainless steel. The results are calculated using laser scanning speed of 1,000 mm/min, sheet thickness of 1.0 mm, and laser spot diameter of 1.0 mm (Reprinted from Cheng and Lin 2000a. With permission. Copyright Elsevier.)

Using this equation of transformed temperature, Θ, the temperature, T, can be determined above in Eq. (8.2). Typical calculated temperature distributions in three directions are presented in Figs. 8.13 and 8.14 for laser forming of 304 stainless steel. The results also show the effect of laser power and laser scanning speed on the temperature distributions. In general, increasing laser power raises the temperature distribution with increasingly higher peak temperature in the distribution, whereas increasing laser scanning velocity decreases the temperature distribution due to decreasing laser energy (Cheng and Lin 2000a).

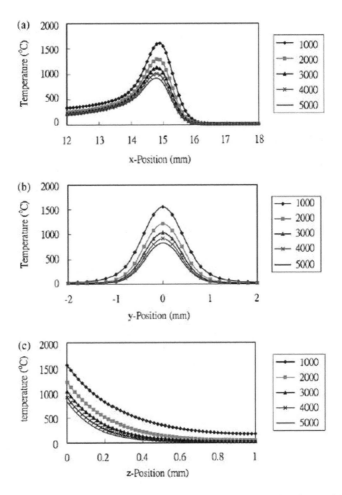

Fig. 8.14 Calculated temperature distributions in (**a**) x direction, (**b**) y direction, and (**c**) z direction as a function of laser scanning speed during laser forming of 304 stainless steel. The results are calculated using laser power of 300 W, sheet thickness of 1.0 mm, and laser spot diameter of 1.0 mm (Reprinted from Cheng and Lin 2000a. With permission. Copyright Elsevier.)

8.3.1.2 Determination of Bending Angle

For calculation of bending angle from the temperature distribution, the following assumptions are made (Cheng and Lin 2001):

1. Heat transfer in longitudinal direction is neglected such that generalized plane deformation conditions can be considered. These conditions are reasonable particularly for the high-speed laser scanning and the longer sheets in longitudinal direction.

2. Angle of bending is determined by both the heating and the cooling phases during laser forming. The angle of bending during heating phase is a function of the temperature distribution when the maximum temperature is reached during heating, whereas the angle of bending during cooling phase is a function of the surrounding temperature.
3. Initial sheet is free from preexisting creep and residual stresses.
4. The effect of gravitational forces is neglected.
5. Compressive and tensile yield stresses are equal in magnitude.
6. The plastic deformation follows von Mises yield criterion.

Since heat transfer is neglected in longitudinal direction, the temperature and stress distribution are expected to be symmetric with respect to y-axis. Hence, only half a section of the sheet is analyzed in this model and is shown by hatched cross section in Fig. 8.12. If the plastic strain is neglected, the total thermoelastic strain, ε_y, can be expressed as a sum of the elastic strain due to external force and the thermal strain:

$$\varepsilon_y(y,z) = \frac{\sigma_y(y,z)}{E(y,z)} + \alpha(y,z)T(y,z), \tag{8.6}$$

where E is the modulus of elasticity and α is the coefficient of thermal expansion. The strain can also be expressed as in-plane strain components as:

$$\varepsilon_y(y,z) = \varepsilon_y^0(y) + z\kappa_y(y), \tag{8.7}$$

where $\varepsilon_y^0(y)$ is the mid-surface strain and the $\kappa_y(y)$ is the curvature of the sheet. Solving Eqs. (8.6) and (8.7), the total thermoelastic stress, ρ_y, is expressed as:

$$\sigma_y(y,z) = E(y,z)\left[\varepsilon_y^0(y) + z\kappa_y(y) - \alpha(y,z)T(y,z)\right] \tag{8.8}$$

In the absence of external force, moment, and external restraint, the force and the moment at any cross section must be zero. The conditions are mathematically expressed as:

$$\sum F_y = 0, \int \sigma_y(y,z)\,dz = 0 \tag{8.9}$$

$$\sum M = 0, \int \sigma_y(y,z)z\,dz = 0 \tag{8.10}$$

Using Eq. (8.8) in Eqs. (8.9) and (8.10), the expressions for force and moment balance can be written as:

$$\int_{-c}^{c} E(y,z)\varepsilon_y^0(y)\mathrm{d}z + \int_{-c}^{c} E(y,z)z\kappa_y(y)\mathrm{d}z - \int_{-c}^{c} E(y,z)\alpha(y,z)T(y,z)\mathrm{d}z = 0 \qquad (8.11)$$

$$\int_{-c}^{c} E(y,z)z\varepsilon_y^0(y)\mathrm{d}z + \int_{-c}^{c} E(y,z)z^2\kappa_y(y)\mathrm{d}z - \int_{-c}^{c} E(y,z)\alpha(y,z)T(y,z)z\mathrm{d}z = 0$$

$$(8.12)$$

Expressing Eqs. (8.11) and (8.12) in a matrix form gives:

$$\begin{bmatrix} \int_{-c}^{-c} E(y,z)\mathrm{d}z & \int_{-c}^{-c} E(y,z)z\mathrm{d}z \\ \int_{-c}^{-c} E(y,z)z\mathrm{d}z & \int_{-c}^{-c} E(y,z)z^2\,\mathrm{d}z \end{bmatrix} \begin{bmatrix} \varepsilon_y^0(y) \\ \kappa_y(y) \end{bmatrix} = \begin{bmatrix} \int_{-c}^{-c} E(y,z)\alpha(y,z)T(y,z)\mathrm{d}z \\ \int_{-c}^{-c} E(y,z)\alpha(y,z)T(y,z)z\mathrm{d}z \end{bmatrix} \qquad (8.13)$$

Solving for mid-surface strain, $\varepsilon_y^0(y)$, and curvature, $\kappa_y(y)$, gives:

$$\begin{bmatrix} \varepsilon_y^0(y) \\ \kappa_y(y) \end{bmatrix} = \begin{bmatrix} \int_{-c}^{-c} E(y,z)dz & \int_{-c}^{-c} E(y,z)zdz \\ \int_{-c}^{-c} E(y,z)zdz & \int_{-c}^{-c} E(y,z)z^2dz \end{bmatrix}^{-1} \times \begin{bmatrix} \int_{-c}^{-c} E(y,z)\alpha(y,z)T(y,z)dz \\ \int_{-c}^{-c} E(y,z)\alpha(y,z)T(y,z)zdz \end{bmatrix} \qquad (8.14)$$

The thermoelastic stress expressed in Eq. (8.8) can then be calculated using Eq. (8.14). When this thermoelastic stress exceeds the yield stress of the material, plastic strains are produced. The plastic yielding can be predicted using the von Mises criterion. The modified force and moment balance equation can then be rewritten as:

$$\begin{bmatrix} \int_{h}^{h'} E(y,z)\mathrm{d}z + \int_{b'}^{b} E(y,z)\mathrm{d}z & \int_{h}^{h'} E(y,z)z\mathrm{d}z + \int_{b'}^{b} E(y,z)z\mathrm{d}z \\ \int_{h}^{h'} E(y,z)z\mathrm{d}z + \int_{b'}^{b} E(y,z)z\mathrm{d}z & \int_{h}^{h'} E(y,z)z^2\mathrm{d}z + \int_{b'}^{b} E(y,z)z^2\mathrm{d}z \end{bmatrix} \begin{bmatrix} \varepsilon_y^0(y) \\ \kappa'_y(y) \end{bmatrix} =$$

$$\begin{bmatrix} \int_{h}^{h'} E(y,z)\alpha(y,z)T(y,z)\mathrm{d}z + \int_{b'}^{b} E(y,z)\alpha(y,z)T(y,z)\mathrm{d}z - \int_{-c}^{b} Y(y,z)\mathrm{d}z - \int_{h'}^{h} Y(y,z)\mathrm{d}z - \int_{b'}^{b'} Y(y,z)\mathrm{d}z \\ \int_{h}^{h'} E(y,z)\alpha(y,z)T(y,z)z\mathrm{d}z + \int_{b'}^{b} E(y,z)\alpha(y,z)T(y,z)z\mathrm{d}z - \int_{-c}^{h} Y(y,z)z\mathrm{d}z - \int_{h'}^{b'} Y(y,z)z\mathrm{d}z - \int_{b'}^{c} Y(y,z)z\mathrm{d}z \end{bmatrix}$$

$$(8.15)$$

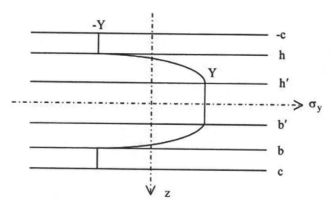

Fig. 8.15 Schematic of the area of plastic zone (Reprinted from Cheng and Lin 2001. With permission. Copyright Elsevier.)

where h, h', b, and b' represent the area of the plastic zone as indicated in Fig. 8.15.

Solving Eq. (8.15) for mid-surface plastic strain, $\varepsilon_y^{0\prime}(y)$, and plastic curvature, $\kappa_y'(y)$, gives:

$$
\begin{bmatrix} \varepsilon_y^{0\prime}(y) \\ \kappa_y'(y) \end{bmatrix} = \begin{bmatrix} \int_h^{h'} E(y,z)\,dz + \int_{b'}^{b} E(y,z)\,dz & \int_h^{h'} E(y,z)\,z\,dz + \int_{b'}^{b} E(y,z)\,z\,dz \\ \int_h^{h'} E(y,z)\,z\,dz + \int_{b'}^{b} E(y,z)\,z\,dz & \int_h^{h'} E(y,z)\,z^2\,dz + \int_{b'}^{b} E(y,z)\,z^2\,dz \end{bmatrix}^{-1} \times
$$

$$
\begin{bmatrix} \int_h^{h'} E(y,z)\,\alpha(y,z)\,T(y,z)\,dz + \int_{b'}^{b} E(y,z)\,\alpha(y,z)\,T(y,z)\,dz - \int_{-c}^{h} Y(y,z)\,dz - \int_{h'}^{b'} Y(y,z)\,dz - \int_{b'}^{c} Y(y,z)\,dz \\ \int_h^{h'} E(y,z)\,\alpha(y,z)\,T(y,z)\,z\,dz + \int_{b'}^{b} E(y,z)\,\alpha(y,z)\,T(y,z)\,z\,dz - \int_{-c}^{h} Y(y,z)\,z\,dz - \int_{h'}^{b'} Y(y,z)\,z\,dz - \int_{b'}^{c} Y(y,z)\,z\,dz \end{bmatrix}
$$

$$(8.16)$$

Finally, the angle of bending, α_b, during heating phase can be evaluated as:

$$\alpha_b = 2\int \kappa_y'\,dy \qquad (8.17)$$

Using a similar approach, the angle of bending during cooling phase can be calculated. The bending angle during the laser forming is considered as a summation of the bending angles during heating and cooling phases. Calculated bending angles as a function of laser power and scanning speed during laser forming of 304 stainless steel are presented in Figs. 8.16 and 8.17. The results are compared with the experimental values and the predictions from the previous models (Cheng and Lin 2001).

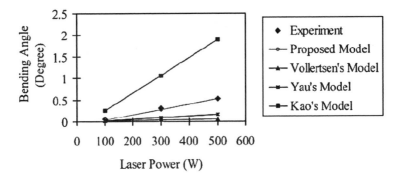

Fig. 8.16 Calculated bending angle as a function of laser power during laser forming of 304 stainless steel. The results are calculated using laser scanning speed of 5,000 mm/min, sheet thickness of 1.0 mm, and laser spot diameter of 0.5 mm (Cheng and Lin 2001; Vollertsen 1994; Yau et al. 1998; Kao 1996) (Reprinted from Cheng and Lin 2001. With permission. Copyright Elsevier.)

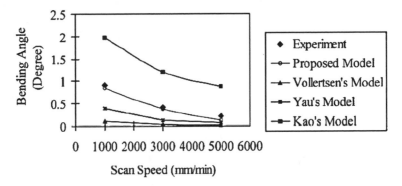

Fig. 8.17 Calculated bending angle as a function of laser scanning speed during laser forming of 304 stainless steel. The results are calculated using laser power of 500 W, sheet thickness of 1.5 mm, and laser spot diameter of 1.0 mm (Cheng and Lin 2001; Vollertsen 1994; Yau et al. 1998; Kao 1996) (Reprinted from Cheng and Lin 2001. With permission. Copyright Elsevier.)

8.3.2 Buckling Mechanism

Vollertsen proposed a simple analytical model of laser forming by BM. In this model the buckling is considered as a two-dimensional bending process. Figure 8.18 shows the geometrical cross section of the buckled sheet. As explained earlier, the buckling is associated with a combination of elastic and plastic strains. The top region of the sheet is directly heated by the laser beam and hence is at a higher temperature. At this region, the temperature-dependent flow stress

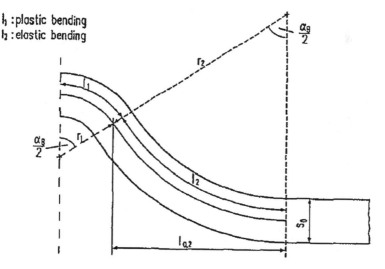

Fig. 8.18 Cross-sectional geometry of a sheet assumed for modeling laser forming by buckling mechanism (Reprinted from Vollertsen et al. 1995. With permission. Copyright Institute of Physics.)

is low and the deformation is nearly pure plastic (region 1). The surrounding region of the sheet (region 2) is not directly heated by the laser beam and is at a lower temperature. The deformation is nearly pure elastic in this surrounding region of the sheet. In the present model both the elastic and plastic deformations are approximated as the arcs of circles. At the end of the laser scan, the elastic region straightens, resulting in the development of a well-defined angle of bending. The angle of bending is primarily defined by the region of plastic deformation (Vollertsen et al. 1995).

The bending angle, α_b, can be calculated from the geometry of the elastic and plastic regions presented in Fig. 8.18 and is given by:

$$\frac{\alpha_b}{2} = \frac{l_1}{r_1} = \frac{l_1}{r_2}$$

(8.18)

where l and r are the length and the (bending) radius of circular arcs respectively. Subscripts 1 and 2 represent the plastic and elastic regions of the buckled sheet respectively.

In region 1, the plastic bending moment is given by:

$$M_{pl} = \frac{1}{4}k_f(T_1)bs_0^2$$

(8.19)

In region 2, the elastic bending moment is given by:

$$M_{el} = \frac{Ebs_0^3}{12r_2},$$ (8.20)

where E is the modulus of elasticity, b is the width of the sheet, S_0 is the thickness of the sheet, and $k_f(T_1)$ is the temperature-dependent flow stress. Equating the elastic and plastic moments at the interface of two regions yields the following expression for r_2:

$$r_2 = \frac{Es_0}{3k_f(T_1)}.$$ (8.21)

The length of the bent sheet, l_2, is given by the sum of the original length of the sheet, $l_{0,2}$, and the total thermal expansion, Δl:

$$l_2 = l_{0,2} + \Delta l.$$ (8.22)

Also,

$$l_{0,2} = r_2 \sin(\alpha_b / 2),$$ (8.23)

$$\Delta l = \alpha_{th} \overline{\Delta T} l_h,$$ (8.24)

where α_{th} is the coefficient of thermal expansion and is $\overline{\Delta T}$ the average temperature in the heated region of length l_h. The temperature can be calculated from the laser processing parameters (such as absorbed laser power, Ap_1, and laser traverse velocity, v_1) and material properties (such as specific heat, c_p, and mass density, ρ):

$$\overline{\Delta T} = \frac{Ap_1}{2c_p \rho s_0 v_1 l_h}$$ (8.25)

Solving the above equation for Δl gives:

$$\Delta l = \frac{\alpha_{th} Ap_1}{2c_p \rho s_0 v_1}$$ (8.26)

The above equation based on simplified analysis gives the thermal expansion of the heated layers neglecting the effects of three-dimensional conduction. Thus, this equation is particularly useful for the laser processing conditions for which the lateral conduction is negligible.

Also, by combining the above equations, the final length of the buckled sheet, l_2, is given by:

$$l_2 = r_2 \sin(\alpha_b / 2) + f' \alpha_{th} A p_1 / 2 c_p \rho s_0 v_1,$$
(8.27)

where, f' in the above equation is the fraction of the thermal expansion that results in the elongation of region 2. The value of f' ranges from 0.5 to 0.9. However, for simplified analysis, the value of $f' = 0.5$ is incorporated in the present model.

The equation of bending angle can then be rewritten as:

$$\alpha_b / 2 = \sin(\alpha_b / 2) + 3 \alpha_{th} A p_1 k_f (T_1) / 4 E s_0^2 c_p \rho v_1.$$
(8.28)

The following series expansion for sine function can be used to simplify the expression of bending angle:

$$\sin x = \sum_{n=1}^{\infty} (-1)^{n+1} \frac{x^{2n-1}}{(2n-1)!}.$$
(8.29)

The simplified expression of bending angle (radian) can then be given as:

$$\alpha_b = \left[36 \frac{\alpha_{th} k_f (T_1)}{c_p \rho E} \frac{A p_1}{v_1} \frac{1}{s_0^2} \right]^{1/3}.$$
(8.30)

The above equation of bending angle is valid for laser forming by BM. As described in Section 8.3.1.2, bending mechanism is operative particularly for sheets geometries (such as thin sheets) and laser processing parameters (such as high laser traverse velocity), where the temperature gradient in the thickness direction is negligibly small. This is also favored by high ratio of high thermal conductivity to sheet thickness.

Compared to the analytical modeling of laser forming process, relatively fewer efforts were concentrated on the numerical modeling. Also, most of these efforts deal with the TGM of laser forming (Vollertsen et al 1993; Alberti et al. 1994, 1997; Hsiao et al. 1997). Laser forming is a complex process where sheet temperature, dimensions, and properties change with time and space. Such a complexity of process is expected to be dealt with more efficiently by numerical approaches, especially in view of the recent developments in the computational techniques.

Recently, a three-dimensional numerical model was developed using finite element codes "ANSYS" for laser bending and forming of sine-shaped plates (Kyrsanidi et al. 1999). The algorithm used for the model includes a nonlinear

transient coupled thermostructural analysis and takes into account the temperature
dependency of the thermal and mechanical properties of the material. The nonlinear
heat transfer equation used for calculating temperature and its dependence of time
is expressed as:

$$[C(T)]\{\dot{T}(t)\} + [K(T)]\{T(t)\} + \{Q(t)\} = 0, \tag{8.31}$$

where $C(T)$ denotes the temperature-dependent specific heat matrix, $K(T)$ the temperature-dependent conductivity matrix, $Q(T)$ the heat flux vector, and T(t) and $\dot{T}(t)$
and the time-dependent nodal temperature and time derivative of the nodal temperature vector, respectively. A nonlinear transient dynamic structural equation for
derivation of stresses and strains was given by:

$$[M(T)]\{\ddot{u}(t)\} + [C(T)]\{\dot{u}(t)\} + [K(T)]\{u(t)\} + \{F(t)\} + F^{\mathrm{th}}(t) = 0, \tag{8.32}$$

where $M(T)$, $C(T)$, and $K(T)$ are the temperature-dependent mass, and the damp-
ing and stiffness matrices, respectively; $F(T)$ is the external load vector; and
$\{u(t)\}$, $\{\dot{u}(t)\}$, $\{\ddot{u}(t)\}$ are the displacement, velocity, and acceleration vectors,
respectively. The equations were solved using Newton–Raphson procedure and
Newmark integration method. The calculated results of the time-dependent temperature distribution, stresses, strains and angular distortion, and final shape are
presented in Fig. 8.19. As indicated in the figure, the temperature distributions at
the surface and the bottom of the sheet differ significantly resulting in setting up
of steep temperature gradients (Fig. 8.19a and b). Also, distortion of the sheet,
and hence its final shape, is primarily due to the maximum plastic compressive
strain formed immediately after the laser irradiation (Fig. 8.19d–f). The time-
dependent distortion results of this numerical model are found to be in good
agreement with the experimental data obtained by linear variable differential
transformer (LVDT) devices (Fig. 8.20).

8.4 Practical Considerations

8.4.1 Processing Parameters

8.4.1.1 Effect of Laser Parameter

The bending angle formed during laser forming process is greatly influenced by the
laser processing parameters such as laser power, laser scanning speed, beam diameter, number of laser scans, holding time between the successive scans, and geometry of scanning path. The total bending angle is determined by the complex

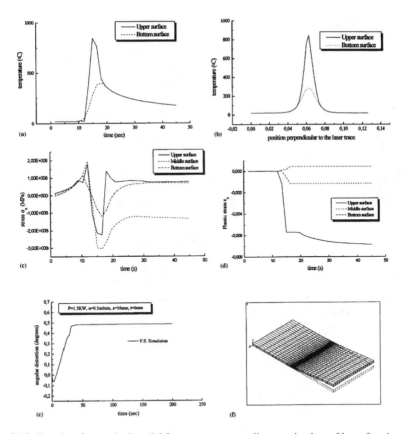

Fig. 8.19 Results of numerical model for temperature gradient mechanism of laser forming: (**a**) temperature distribution with time, (**b**) temperature distribution with position at upper and bottom surface of sheet, (**c**) stress distribution with time, (**d**) plastic strain distribution with time, (**e**) angular distortion with time, and (**f**) final shape after one scan. The results are obtained with 300 × 150 mm plate of 6 mm thickness using laser power of 1.5 kW, laser scan speed of 0.3 m/min and beam diameter of 16 mm (Reprinted from Kyrsanidi et al. 1999. With permission. Copyright Elsevier.)

combination of such parameters. Here general trends regarding the effect of these parameters on bending angle are discussed.

The typical effects of various laser processing parameters on the final bending angle are presented in Figs. 8.21–8.23 for the case of 0.8 mm thick mild steel plate (Lawrence et al. 2001). Figure 8.21 shows the effect of laser power on bending angle obtained at various scanning speeds after 40 scans and with zero holding time between the successive passes. Two important observations can be made regarding variation of bending angle with laser power. First, there exists a threshold value of laser power below which no bending occurs. At lower laser power, the surface temperature is not expected to reach high enough to establish steep temperature gradients for bending action. Second, there exists a maximum power beyond which

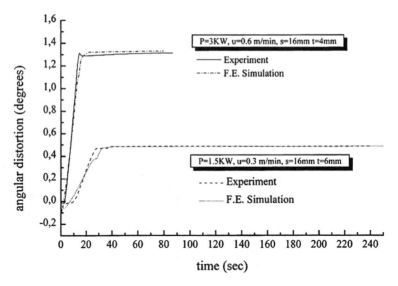

Fig. 8.20 Comparison between calculated and experimental angular distortion formed during laser forming process (Reprinted from Kyrsanidi et al. 1999. With permission. Copyright Elsevier.)

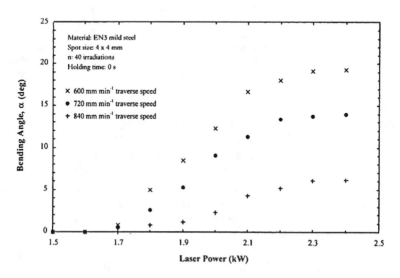

Fig. 8.21 Variation of bending angle with laser power during laser forming of 0.8 mm thick mild steel plate (Reprinted from Lawrence et al. 2001. With permission. Copyright Elsevier.)

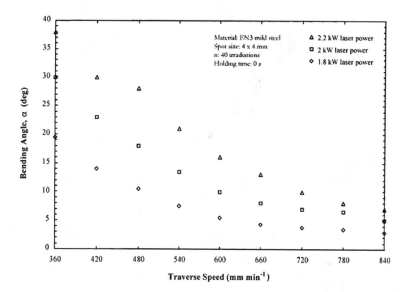

Fig. 8.22 Variation of bending angle with laser traverse speed during laser forming of 0.8 mm thick mild steel plate (Reprinted from Lawrence et al. 2001. With permission. Copyright Elsevier.)

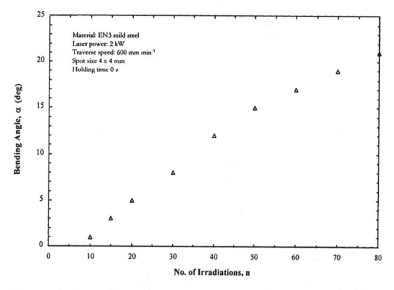

Fig. 8.23 Variation of bending angle with a number of irradiations during laser forming of 0.8 mm thick mild steel plate (Reprinted from Lawrence et al. 2001. With permission. Copyright Elsevier.)

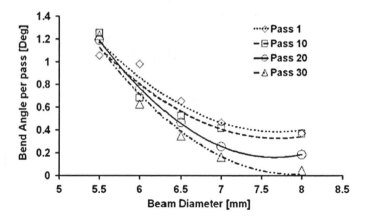

Fig. 8.24 Variation of bending angle with laser beam diameter during laser forming of 1.5 mm mild steel plate with laser power of 760 W and traverse velocity of 30 mm/s (Reprinted from Edwardson et al. 2006. With permission. Copyright Institute of Physics.)

no significant increase in bending angle is achieved. Further increase in the power may result in melting of the workpiece, which may have undesirable effect on the bend quality (Lawrence et al. 2001).

Figure 8.22 presents the effect of laser traverse speed on the final bending angle obtained at various laser powers after 40 scans and with zero holding time between the successive passes. The figure indicates that the bending angle decreases with increasing laser traverse speed. This can again be related with the effect of laser traverse speed on the surface temperature. At faster laser scanning speed, the irradiation time (irradiation time being inversely proportional to traverse velocity) of laser is decreased, resulting in lower surface temperature (Cheng and Lin 2000a). Thus, as the laser scanning velocity increases, the temperature gradient decreases, resulting in lower bending angle.

A typical effect of the number of irradiations (number of scans) on the bending angle is presented in Fig. 8.23. In general, the bending angle increases with the number of scans. However, the variation is not linear, suggesting the complexity of thermomechanical effects taking place in the sheet after each laser scan (Lawrence et al. 2001). Bending angle is also expected to be influenced by the diameter of the laser beam.

In general, the bending angle per pass decreases with increasing beam diameter at constant laser power and velocity as indicated in Fig. 8.24 (Edwardson et al. 2006). This is expected as the laser energy density decreases with increasing beam size, resulting in spreading of temperature distributions with moderate temperature gradients.

Figure 8.25 shows the typical influence of holding time (dwell time) between the successive paths on the variation of the bending angle with a number of passes during laser forming of 0.2 mm mild steel sheet. As indicated in the figure, longer the

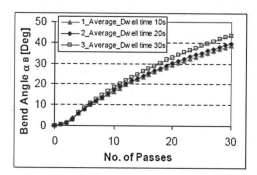

Fig. 8.25 Effect of dwell time between the successive passes on the variation of bending angle with the number of passes during laser forming of 0.2 mm mild steel. Laser forming was carried out with beam power of 26 W, spot size of 0.2 mm, and laser traverse speed of 30 mm/s (Reprinted from Bartkowiak et al. 2004. With permission. Copyright Laser Institute of America.)

holding time, higher is the bending angle, keeping the other parameters such as number of passes, laser power, and traverse velocity constant. The effect is more pronounced at higher number of passes. The increase in bending angle seems to be due to effective temperature gradients that can be achieved after each laser scanning pass if the holding time between the successive passes is increased (Bartkowiak et al. 2004).

Sometimes, it is convenient to express the dependence of bending angle on a parameter called line energy, which combines the effects of laser power and traverse velocity. Line energy (LE) is expressed as a ratio of laser power, P, and traverse speed, V, such that $LE = P/V$. It represents the laser energy input per unit length along the scanning path. At constant line energy, i.e., at constant energy input to the surface, the bending angle increases with increasing laser traverse speed as indicated in Fig. 8.26 (Li and Yao 2001a; Magee et al. 1997). This suggests that temperature gradient and efficiency of the process can be enhanced at higher speeds.

Most of the experimental laser forming studies reported in the literature are concentrated on the simple bending of sheets using linear (straight line) scans of laser beam on the sheet surface. The bending angle of the sheet is expected to be influenced by the scanning path of the laser beam. In one of the studies it is reported that the bending angle during laser forming decreases with increasing the curvature of the laser scanning path (Fig. 8.27) (Chen et al. 2004).

8.4.1.2 Effect of Sheet Geometry

In addition to the laser processing parameters, the bending angle during laser forming is also influenced by the geometry of the sheet. This includes

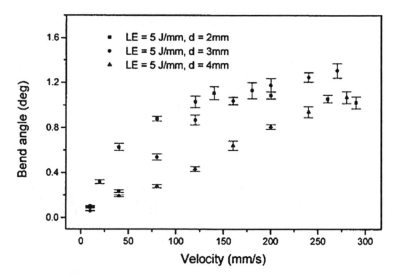

Fig. 8.26 Variation of bending angle with laser traverse velocity at constant line energy during laser forming of 0.89 mm low carbon steel sheet (Reprinted from Li and Yao 2001a. With permission. Copyright American Society of Mechanical Engineers.)

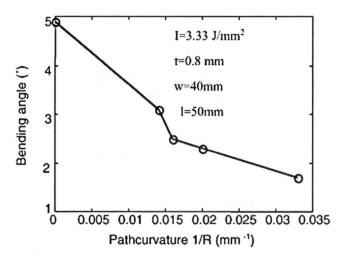

Fig. 8.27 Influence of laser scanning path curvature on the bending angle during laser forming of 0.8 mm Ti-6Al-4V sheets (Reprinted from Chen et al. 2004. With permission. Copyright Elsevier.)

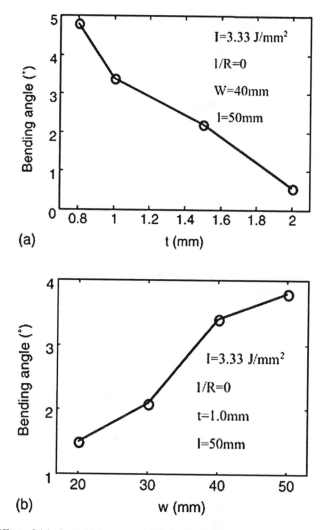

Fig. 8.28 Effect of (**a**) sheet thickness, and (**b**) sheet width on the bending angle during laser forming of Ti-6Al-4V sheets with laser intensity of 3.33 J/mm² and sheet length of 50 mm (Reprinted from Chen et al. 2004. With permission. Copyright Elsevier)

parameters like sheet thickness, width, and length. The effects of sheet width and thickness on the bending angle during laser forming of Ti-6Al-4V alloy sheets are presented in Fig. 8.28. The figure indicates that bending angle decreases with increasing sheet thickness, whereas the bending angle increases with increasing sheet width (Chen et al. 2004). The bending angle is not found to be significantly influenced by the length of the sheet (Shichun and Jinsong

2001). The dependency of the bending angle on the sheet thickness is due to the sectional moment which is proportional to the square of the sheet thickness (Vollertsen 1994a).

8.4.1.3   Effect of Material Properties

Since the bending phenomenon is associated with the temperature gradients and the corresponding thermal stresses established in the material, the thermophysical and thermomechanical properties of the material are expected to influence the bending behavior of the sheet during laser forming.

Effect of Thermophysical Properties

Early investigations on the effect of thermophysical properties on the bending angle were conducted by Vollertsen (1994b). The following relationship between the bending angle and the thermophysical properties was proposed:

$$\alpha_B = C\frac{\alpha_{th}}{c_p\rho},\tag{8.33}$$

where C is the factor depending on processing and geometric parameters, α_{th} is the coefficient of thermal expansion, c_p is the specific heat, and ρ is the density. The relationship is illustrated for laser forming of various materials with widely varying thermophysical properties (Fig. 8.29). General observations from the figure are:

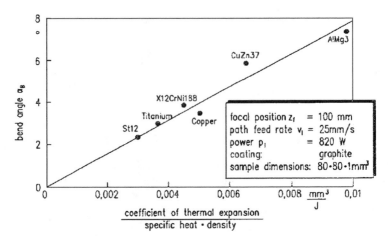

Fig. 8.29 Effect of thermophysical properties of materials on bending angle (Reprinted from Vollertsen 1994b. With permission. Copyright Old City Publishing.)

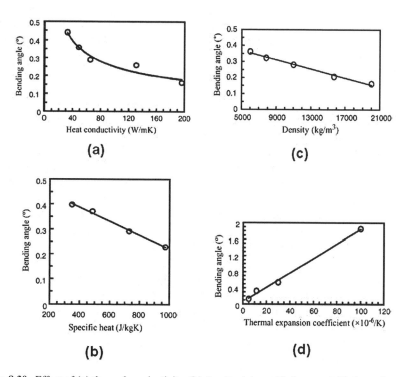

Fig. 8.30 Effect of (**a**) thermal conductivity, (**b**) density, (**c**) specific heat, and (**d**) thermal expansion coefficient on the bending angle during laser forming (Reprinted from Yanjin et al. 2005. With permission. Copyright Elsevier.)

1. The angle of bending increases with thermal expansion coefficient. As the thermal expansion coefficient approaches zero, the bending angle tends to zero. The relationship is a direct consequence of the larger compressive plastic deformation during heating the material with higher coefficient of thermal expansion.
2. The bending angle increases with decreasing values of product $c_p \rho$. This factor determines the temperature increase at the surface. For the same amount of energy incident on the surface, temperature increase is higher for a material with smaller value of specific heat (Vollertsen 1994b).

Recently, Yanjin et al. (2005) proposed a three-dimensional coupled thermomechanical finite element model to establish the relationship between the bending angle and the thermophysical properties of material. The simulations were carried out for laser forming using laser power of 1 kW, laser scanning speed of 1.5 m/min, laser spot diameter of 4 mm, and initial temperature of 20 °C. The results of these modeling efforts are summarized in Fig. 8.30 (Yanjin et al. 2005). The effect of thermal expansion coefficient, specific heat, and density on the bending angle is consistent with the experimental data of Vollertsen (1994b). In addition, Fig. 8.30 shows the influence of thermal conductivity and density on the bending angle.

Higher the thermal conductivity of a material, the more will be the heat conducted away from the surface, resulting in lower surface temperature and relatively moderate temperature gradients, and thus in smaller bending angle for materials with higher thermal conductivity.

Effect of Mechanical Properties

During laser forming, plastic deformation occurs when the thermal stress exceeds the temperature-dependent yield stress of the material. Hence, the bending angle during laser forming is expected to be influenced by the mechanical properties, particularly yield strength, of the material. The previously mentioned modeling efforts of Yanjin et al. (2005) were extended to study the influence of mechanical properties on the bending angle. Figure 8.31 shows the influence of yield strength and the elastic modulus on the bending angle. The figure indicates that the bending angle decreases with increase in yield strength. This means that the material with lower yield strength can be plastically deformed more easily than the high-strength material, resulting in higher bending angle at lower yield strength. In contrast to these numerical results, the experimental results of Vollertsen (1994b) indicated that room temperature yield strength of the material does not significantly influence the bending angle. Vollertsen argued that during laser forming, the strength of the heated region is always lower than the strength of the surrounding material irrespective of the room temperature strength.

Figure 8.31 also indicates that bending angle decreases with increasing Young's modulus. However, the influence of Young's modulus on bending angle is relatively less pronounced compared to that of yield strength (Yanjin et al. 2005).

As explained above, the laser forming process is complex and involves a number of laser processing, geometric, and material parameters. Hence, in order to predict the bending angle during laser forming, it is necessary to identify the parameters with significant effect on the bending angle. In this context, application of neural networks is attracting significant interest. In one of the recent studies, three supervised neural networks (radial basis function neural network (RBFN), back propagation model using a logistic function as activation function (BPLF), and back propagation model using a hyperbolic tangent function as an activation function (BPHTF)) were used to estimate the bending angle during laser forming (Cheng and Lin 2000b). Various processing and geometric parameters used as inputs to the neural networks were laser power, scan speed, spot diameter, sheet thickness, and sheet length. Figure 8.32 presents a comparison between the estimated bending angle using neural networks with the experimental bending angles during laser forming. As indicated in the figure, the RBFN model is superior to the other models in predicting the bending angle. The application of such an approach of neural networks to a complex process like laser forming is associated with significant advantages such as superior learning, noise suppression, and parallel computation abilities (Cheng and Lin 2000b).

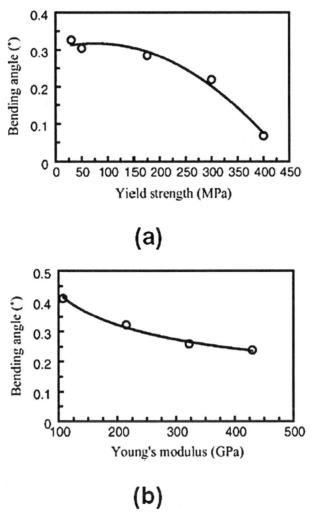

Fig. 8.31 Effect of (**a**) yield strength, and (**b**) Young's modulus on the bending angle during laser forming (Reprinted from Yanjin et al. 2005. With permission. Copyright Elsevier.)

8.4.2 Bending Rate and Edge Effects

8.4.2.1 Bending Rate

Extensive experimental studies have been reported in the literature regarding the increase in bending angle with the number of laser scanning passes. However, very few studies have discussed the detailed development of bending angle with the number of passes. Significant analysis of the development of bending angle after each laser scanning pass was recently conducted by the Laser Group at the

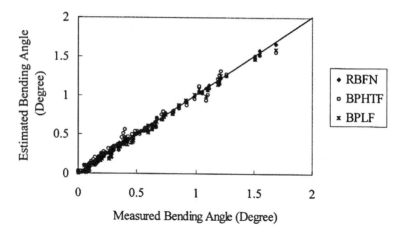

Fig. 8.32 Comparison of the estimated bending angle using neural networks with the experimental bending angles during laser forming (Reprinted from Cheng and Lin 2000b. With permission. Copyright Elsevier.)

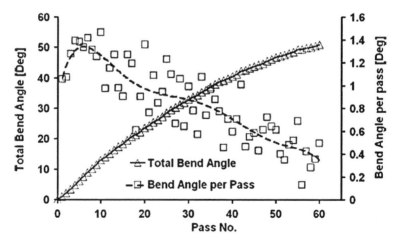

Fig. 8.33 Variation of bending angle and bending angle per pass with the number of laser scanning passes during laser forming of 1.5 mm mild steel (Reprinted from Edwardson et. al 2006. With permission. Copyright Institute of Physics.)

University of Liverpool. One of the most important and interesting observations made during these studies is that the bending angle per pass during multipass laser bending decreases with an increase in the number of passes. This is shown in Fig. 8.33 for 1.5 mm mild steel using laser power of 760 W, laser scanning velocity of 30 mm/s, and laser beam diameter of 5.5 mm (Edwardson et al. 2006). Such a decrease in bending angle per pass can be attributed to a number of laser forming parameters such as strain hardening, section thickening, variation in absorption, and thermal effects. These effects are briefly discussed in the following sections.

Strain hardening of materials is associated with the increase in the strength of the material with further plastic deformation. As the material is plastically deformed, the dislocation density increases and further motion of the dislocations becomes difficult due to dislocation entanglements. This amounts to the increase in the strength and hardness of the plastically deformed material. In the context of laser bending, each laser scan produces some plastic deformation in the sheet, making it stronger and difficult to be deformed in the subsequent passes. This is translated in the decrease in bending angle per pass after each subsequent pass. The effect has been verified by the observation that microhardness of the sheet increases substantially after laser bending (Edwardson et al. 2006; Thomson and Pridham 2001). Also, microstructural characterization of the laser bent sheets indicate that the increase in hardness of the aluminum alloy AA1050 is associated with the increased dislocation density in the microstructure after laser bending (Fig. 8.34). This suggests the role of strain hardening in influencing the bending angles during laser forming (Merklein et al. 2001). The effect of strain hardening is more pronounced for thick sheets and materials with large strain hardening coefficients (Magee et al. 1997).

During laser bending, the localized plastic deformations tend to cause the thickening (defined as percentage increase in thickness with respect to initial thickness of sheet) of the bent section. This is due to the need of conservation of volume. The thickening of the section increases after each pass, thus making it progressively difficult to deform the sheet after each pass. This results in the progressively decreasing bending angle per pass during multipass bending. Figure 8.35 indicates the progressively decreasing bending rate (rate of change of bending angle with the number of scans, i.e., slope of Fig. 8.35a, and the accompanied increasing sheet thickening with the number of passes during laser forming of titanium alloy sheet (Marya and Edwards 2001).

The decrease in bending angle per pass with increasing number of passes has also been partly attributed to the decrease of the absorptivity of the material with the number of passes. The variation of absorptivity with the number of passes may be due to damage of the absorptive coatings with repeated irradiations, complex surface darkening effects, etc. Also, the decreasing bending angle with the number of passes can be explained on the basis of thermal effects taking place during laser forming. After each scan of the laser beam, the temperature gradients established in the sheet material causes the plastic deformation, resulting in a contribution to total bending angle. However, temperature distribution during subsequent passes is influenced by the temperature history during previous passes. If the second pass of laser scanning is carried out immediately (i.e., zero holding time) after the first pass, temperature distributions in the material at the end of first pass influence the temperature distributions during the second pass. The effect is primarily on reducing the temperature gradients in the subsequent passes, thus resulting in reducing bending rate with the number of passes. This also means that if the holding time between the two passes is increased suffi-

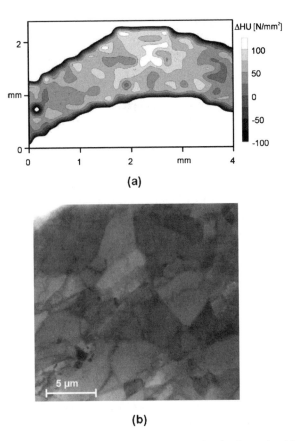

(a)

(b)

Fig. 8.34 (a) Hardness profile indicating increase in microhardness (positive ΔHU), and (b) microstructure indicating subgrain structure in laser-formed AA1050 alloy (Reprinted from Merklein et al. 2001. With permission. Copyright Elsevier.)

ciently such that temperature distributions of the first pass do not influence the temperature distribution of the second pass, the decreasing trend of bending rate with the number of passes may be avoided. However, all these parameters influence the bending rate in a complex way such that one or more parameters may be more dominant than the others.

Recently, the variation of bending rate with the number of passes has been explained based on the geometric effects during laser forming (Edwardson et al. 2006). During laser forming, as the bending angle increases, the area irradiated with the laser beam increases, resulting in reduced energy density at higher bending angle. Such lowered laser energy fluence at higher bending angles results in the smaller bending angle per pass with the increasing number of scans. This effect is illustrated in Fig. 8.36. As indicated in the figure, the semicircular area of the laser

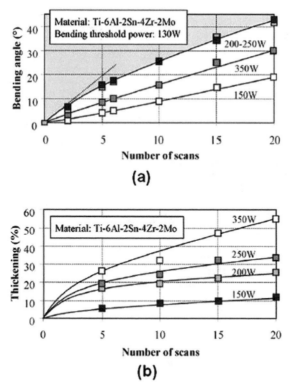

Fig. 8.35 Variation of (**a**) bend angle, and (**b**) percentage thickening during laser forming of a titanium alloy sheet (Reprinted from Marya and Edwards 2001. With permission. Copyright Elsevier.)

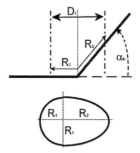

Fig. 8.36 Schematic geometry of the area irradiated with laser beam at high bending angle (Reprinted from Edwardson et al. 2006. With permission. Copyright Institute of Physics.)

beam gets spread into semielliptical shape, resulting in increased area of the incident beam (Fig. 8.37a). For a given incident laser power, increasing laser beam area with bending angle causes progressive decrease in laser fluence (energy per unit area) (Fig. 8.37b) (Edwardson et al. 2006).

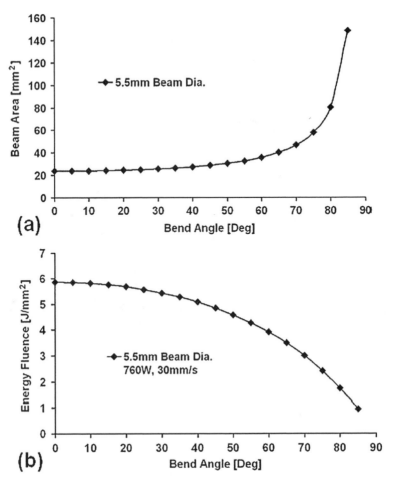

Fig. 8.37 Effect of bending angle on **(a)** the incident beam area, and **(b)** the energy fluence during laser forming of sheet (Reprinted from Edwardson et al. 2006. With permission. Copyright Institute of Physics.)

8.4.2.2   Edge Effects

Laser bending is often associated with geometric irregularities. During straight line bending, the bending angle shows variation along the bending edge of the sheet such that the maximum bending angle is located somewhere near the center of the edge. Also, the angle near the exit edge is generally larger than that near the entrance edge. This is referred to as edge effect. The edge effect causes the development of curves on the edges of the sheet. The schematic of the geometry of the sheet with and without edge effects is presented in Fig. 8.38. The possible reasons for the edge effects are (Magee et al. 1997; Bao and Yao 2001):

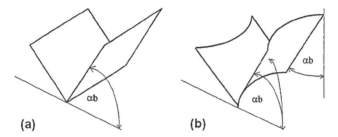

Fig. 8.38 Schematic of the geometry of sheet (**a**) without, and (**b**) with edge effects during laser forming of sheet (Reprinted from Magee et al. 1997. With permission. Copyright Laser Institute of America.)

1. Variation of temperature distribution (peak temperature, temperature gradient, etc.) along the bending edge during forming process
2. Changing mechanical resistance along the bending edge to hinder the free thermal expansion of material during laser forming
3. Contraction of the material along the bending edge (i.e., along the direction of laser scanning)

The tendency to cause edge effects can be minimized by varying the processing parameters along the bending edge such that nearly equal bending angle is achieved along the bending edge. One such approach is based on varying the laser scanning speed along the bending angle. Since the bending angle near the entrance and the exit are smaller than that at the center, slower scanning speed at the entrance and exit edges is expected to enhance the yielding such that nearly uniform bending angle is achieved along the bending edge. This is shown in Fig. 8.39 for the laser forming of titanium alloy. The top curve in this figure corresponds to the constant laser scanning velocity of 30 mm/s and indicates the maximum variation of bending angle along the edge to be 1.15°. The bottom curve is obtained with a varying scanning velocity along the bending edge and indicates the maximum variation of bending angle along the edge to be 0.4°. The variation of the laser scanning velocity counters the effect of varying mechanical restraint along the bending edge and minimizes edge effects (Magee et al. 1997).

8.4.3 Laser Forming Strategies and Control

This section discusses the various laser scanning strategies for achieving the desired shape during laser forming, measurement of the bending angle, and the control during the laser forming process.

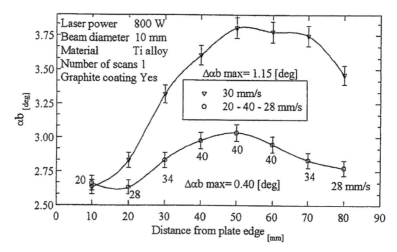

Fig. 8.39 Variation bending angle along the bending edge of the sheet showing the edge effects associated with laser forming of sheet (Reprinted from Magee et al. 1997. With permission. Copyright Laser Institute of America.)

8.4.3.1 Scanning Strategies

Laser scanning strategies refer to the designs of laser scanning paths for achieving the desired geometry of laser-formed shapes. For the given laser processing parameters (power, scanning speed, beam diameter, etc.) and the workpiece material (with corresponding geometric and thermophysical properties), the laser scanning paths can influence the bending angle, edge effects, and final shape of the sheet. The following section primarily discusses the influence of scanning path on the bending angle and the shape of the final sheet.

Scanning Strategies for Two-dimensional Laser Forming

This is the simplest type of laser bending operation consisting of linear laser scanning to produce the two-dimensional out-of plane bends. Three of the possible scanning strategies for two-dimensional laser forming presented in Fig. 8.40 are single direction, alternating direction, and dashed line. In single direction strategy, each laser scanning pass is directed in the same direction along the bending path, whereas in alternating direction strategy, each laser scanning pass is directed in the direction opposite to that of the previous pass. The dashed line strategy consists of dashed straight-line scans of opposite direction and sequence (Bartkowiak et al.2004, 2005).

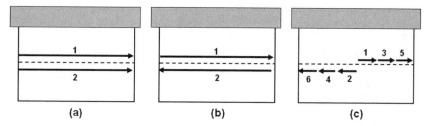

Fig. 8.40 Laser scanning strategies for two-dimensional laser forming: **(a)** Single direction, **(b)** alternating direction, and **(c)** dashed line strategies (Bartkowiak et al. 2005)

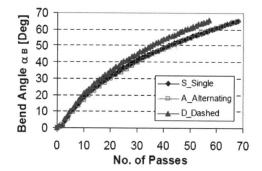

Fig. 8.41 Variation of bending angle with the number of scanning passes for three laser scanning strategies during laser forming of mild steel. Laser forming was carried out with laser power of 25.3 W, laser scanning speed of 30 mm/s, laser spot diameter of 0.2 mm, and dwell time of 10 s (Reprinted from Bartkowiak et al. 2004. With permission. Copyright Laser Institute of America.)

Figure 8.41 presents the variations of the bending angle with the number of scanning passes for three different laser scanning strategies. The figure indicates that dashed line strategy produces a larger bending angle when compared with the single direction and alternating direction scanning strategies (Bartkowiak et al. 2004). Even though this effect is not well understood, possible reason for such variation can be the "heat corralling" effect due to outwardly progressing dashed line sequence resulting in reduced cooling cycle phase per pass (Dearden et al. 2003). Also, the waving motion of the sample edge associated with the dashed line laser scanning is significantly less when compared with the single direction and alternating direction scanning strategies (Reeves et al. 2003).

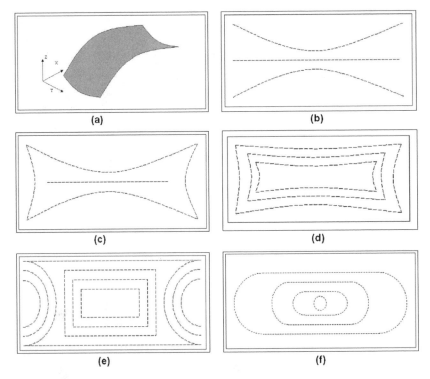

Fig. 8.42 (a) Ideal saddle shape and (b)–(f) various laser scanning strategies for forming saddle shape (Reprinted from Edwardson et al. 2001. With permission. Copyright Springer.)

Scanning Strategies for three-dimensional laser forming

Three-dimensional laser forming consists of producing a combination of multi-axis two-dimensional out-of-plane bends and in-plane shortenings to form three-dimensional shapes (such as saddle, dome, or pillow shape). The scanning strategies for three-dimensional forming are complex and often involve laser scanning in continuous or separate combinations of straight and curved paths. Most of the early experiments of three-dimensional laser forming are based on empirical approaches. Such laser scanning strategies are derived from the knowledge of laser forming mechanisms such as TGM, BM, and UMs that are necessary to form the desired shapes (Edwardson et al. 2001; Bartkowiak et al. 2004). For example, Fig. 8.42 presents the empirical plan for the development of the desired laser scanning strategy to form a saddle shape. To form a saddle shape from the flat rectangular piece of sheet, the material needs to be shortened in the diagonal, length, and width directions. Also, to form a symmetric saddle shape, the laser scanning patterns needs to be symmetric along the length and width of the sheet. The design starts with the simplest laser scanning strategy as indicated in Fig. 8.42b. The results (laser-formed shapes) obtained with this simplest strategy are then compared with the desired shapes. If the results of this strategy are within tolerance, it may be adopted as a desired strategy. Else, further development needs to be made until desired laser scanning strategy is evolved. It was found that this simplest scanning strategy results in a distorted saddle with little or no forming along the shorter edges. The strategy is then further developed until the desired shape is formed within the specified tolerance. Among the scanning strategies for three-dimensional saddle shapes presented in Fig. 8.42, the last strategy (Fig. 8.42f) forms the better symmetric saddle shape. Laser processing parameters can also be varied along the scanning paths to achieve the desired shape (Edwardson et al. 2001).

Recently, predictive and adaptive approaches for laser scanning strategies were used such that repeatable, closed loop-controlled laser forming processes can be developed. Figure 8.43 presents the steps for developing the three-dimensional laser forming strategy for a pillow shape. These steps are (Edwardson et al. 2005a):

1. Representation of the desired surface as "Bezier surface patch" which is defined by a grid of evenly spaced control points (MATLAB functions)
2. Calculation of localized gradients ($\partial z / \partial x$ and $\partial z / \partial y$) in x and y directions and contour plots of constant gradient values over the surface
3. Overlapping of contour plots to arrive at gradient-based scan strategy (strategy 1) which can be directly converted into CNC codes for forming the desired shapes
4. Calculation of the resultant gradient vector and magnitude (Quiver plot) and contour plot of pillow shape. Various different patterns (such as radial and concentric) can be obtained depending on how the contours are drawn through the resultant gradient vector. The figure shows the concentric pattern obtained by rotating the gradient vectors through $90°$ (strategy 2).
5. Calculation of height contour plots where the resultant gradient vector magnitude is calculated for points along the contour lines of constant height for a pillow shape.

Figure 8.44 presents the comparison of the desired shape and the laser-formed shape (with predicted scan strategy) after first pass and the corresponding error plot. The error plot helps in scanning strategy prediction for the next pass. The laser processing parameters such as laser scanning speed can be varied along the laser scanning paths to get the desired shape. Thus, iterative approach can be used for controlled three-dimensional laser forming of continuous surfaces (Edwardson et al. 2005a).

8.4.3.2 Process Control

Laser forming is a complex process involving a number of parameters corresponding to laser, material, geometry, processing system, etc. In addition, each

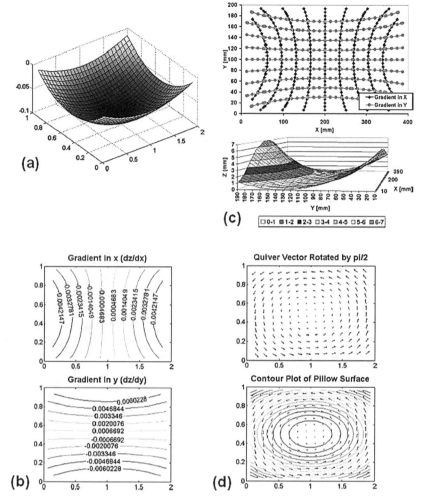

Fig. 8.43 Prediction of three-dimensional laser scanning strategy for pillow shape: (**a**) Bezier surface patch, (**b**) surface gradient contour plots, (**c**) overlaid gradient-based scan strategy with corresponding forming result, and (**d**) Quiver plot and corresponding contour plot of resultant gradient vector (Reprinted from Edwardson et al. 2005a. With permission. Copyright Laser Institute of America.)

of these parameters is subject to fluctuation, which tends to impart process variability with nominally identical test pieces and process conditions. Hence, process control of laser forming becomes important to minimize the process variability and achieve the high degree of repeatability. Two approaches for process control are predictive open loop and feedback control (Thomson and Pridham 1997a).

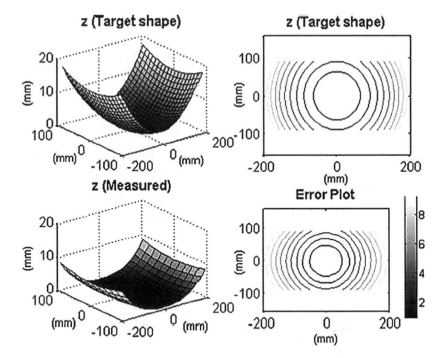

Fig. 8.44 Comparison between the desired shape and formed shape after one pass and corresponding error plot (Reprinted from Edwardson et al. 2005a. With permission. Copyright Laser Institute of America.)

In predictive control approach, the processing parameters for a given experiment are determined using the previous empirical results and/or theoretical methods. This involves laser forming trials on a large number of materials and thicknesses and recording the results corresponding to processing parameters and bending angle in a database. These results are then consulted while working on a new job. However, it is not always the case where the close match of the new job is found in the database. In such cases, interpolation techniques can be used to arrive at processing parameters. Due to the complexity of laser processing parameters, the predictive approach is generally associated with large process variability (Thomson and Pridham 1997a).

To achieve a high degree of repeatability of the process, feedback control is generally recommended. In feedback control, in-process deformations are measured with the sensors and results are compared with the target value such that processing parameters are continuously adjusted. This ensures that the deformations reach the desired value without overshooting. A schematic of the feedback control approach for laser forming is presented in Fig. 8.45. In laser forming, the

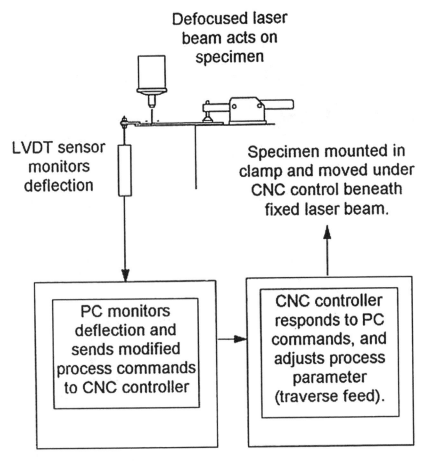

Fig. 8.45 Feedback approach for process control in laser forming (Reprinted from Thomson and Pridham 1998. With permission. Copyright Elsevier.)

targeted bending angle is achieved by scanning the number of laser passes on the workpiece. For rapid processing, the total number of laser passes for achieving the targeted deformation needs to be minimized. This requires large deformation per pass. However, this may result in exceeding the targeted deformation after a certain number of passes. Hence, to balance between the rapid processing and accuracy to achieve the targeted deformation, the feedback control system necessitates the selection of processing parameters such that large incremental deformations are achieved in the early stages and small incremental deformations are achieved as the deformation approaches targeted deformation. This is illustrated in Fig. 8.46 (Thomson and Pridham 1998).

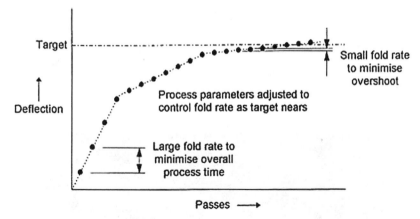

Fig. 8.46 Process control during laser forming process to minimize the processing time and to achieve the targeted deformation accurately (Reprinted from Thomson and Pridham 1998. With permission. Copyright Elsevier.)

8.4.4 Microstructure and Properties of Laser-Formed Components

Even though extensive experimental and modeling studies have been conducted on the formability of the materials during the laser forming process, relatively fewer studies are devoted toward the evaluation of microstructure and mechanical properties of laser-formed components. Evaluation of microstructure and mechanical properties is important for reliable utilization of laser-formed components in practical applications.

Typical cross-sectional microstructure of the laser-formed steel sheet is presented in Fig. 8.47. The figure clearly indicates the three microstructural regions: the heat affected zone (HAZ); the unaffected zone; and a boundary between HAZ and the unaffected zone (Zhang et al. 2005). Such microstructural changes associated with the laser-formed components are expected to influence the properties of the components.

Figure 8.48 shows the variation of microhardness in the depth of laser-formed mild steel components. The figure clearly indicates the increase in microhardness near the surface, which is most probably due to wok hardening effect during laser forming (Bartkowiak et al. 2004).

Thomson and Pridham (2001) conducted tensile tests and straightening tests on the laser-formed steel sheets. The tensile test specimen geometry and the test results for laser-formed mild steel sheets are presented in Fig. 8.49. The results indicate that as the laser traverse speed decreases (i.e., input laser energy increases), the yield stress increases and the corresponding deflection at break (which is a measure of ductility) decreases. These tensile deforma-

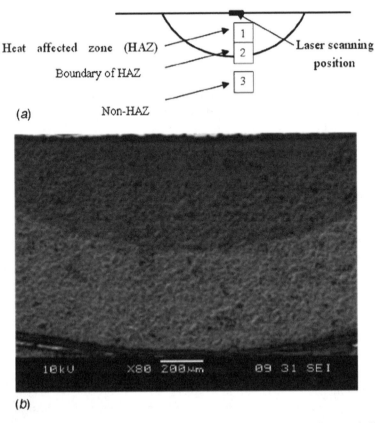

Fig. 8.47 Typical cross-sectional microstructure of the laser-formed components indicating heat affected zone, unaffected zone, and a boundary between HAZ and unaffected zone (Reprinted from Zhang et al. 2005. With permission. Copyright American Society of Mechanical Engineers.)

tions of the laser-formed sheets are also associated with a degree of strain hardening indicated by discontinuous yielding in the load-deformation curves. The straightening test results presented in Fig. 8.50 indicates that conventionally formed and laser-formed specimens do not show significant difference with respect to load during straightening or the ultimate load (Thomson and Pridham 2001).

To investigate the effect of microstructure on the fracture toughness, Cheng et al. (2005) conducted the experimental computer simulation of fracture toughness of sheet metal (low carbon steel) after laser forming. It has been reported that the fracture toughness at the center of the laser track is the highest (637.3 MPa m$^{1/2}$). whereas it is the lowest (71.6 MPa m$^{1/2}$) at the boundary near HAZ compared to the initial fracture toughness of the sheet (140 MPa m$^{1/2}$). The

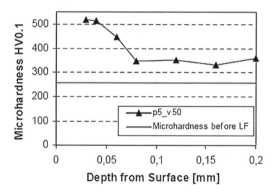

Fig. 8.48 Hardness profile in the depth of laser-formed mild steel component. The forming was carried out with laser power of 30.1 W, scan speed of 50 mm/s, spot size of 0.2 mm, dwell time of 10 s, and 5 passes (Reprinted from Bartkowiak et al. 2004. With permission. Copyright Laser Institute of America.)

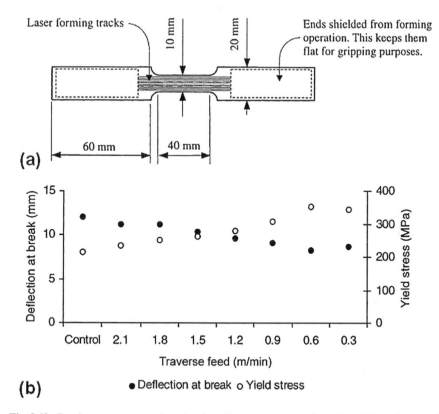

Fig. 8.49 Specimen geometry and results of tensile test on the laser-formed steel sheet (Reprinted from Thomson and Pridham 2001. With permission. Copyright Elsevier.)

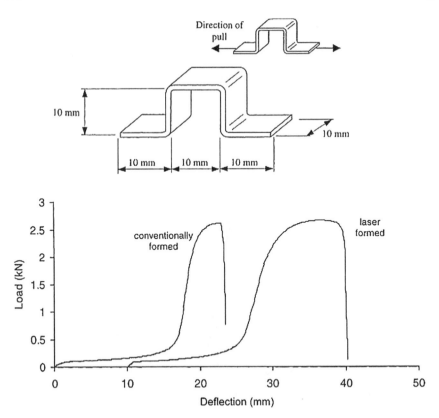

Fig. 8.50 Specimen geometry and results of straightening test on the conventionally formed and laser-formed steel sheet (Reprinted from Thomson and Pridham 2001. With permission. Copyright Elsevier.)

HAZ in the laser-formed sheet primarily consisted of martensite, whereas the unaffected region was composed of ferrite and pearlite. The increase in the fracture toughness at the center of the laser track is primarily due to refinement in the microstructure. The inhomogeneity of the microstructure at the boundary near HAZ resulted in the stress concentrations responsible for the lowest fracture toughness at the boundary (Cheng et al. 2005). Similar investigations have been conducted to evaluate the fatigue response of the laser-formed components. The experimental and modeling results indicated that fatigue life of the laser-formed steel components (~29,000 cycles) is deteriorated compared to that of initial sheet (~37,000 cycles). The deterioration of fatigue life of the laser-formed components is primarily due to nonhomogeneous microstructure across the HAZ and the tensile residual stress around the boundary of HAZ. Improvements in the fatigue life can be obtained by controlling the microstructure formation by various methods such as alloying, tempering, or application of an additional laser beam to control the thermal effects such that martensite formation in the HAZ is prevented (Zhang et al. 2005).

8.5 Laser Forming Applications

Laser forming of materials is increasingly being used in industrial forming applications. New applications are emerging where laser forming can be potentially used. This section provides a brief overview of the current and potential applications of laser forming.

8.5.1 Correction of Bending Angles

Mechanical forming of the sheets is often associated with the spring-back effects. This requires the overbending of the sheet to achieve the desired geometric accuracy. For complex shapes with variation in thicknesses and having a number of folds, mechanical forming often results in collision or unwanted gaps. Attempts have been made to correct the spring-back using laser forming as a final stage. The laser scanning can be carried out at the centerline or the bottom of the existing folds to close the gaps. Furthermore, the same laser can be used for welding the gaps or for cutting the special features in the components. Thus, laser forming offers a great potential to achieve geometric accuracy and rapid processing by combining with the conventional forming techniques (Magee and De Vin 2002).

8.5.2 Laser Forming of Complex Shapes

One of the important suggested applications of laser forming is in shipbuilding. The mechanical forming is a primary technique for forming thick plates in shipbuilding applications. Laser forming has the potential to be applied as a primary forming method for forming accurate shapes or as a secondary forming method for correcting the distortions introduced by conventional forming processes. However, successful application of the laser forming in shipbuilding applications is limited due to high equipment costs, safety requirements, etc. (Vollertsen 1994a; Dearden and Edwardson 2003a).

Laser forming has demonstrated a potential to be used in the forming of complex shapes from sheets and tubes. Figure 8.51 presents some of the artistic and component shapes produced by two-dimensional and three-dimensional laser

Fig. 8.51 Complex shapes formed by three-dimensional and two-dimensional laser forming (Reprinted from Dearden and Edwardson 2003b. With permission. Copyright Institute of Physics.)

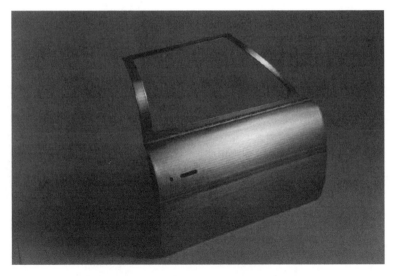

Fig. 8.52 Laser-formed part for car rear door panel (Reprinted from Thomson and Pridham 1998. With permission. Copyright Elsevier.)

forming (Dearden and Edwardson 2003b). Useful part shapes such as rear door panels (Fig. 8.52) have also been demonstrated using laser forming (Thomson and Pridham 1998). The production of such shapes requires the development of laser forming strategies derived from the understanding of laser–material interaction.

Fig. 8.53 Cover plates formed by a combination of laser forming and cutting processes (Reprinted from Thomson and Pridham 1997b. With permission. Copyright Emerald.)

8.5.3 Rapid Prototyping

Laser forming has demonstrated a great potential to be used in rapid prototyping of sheet metal parts. Rapid prototyping with laser forming does not require any hard-tooling. Also, it is associated with advantages such as high tolerances, and economic and rapid manufacturing. Earlier studies have reported the fabrication of prototypes of lamp housing and spoon shapes (Vollertsen 1995). Recently, parts such as cover plates and car door panels have been produced using a combination of laser forming and cutting (Fig. 8.53). The process starts with a cutting of correct size of blank from a sheet followed by a sequence of laser bending and cutting (Thomson and Pridham 1995, 1997). Major developments in the control and measurement systems are needed to fully exploit the potential laser forming for the rapid prototyping of industrially useful parts.

8.5.4 Flexible Straightening of Car Body Shells

Laser forming offers a potential for correcting the distortions introduced in automotive parts during the various welding processes. Geiger et al. demonstrated this application of laser forming for the straightening of laser-welded car body shells. Laser welding of such visible automotive parts often introduces the distortions which need to be removed to improve the aesthetic and functional quality of the parts. Laser forming for straightening is particularly useful because it can be sequenced with the laser welding operation for fully automated production system. Due to the complexity of the processing laser processing parameters, one step straightening is difficult. Hence, a closed loop straightening system is suggested (Geiger et al. 1993).

8.5.5   Microfabrication

Recently, significant interests have been attracted toward application of laser forming in microfabrication, particularly for precision adjustment of components by shortening or UMs. An excellent review of recent applications of laser forming in microfabrication is given by Dearden and Edwardson (2003b). Some of these applications are laser adjustment of digital audio head mounting frames, laser adjustment of reed switches, and laser adjustment of curvature of magnetic head sliders in disc drives (Dearden and Edwardson 2003b).

8.5.6.   Laser Forming in Space

One of the earliest suggested applications of laser forming is the forming of sheets in space. Such an application can minimize the space limitations associated with the transportation of bulky structures or a number of substructures to the space. The idea proposed was to convey the coiled sheets into space by a space shuttle and to form the sheets to desired dimensions in space using a laser beam generated on earth. Laser forming was also suggested for repairing jobs in space (Namba 1986).

8.6   Advances in Laser Forming

Most of the previous experimental and modeling studies on laser forming were focused on the bending of sheets to produce simple out-of plane deformations. Recently, significant efforts have been directed toward extending the laser bending into forming of complex shapes such that direct fabrication of critical components is realized. Some of these advanced topics are briefly discussed in this section.

8.6.1   Laser Forming of Tubes

Laser forming can be applied to the bending of tubes with a variety of cross-sectional shapes such as circular and square cross sections. However, most of the recent efforts were directed toward bending of tubes with circular cross sections. The bending of tubes is primarily realized by two laser scanning schemes: circumferential and axial. The schematic of these schemes are shown in Fig. 8.54. In circumferential scanning, a circular beam of laser is scanned along the circumference of the tube through a prescribed scanning angle (up to 180°). The subsequent laser scans may be carried out at the same axial location (i.e., fully overlapping multiple scans along the same path) or at different axial locations (i.e., partially overlapping or separate multiple scans). The bending of the tube can also be achieved by scanning a laser beam along the axial direction of the tube. Single axial scanning with a line source is expected to give much higher bending angle compared to multiple circumferential scanning with point laser source. Many variations of these primary

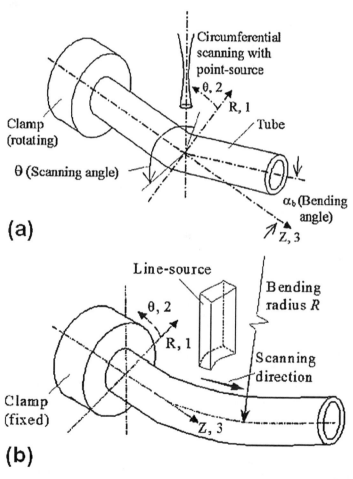

Fig. 8.54 Laser bending of tubes: (a) circumferential scanning and (b) axial scanning (Reprinted from Zhang et al. 2006. With permission. Copyright Elsevier.)

laser schemes are possible depending on the source of laser (point or line source, and continuous or pulsed beam) and application of water cooling (with or without water cooling) (Zhang et al. 2006).

The primary mechanism of laser tube bending during circumferential scanning is generally considered as UM. During circumferential scanning, the laser beam size is generally larger than the thickness of the tube such that nearly homogeneous heating takes place through the thickness of the tube. When the laser is scanned around the circumference of the tube through a scanning angle (typically 180°), the homogeneous heating through the thickness of the tube causes the compressive plastic deformation in the laser-heated region. The subsequent axial shortening of the tube associated with tube thickening causes the out-of-plane bending of the tube. The circumferential compressive stresses may also contribute to the bending of the tube (Li and Yao, 2001b). The variation of the bending angle during laser bending of tubes follows the trends similar to those obtained during laser bending of sheets. For example, Fig. 8.55 presents the variation of bending angle and curvature radius with a number of laser scanning passes during laser bending of 0.89 mm thick steel tube (with outer diameter of 12.7 mm). As indicated in the figure, the bending angle increases with the increase in the number of scans. Also, the increase in bending angle with the number of passes slows down at larger number

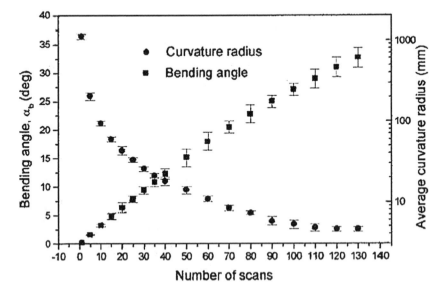

Fig. 8.55 Variation of the bending angle and average curvature radius with the number of laser scans during laser bending of 0.89 mm thick steel tube with 12.7 mm outer diameter (laser power: 780 W; scanning speed: 1.57 rad/s; beam diameter: 11 mm) (Reprinted from Li and Yao 2001b. With permission. Copyright American Society of Mechanical Engineers.)

of passes, suggesting the influence of work hardening on the further plastic deformation (Li and Yao, 2001b).

8.6.2   Laser Forming with Preload

In the typical laser forming operation, one end of the sheet is clamped and the other end of the sheet is allowed to deform freely. Such an experimental arrangement gives the predictable bending direction (positive or negative) in the case of thick sheets. However, in the case of thin sheets, the control and prediction of bending angle becomes difficult. To control the bending direction and deformation of the sheet, laser bending operations under prestresses have been suggested. The application of prestresses during bending can also increase the deformation per laser scanning pass. The deformation during bending can be controlled by the magnitude and direction of the applied load. Both positive (force opposite to the irradiation direction) and negative (force in the irradiation direction) forces can be applied. Figure 8.56 presents the influence of positive prestress on the bending angle. The results were simulated for 2 mm thick steel sheets with laser power of 1 kW, scanning velocity of 1.5 m/min, and spot diameter of 4 mm. The figure indicates that bending angle increases significantly (nearly exponential) with increasing prestress. The applied positive load produces compressive prestresses at the top of the sheet, which in combination with the compressive stresses due to laser heating results in the enhanced deformation of the sheet (Yanjin et al. 2003).

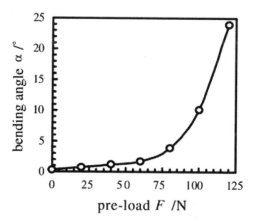

Fig. 8.56 Influence of preload on the bending angle during laser bending of sheet (Reprinted from Yanjin et al. 2003. With permission. Copyright Elsevier.)

Similar studies have been conducted for the bending of tubes with axial preloads. It was observed that the BM dominates the deformation of the tube under axial preload (Hsieh and Lin 2005).

8.6.3   Laser Forming with Two Beams

A typical laser forming process consists of sequential scanning of a laser beam on the surface of the sheet. The subsequent laser scanning passes are carried out either along the same scanning path or parallel paths adjacent to the previous path. Since the passes are carried out in sequence, the thermal effects from the previous passes do not significantly influence the thermal effects during subsequent passes. Recently, it was suggested that significantly higher deformation can be achieved if simultaneous scanning is carried out along the two parallel paths using two laser sources provided the distance between the parallel paths is not large. When the two parallel paths are close, the temperature distributions from each path intersect, resulting in steeper temperature gradient across the thickness of the sheet. Such steeper gradient due to simultaneous sources of laser are expected to provide the enhanced deformation during laser bending (Shen et al. 2006).

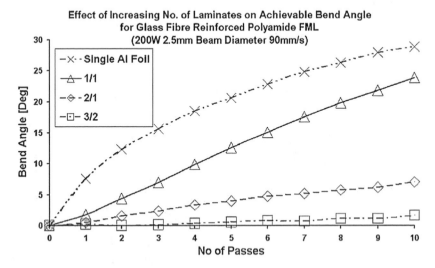

Fig. 8.57 Variation of bending angle with the number of laser scanning passes for fiber metal laminates with different number of layers (Reprinted from Edwardson et al. 2005b. With permission. Copyright Old City Publishing.)

8.6.4 Laser Forming of Composite Materials

Recently, considerable research interests were directed toward laser forming of advanced materials such as laminates and composites (Edwardson et al. 2005b; Chan and Liang 2001). Due to difference in the thermophysical properties among the constitutive phases in such composites, it is interesting to study the response of these materials to laser forming. When the laminates are scanned with a beam of laser, semiuniform temperature gradients established through the thickness of the laminate are expected to result in the out-of-plane deformation. However, careful selection of the parameters is required. For the 2/1 (2 layers of aluminum and 1 layer of glass fiber reinforced polyimide composite) fiber metal laminates (FML), it was observed that it is difficult to set up the TGM conditions throughout the thickness of the laminate, resulting in delamination of the upper aluminum layer. However, if the high temperature gradient is established only across the upper aluminum layer of the laminate, it is possible to produce the significant deformation without damage. Figure 8.57 presents the variation of bending angles with the number of laser scanning passes for laminates with different number of layers. The figure indicates that for the same number of laser scanning passes, larger bending angle can be achieved in the laminate with smaller number of layers (Edwardson et al. 2005b).

References

Alberti N, Fratini L, Micari F (1994) Numerical simulation of the laser bending process by a coupled thermal mechanical analysis. In: Geiger M, Vollertsen F (eds) Laser Assisted Net Shape Engineering, Proceedings of the LANE' 94, Maisenbach Bamberg, Germany, vol. I, pp. 327–336.

Alberti N, Fratini L, Micari F, Cantello M, Savant G (1997) Computer aided engineering of a laser assisted bending processes. In: Geiger M, Vollertsen F (eds) Laser Assisted Net Shape Engineering 2, Proceedings of the LANE' 97, Maisenbach Bamberg, Germany, vol. 2, pp.375–382.

Bao J, Yao YL (2001) Analysis and prediction of edge effects in laser bending. Journal of Manufacturing Science and Engineering 123:53–61.

Bartkowiak K, Dearden G, Edwardson SP, Watkins KG (2004) Development of 2D and 3D forming strategies for thin section materials using scanning optics. Proceedings of the 23rd International Congress on Laser Applications and Electro-optics, ICALEO 2004, Laser Institute of America, Orlando, Florida, Publication number 597, vol. 97.

Bartkowiak K, Edwardson SP, Borowski J, Dearden G, Watkins KG (2005) Laser forming of thin metal components for 2D and 3D applications using a high beam quality, low power Nd:YAG laser and rapid scanning optics. In: Vollertsen F, Seefeld T (eds) International Workshop on Thermal Forming, Bremen, vol. 26.

Chan KC, Liang J (2001) Thermal expansion and deformation behavior of aluminum-matrix composites in laser forming. Composite Science and Technology 61:1265–1270.

Chen DJ, Wu SC, Li MQ (2004) Studies of laser forming of Ti-6Al-4V alloy sheet. Journal of Materials Processing Technology 152:62–65.

Cheng JG, Zhang J, Chu CC, Zhe J (2005) Experimental study and computer simulation of fracture toughness of sheet metal after laser forming. International Journal of Advanced Manufacturing Technology 26:1222–1230.

Cheng PJ, Lin SC (2000a) An analytical model for the temperature field in the laser forming of sheet material. Journal of Materials Processing Technology 101:260–167.

Cheng PJ, Lin SC (2000b) Using neural networks to predict bending angle of sheet metal formed by laser. International Journal of Machine Tools and Manufacture 40:1185–1197.

Cheng PJ, Lin SC (2001) An analytical model to estimate angle formed by laser. Journal of Materials Processing Technology 108:314–319.

Dearden G, Edwardson SP (2003a) Laser assisted forming for ship building. SAIL 2003, Williamsburg, VA, June 2–4.

Dearden G, Edwardson SP (2003b) Some recent developments in two- and three-dimensional laser forming for "macro" and "micro" applications. Journal of Optics A: Pure Applied Optics 5:S8–S15.

Dearden G, Taylor C, Bartkowiak K, Edwardson SP, Watkins KG (2003) An experimental study of laser micro-forming using a pulsed Nd:YAG laser and scanning optics. Proceedings of the 22nd International Congress on Laser Applications and Electro-optics, ICALEO' 2003, Laser Institute of America, Orlando, Florida, Publication No. 595, vol. 95.

Edwardson SP, Watkins KG, Dearden G, Magee J (2001) 3D laser forming of saddle shapes. In: Geiger M, Otto A (eds) Laser Assisted Net Shape Engineering, Proceedings of 3rd International Conference on LANE' 2001, Maisenbach Bamberg, Germany, pp. 559–568.

Edwardson SP, Abed E, French P, Dearden G, Watkins KG, McBride R, Hand DP, Jones JD, Moore AJ (2005a) Developments towards controlled three-dimensional laser forming of continuous surfaces. Journal of Laser Applications 17:247–255.

Edwardson SP, French P, Dearden G, Watkins KG, Cantwell WJ (2005b) Laser forming of fibre metal laminates. Lasers in Engineering 15:233–255.

Edwardson SP, Abed E, Bartkowiak K, Dearden G, Watkins KG (2006) Geometric influences on multi-pass forming. Journal of Physics D: Applied Physics 39:382–389.

Geiger M, Vollertsen F (1993) The mechanisms of laser forming. CIRP Annals 42:301–304.

Geiger M, Vollertsen F, Deinzer G (1993) Flexible straightening of car body shells by laser forming. SAE Special Publication 944:37–44.

Hertzberg RW (1976) Deformation and Fracture Mechanics of Materials. Wiley, New York.

Hsiao YC, Shimizu H, Firth L, Mather W, Masubuchi K (1997) Finite element modeling of laser forming. Proceedings of the International Congress on Laser Applications and Electro-optics, ICALEO'97, Laser Institute of America, Orlando, Florida, Section A, pp. 31–40.

Hsieh H, Lin J (2005) Study of the buckling mechanism in laser tube forming with axial preloads. International Journal of Machine Tools and Manufacture 45:1368–1374.

Hu Z, Kovacevic R, Labudovic M (2002) Experimental and numerical modeling of buckling instability of laser sheet forming. International Journal of Machine Tools and Manufacture 42: 1427–1429.

Kao MT (1996) Elementary study of laser sheet forming of single curvature. Master Thesis, Department of Power Mechanical Engineering, Tsing Hua University.

Kitamura N (1983) Technical report of joint project on materials processing by high power laser. JWES-TP-8302, pp. 359–371.

Kyrsanidi AK, Kermanidis TB, Pantelakis (1999) Numerical and experimental investigation of the laser forming process. Journal of Materials Processing Technology 87:281–290.

Lawrence J, Schmidt MJ, Li L (2001) The forming of mild steel plates with a 2.5 kW high power diode laser. International Journal of Machine Tools and Manufacture 41:967–977.

Li W, Yao YL (2001a) Laser forming with constant line energy. International Journal of Advanced Manufacturing Technology 17:196–203.

Li W, Yao YL (2001b) Laser bending of tubes: mechanism, analysis, and prediction. Transactions of ASME 123:674–681.

Magee J, De Vin LJ (2002) Process planning for laser-assisted forming. Journal of Materials Processing Technology 120:322–326.

Magee J, Watkins KG, Steen WM, Calder NJ, Sidhu J, Kirby J (1997) Laser forming of aerospace alloys. Proceedings of the International Congress on Laser Applications and Electro-optics, ICALEO'97, Laser Institute of America, Orlando, Florida, vol. 83, pp. 156–165.

Magee J, Watkins KG, Steen WM (1998) Advances in laser forming. Journal of Laser Applications 10:235–246

Marya M, Edwards GR (2001) A study on the laser forming of near-alpha and metastable beta titanium alloy sheets. Journal of Materials Processing Technology 108:376–383

Merklein M, Hennige T, Geiger M (2001) Laser forming of aluminum and aluminum alloys-microstructural investigation. Journal of Materials Processing Technology 115:159–165.

Moshaiov A, Vorus W (1987) The mechanics of the flame bending process, theory and applications. Journal of Ship Research 31:269–281

Namba Y (1986) Laser forming in space. In: Wang CP (ed) Proceeding of the International Conference on Lasers'85, STS Press, McLean, pp. 403–407.

Ready JF (1997) Industrial Applications of Lasers. Academic Press, San Diego.

Reeves M, Moore AJ, Hand DP, Jones JD, Cho JR, Reed RC, Edwardson SP, Dearden G, French P, Watkins KG (2003) Dynamic distortion measurements during laser forming of Ti-6Al-4V and their comparison with a finite element model. Journal of Engineering Manufacture 217:1685–1696.

Semiatin SL (1988) Introduction to forming and forging processes. ASM Handbook, vol. 14, pp. 15–21.

Shen H, Shi Y, Yao Z (2006) Numerical simulation of the laser forming of plates using two simultaneous scans. Computational Materials Science 37:239–245.

Shichun W, Jinsong Z (2001) An experimental study of laser bending for sheet metals. Journal of Materials Processing Technology 110:160–163.

Thomson GA, Pridham M (1995) Laser forming. Manufacturing Engineer 74:137–139.

Thomson G, Pridham M (1997a) A feedback control system for laser forming. Mechatronics 7:429–441.

Thomson G, Pridham MS (1997b) Controlled laser forming for rapid prototyping. Rapid Prototyping Journal 3:137–143.

Thomson G, Pridham M (1998) Improvements to laser forming through process control refinements. Optics and Laser Technology 30:141–146.

Thomson G, Pridham M (2001) Material property changes associated with laser forming of mild steel component. Journal of Materials Processing Technology 118:40–44.

Vollertsen F, Komel I, Kals R (1995) The laser bending of steel foils for microparts by the buckling mechanism-a model. Modelling and Simulation in Materials Science and Engineering 3:107–119.

Vollertsen F (1994a) Mechanisms and models for laser forming. In: Geiger M, Vollertsen F (eds) Laser Assisted Net Shape Engineering, Proceedings of the LANE' 94, Maisenbach Bamberg, Germany, vol. I, pp. 345–360.

Vollertsen F, Rodle M (1994) Model for the temperature gradient mechanism of laser bending. In: Geiger M, Vollertsen F (eds) Laser Assisted Net Shape Engineering, Proceedings of the LANE' 94, Maisenbach Bamberg, Germany, vol. I, pp. 371–378.

Vollertsen F (1994b) An analytical model for laser bending. Lasers in Engineering 2:261–276.

Vollertsen F, Geiger M, Li WM (1993) FDM and FEM simulation of laser forming. In: Wang ZR, He Y (eds) Advanced Technology of Plasticity, vol. III, pp. 1793–1798.

Vollertsen F (1995) Applications of lasers for flexible shaping processes. In: Geiger M (ed) Schlusseltechnologie Laser: Herausforderung an die Fabrik 2000, Proceedings of the 12th International Congress (LASER95), Maisenbach Bamberg, Germany, pp. 151–162.

Yanjin G, Sheng S, Guoqun Z, Yiguo L (2003) Finite element modeling of laser bending of pre-loaded sheet metals. Journal of Materials Processing Technology 142:400–407.

Yanjin G, Sheng S, Guoqun Z, Yiguo L (2005) Influence of material properties on the laser-forming process of sheet metals. Journal of Materials Processing Technology 167:124–131.

Yau CL, Chan KC, Lee WB (1998) Laser bending of leadframe materials. Journal of Materials Processing Technology 82:117–121.

Zhang J, Cheng P, Zhang W, Graham M, Jones J, Jones M, Yao YL (2006) Effects of scanning schemes on laser tube bending. Transactions of the ASME 128:20–33.

Zhang J, Pirzada D, Chu CC, Cheng GJ (2005) Fatigue life prediction after laser forming. Journal of Manufacturing Science and Engineering 127:157–164.

Chapter 9
Laser-Based Rapid Prototyping Processes

9.1 Introduction

In the present competitive manufacturing environments, the product manufacturing industries are faced with numerous challenges such as increasing variations of products and product complexity, and decreasing delivery time. In response to these challenges, major industrial and research efforts have been concentrated on rapid development of prototypes, rapid tooling, and rapid manufacturing. Lasers have been extensively used in various rapid prototyping (RP) technologies. This chapter presents the background of the RP processes followed by a detailed discussion on various laser-based RP processes.

9.2 Basics of Rapid Prototyping Processes

RP is a set of technologies used for rapidly creating 3D physical models of the product or prototype directly from the computer-aided design (CAD) data. RP technologies are also referred to as solid free-form fabrication (SFF), layered manufacturing (LM), Automated Fabrication, 3D Printing, etc. RP facilitates the rapid development of the product by decreasing the development time, reducing the costly mistakes, and effective incorporation of appropriate design changes.

There are various RP techniques used in the industrial practices. Figure 9.1 presents a schematic of the common methodology for the implementation of RP processes. The important steps in these processes can be summarized as:

1. *Construction of a computer-aided design (CAD) model*: In most of the design projects, the CAD models of the product to be built up are generally available and can be directly subjected to the implementation of RP. However, in many cases, particularly during the new product design, the CAD models of the product to be built up are not available. In such cases, models are constructed from the physical models by implementing reverse engineering (RE), computer tomography (CT), or (magnetic resonance imaging (MRI) techniques. CAD models are generally constructed by the design team.

2. *Conversion of CAD model into STL file*: The true CAD models of the complex objects with curved surfaces are generally large and difficult for implementation of RP analysis. Hence, the RP industry has developed stereolithography (STL) format in which 3D surface of the object is represented by an assembly of nonoverlapping, planar triangles. This is referred to as tessellation of CAD data. The STL file contains the coordinates of the vertices and the direction of the outward normal of each triangle. Since the STL files use planar elements to construct the surfaces, the tessellated models are too coarse to accurately represent the curved surfaces. The accuracy of the model can be increased by increasing the number of triangular elements; however, this increases the size of the STL file making it difficult to process. Hence balance must be reached between the accuracy of the model and the manageability to produce useful STL files.

3. *Creation of the model*: The STL file prepared in the second step is then processed to slice the model into a number of layers with thickness usually less than 0.7 mm. The buildup time of the object, among various parameters, depends on the orientation of the object selected for the buildup. The buildup time can be shortened by constructing the shortest dimension in the vertical axis such that the number of layers gets reduced. The preprocessing programs can be used to determine the buildup time of the object. Since the structures are built layer by layer, features such as overhangs and internal cavities need to be supported. These auxiliary supports can also be generated by the preprocessing programs. The actual construction of the model is carried out by stacking the layers of polymer or powder metal where each layer corresponds to the cross section of the actual model. A voxel-based approach can also be used where each layer is subdivided into a number of voxels (3D volume elements). Such an approach provides close resemblance between the voxel model of an object and the object fabricated by LM (Chandru et al. 1995).

4. *Postprocessing*: Finally, the structure is removed from the platform and the auxiliary supports on which it was built. If necessary, the prototype may be subjected to surface preparation operations such as cleaning, sealing, surface painting, etc.

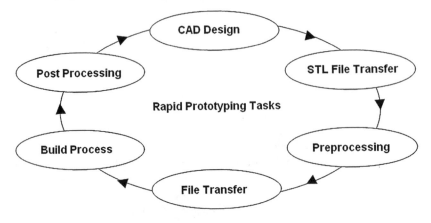

Fig. 9.1 Schematic of methodology for implementing rapid prototyping process

From the above discussion of the various steps in the RP processes, two important characteristics of these processes can be listed:

1. Manufacturing physical objects directly from the digital 3D images obtained using CAD, CT, or MRI
2. Manufacturing by means of layer-by-layer stacking of cross-sectional features of the model from bottom upwards

9.3   Classification of Rapid Prototyping Processes

Various RP processes can be classified in numerous ways; however, most widely used classification of RP processes is based on the nature of the process (additive or subtractive), the state of raw material, and the choice of physical processes for adhesion (Kruth 1991; Xu et al. 1999; Kochan et al. 1999; Kulkarni et al. 2000). Figure 9.2 presents the classification of additive RP processes according to the state of raw material before the prototype formation (Kruth 1991). The raw materials for the prototype formation primarily consist of liquid, powder, or solid sheets. Liquid-based RP methods rely upon the controlled and localized solidification of liquid polymers by exposing it to the light or a sequence of melting, deposition, and resolidification of material. RP methods using powders as raw material are based on selective sintering (using laser) or gluing (using adhesive binder) of powders into desired shapes. In the third class of RP processes, the raw material is in the form of thin solid foils which may be welded or bonded on top of each other using the laser or adhesion. This nonexhaustive classification is intended to provide a brief overview of the RP processes. Some of these processes are still under development. It is beyond the scope of this chapter to provide the complete listing and detailed description of all the available RP processes.

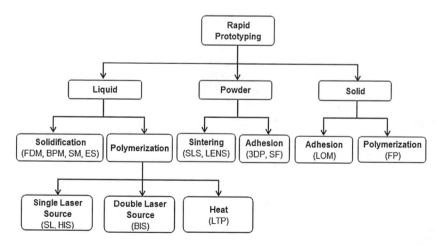

Fig. 9.2 Classification of additive rapid prototyping processes (Kruth 1991)

9.3.1 Liquid-Based RP Processes

9.3.1.1 Stereolithography (SL)

SL is based on selective curing of photosensitive liquid polymers using laser as a source of energy. The laser beam traces the slice of the CAD model point by point such that each layer of solidified photopolymer is formed from the assembly of voxels (3D volume elements). The process is repeated to create all the layers to build the complete structure.

9.3.1.2 Holographic Interference Solidification (HIS)

In this process a holographic image is projected on a liquid photosensitive polymer contained in a vat such that the entire surface of the polymer is solidified instead of point-by-point solidification (Kruth 1991).

9.3.1.3 Beam Interference Solidification (BIS)

This process is based on the point-by-point solidification of photosensitive polymers (contained in a transparent vat) at the intersection of two laser beams having different wavelengths. The first laser excites the liquid polymer to the reversible metastable state which is subsequently polymerized by the radiations from the second laser. The process is associated with various technical limitations such as insufficient absorption of laser intensity at higher depths, shadowing effect of already solidified material, and diffraction of laser lights leading to difficulties in obtaining the precise intersection of the beam (Kruth 1991).

9.3.1.4 Liquid Thermal Polymerization (LTP)

LTP process is quite similar to SL in a way that the part is built by solidification of successive layers of liquid polymer. However, the polymers used in LTP process are thermosetting polymers instead of photopolymers and hence the solidification is induced by thermal energy rather than light energy. The thermal nature of the process is expected to make the control of the size of the voxels difficult due to dissipation of heat (Kruth 1991).

9.3.1.5 Fused Deposition Modeling (FDM)

FDM is one of the most common RP processes in which the part is built by solidi-fication of the molten material on the substrate. Typical FDM apparatus consists of an extrusion head where the filament of material (generally thermoplastic polymer) is fed. The filament material is heated slightly (~1 °C) above the melting point such that the deposited molten material solidifies quickly (~<0.1 s) on the substrate or the previously solidified layer.

9.3.1.6   Ballistic Particle Manufacture (BPM)

The BPM process involves a stream of molten droplets ejected from the piezoelectric ink-jet printing nozzles to deposit on the target substrate. The process still uses the 3D data of the solid model to position the stream of material on the substrate. Since the process is based on the melting of the material, it is particularly suited for the materials that easily melt and solidify such as thermoplastics and metals.

9.3.1.7   Shape Melting

The SM process is similar to FDM and is used for producing the near net-shape metallic parts. In this process, a thread of metal is melted and deposited by arc welding. In contrast to the other RP processes, the parts produced by SM are strongly limited due to poor achievable accuracy (~1 mm). Various materials used with SM technique include Inconel, tungsten carbide, and other alloys (Kruth 1991).

9.3.1.8   Electrosetting

The process is based on an approach in which the electrodes printed by the standard pen plotter or laser printer on the conductive material are stacked and immersed in a bath of thermosetting fluid. When the electrodes are subjected to intense electric charge, the solidification of the fluid is induced by the biasing electric field through the fluid.

9.3.2   Powder-Based RP Processes

9.3.2.1   Selective Laser Sintering (SLS)

In SLS, a laser beam is scanned over a layer of powder such that the powder is sintered point by point into a cross section of 3D model. The process is repeated by spreading, leveling, and sintering of additional layers of powder to produce a complete structure of the prototype. The powder is generally preheated to a temperature slightly less than the melting point so that laser interaction rapidly increases the temperature of the powder to the sintering temperature. The process is most widely used for thermoplastic polymers such as nylon, PVC, and polystyrene; however, the process can also be used with metallic powders of titanium, brass, etc.

9.3.2.2 Laser Engineered Net Shaping (LENS)

LENS is a similar process to the SLS except for the way in which the powder is delivered for the construction of the part. In LENS, the laser beam is focused on the part bed through a deposition head and the material powder is delivered and distributed around the circumference of the head. The powder may be delivered by gravity forces or by using high-pressure carrier gas. The laser causes the melting of the powder and subsequent deposition to produce a layer corresponding to the cross section of the desired part. The part is built directly from the CAD solid model by depositing several layers corresponding to all the cross sections of the geometry. The process can be used for a variety of metals, alloys, and composite materials.

9.3.2.3 Three-Dimensional Printing (3DP)

The 3DP process is based on the selective bonding of the powder using a binder. A nozzle similar to the ink-jet printing head projects the fine droplets of binder on a powder layer at selective places corresponding to the cross-sectional geometry of the part to be built. The complete part is constructed by bonding the successive layers of powder spread on previous layers. Once the part is built, excess unbound powder is removed after adequate heat treatments. The process has been used for the fabrication of ceramic molds, cores, and porous preforms. However, the process is flexible and can be used for a variety of other material systems.

9.3.3 Solid-Based RP Processes

9.3.3.1 Laminated Object Manufacturing

The process involves laser cutting of the profiles from thin foil (supplied by a feed roll) which are subsequently glued to previous layers, thus building a part by stacking. The height of the stack is adjusted after each layer such that the laser cuts the profile in a thickness exactly equal to the thickness of the foil. The gluing is facilitated by a hot roller which activates the heat-sensitive glue on the other side of the foil. The process can be used with a variety of foil materials including papers, plastics, metals, and synthetic materials (Kruth 1991).

9.3.3.2 Solid Foil Polymerization

The process is based on complete polymerization of semipolymerized plastic foils on exposure to suitable light source. The semipolymerized foil is first stacked on the previously solidified part and then illuminated such that bonding is achieved after complete polymerization. The excess foil which is not illuminated can be removed by dissolving into suitable solvent leaving behind the desired part.

9.3.4   Concept Modelers

Concept modelers are a class of RP systems designed to produce quick and cost-effective physical models at any design stage as against the fully functional high-end RP systems which are used at the end of design stages. The concept modelers are also called office modelers and are marketed as CAD peripheral solutions. The concept modelers enable the designers to verify and iterate the design at various stages in office environment so that accurate models for the fully functional RP processes can be designed without high-cost design changes. Typical requirements for the concept modelers are low-cost, fast, office-friendly, fully automatic units. Various commercially available concept modelers are JP System 5, Model Maker II, Z-Corporation Z402, Genisys Xs, Multi-Jet Modeler, etc.

9.4   Laser-Based Rapid Prototyping Processes

Lasers are at the forefront of RP technologies. The applications of lasers in RP are derived from the ability to control the laser beam characteristics and quality such that desired interactions between the laser beam and a variety of materials can be achieved for the creation of final part geometry. With a brief overview of RP processes in the previous sections, the following sections are designed to discuss in detail the important laser-based RP processes.

9.4.1   Stereolithography

SL, also referred to as laser photolithography, is an RP process used for layered fabrication of 3D physical objects by selective polymerization of photosensitive liquid polymer using laser beam. SL is the first commercially developed RP process by 3D Systems, in 1987. Since then the process has undergone several advancement in terms of hardware, software, and materials to emerge as one of the most widely used RP processes. The advanced photopolymers for the SL products at 3D Systems are codeveloped with Ciba Specialty Chemicals. The following sections deal with the process details and the critical issues relating to the SL process.

9.4.1.1   Stereolithography Process

The process of SL primarily consists of three main operations: preprocessing, buildup process, and postprocessing (Dutta et al. 2001).

Preprocessing: The process begins with the creation of a suitable CAD model of the part to be built. The CAD data is tessellated into STL format which is standard for the SL systems. The important steps in the preprocessing are determination of optimum orientation to reduce the build time and the generation of supports required for building the overhangs/undercuts of the part. The STL file is then modified to incorporate the supports and sliced into a number of horizontal cross sections.

Buildup process: The schematic of the apparatus for the SL is shown in Fig. 9.3. The important units of the machine are:

1. A removable vat which holds the liquid photosensitive polymer
2. A base plate mounted on an elevator frame with devices for controllable vertical movement
3. Devices for automated resin-level sensing
4. A recoater blade with a horizontal movement for smoothing action to establish uniformly thick layer of liquid above the previously cured layer
5. A laser with its optics

The entire system is enclosed in a build chamber to prevent the fumes generated during operation from escaping into the surrounding. The building of the part starts with the positioning of the base plate just below the resin surface such that the first layer of the resin of controlled thickness is cured by scanning a desired geometry corresponding to the cross section of the part by a focused laser beam. Subsequently, the elevator is incrementally lowered into the liquid polymer and the recoater blade sweeps a uniform layer of resin over the previously cured layer. The laser scans the subsequent layers of resins corresponding to all the cross sections of the part such that the complete structure is built in a layered fashion directly from the CAD model. One of the variations of SL is solid ground curing (SGC) in which the entire

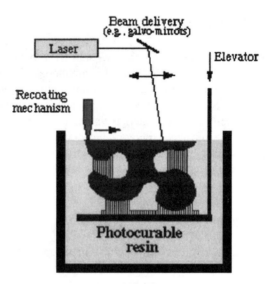

Fig. 9.3 Schematic of stereolithography process (Reprinted from Dutta et al. 2001. With permission. Copyright American Society of Mechanical Engineers.)

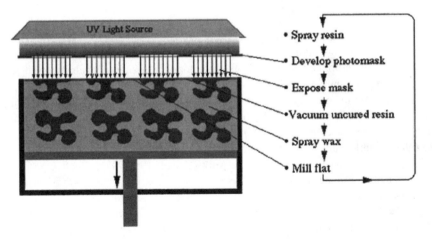

Fig. 9.4 Schematic of Solid Ground Curing process (Reprinted from Dutta et al. 2001. With permission. Copyright American Society of Mechanical Engineers.)

Table 9.1 Specifications of some of the stereolithography systems available from 3D Systems (Courtesy of 3D Systems)

Model	SLA 5000	SLA 7000	Viper Pro
Laser type	Nd:YVO$_4$	Nd:YVO$_4$	Nd:YVO$_4$
Laser wavelength (nm)	354.7	354.7	354.7
Laser power (mW)	216	800	2000
Typical build layer (mm)	0.10	0.10	0.10
Elevator vertical resolution (mm)	0.00177	0.001	0.001
Maximum part weight (kg)	68.04	68	75
Maximum build envelop (mm^3)	$508 \times 508 \times 584$	$508 \times 508 \times 584$	$1500 \times 750 \times 500$
Maximum part drawing speed (m/s)	5.0	9.52	25.0

cross section of the part is built by exposing the resin to UV light through a photomask instead of point-by-point scanning (Fig. 9.4). The method of generating masks is based on electrophotography (xerography). The process was originally developed by Cubital.

Postprocessing: Once the part is built, it is removed from the base plate and washed to remove the excess resin attached to the surfaces of the part. Subsequently, the supports are detached and the part is placed in a UV postcuring apparatus (PCA) to obtain fully cured finished part.

Some of the commercial SL systems along with important specifications available from 3D Systems are listed in Table 9.1.

9.4.1.2 Stereolithography Resins

SL processes are based on the laser-induced curing of liquid resin into solid polymer material. The degree of cure in the area irradiated with laser and the cure time are dependent on the kinetic behavior of the resins. Various resins have been used for the SL process. The first resins used for SL were acrylate-based and involved photopolymerization of acrylate monomers by free radical mechanism. However, epoxy-based resins are increasingly being used. Two photopolymerization mechanisms of SL resins are radical-initiated polymerization and cationic-initiated polymerization.

In radical-initiated polymerization, a photoinitiator is used which, upon irradiation with the laser, absorbs the photon energy and produces free radicals by splitting into its excited state (Fig. 9.5). Since each monomer molecule has at least two acrylate double bonds, the polymerization will propagate in three dimensions to yield a highly cross-linked polymer network (Decker 1999). For effective performance, the photoinitiator must absorb the light at the emission wavelength of the laser. Acrylate-based resins are generally characterized by high reactivity or photospeed; however, the polymerization is accompanied by a high volumetric shrinkage (up to 5%). Also, oxygen inhibition may result in incompletely cured surfaces.

Cationic-initiated polymerization uses cationic photoinitiators such as salts of triarylsulfonium ($Ar_3S^+ X^-$) or diaryliodonium ($Ar_2I^+ X^-$) which produce photonic acid when irradiated with UV radiation in the presence of hydrogen donor molecule (Fig. 9.6). The most attractive characteristics of cationic polymerization for SL are that polymerization is not influenced by the oxygen inhibition and is associated with very limited volume shrinkage. The polymerization reaction

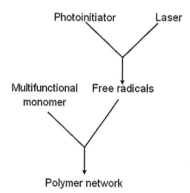

Fig. 9.5 Radical-initiated polymerization (Decker 1999)

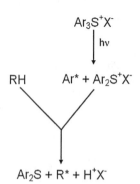

Fig. 9.6 Cationic-initiated polymerization (Decker 1999)

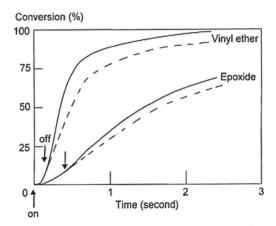

Fig. 9.7 Laser-initiated cationic polymerization (solid lines) and dark polymerization (dashed lines) of diethylene glycol divinyl ether and a cycloaliphatic diepoxide (Reprinted from Decker 1999. With permission. Copyright Elsevier.)

continues even after the laser source has been removed (dark reaction). This continued curing reaction after the laser source is removed may substantially improve the green strength of the SL part (Corcione et al. 2003). Figure 9.7 presents the RTIR (time-resolved infrared spectroscopy) traces representing the laser-initiated cationic polymerization and dark polymerization (dashed lines) of vinyl ether and epoxide using Kr^+ laser beam. The figure indicates that short exposure (as indicated by laser on-and-off arrows) causes extensive dark polymerization (dashed curves). Also, vinyl ethers polymerize faster and more extensively than epoxides due to high reactivity. These are the significant advantages over the acrylate-based resins which show very negligible dark reaction and correspondingly insignificant improvement in strength after the laser has been removed. However, the kinetics of cationic-initiated polymerization is complex and strongly affected by resin formulations (Decker 1999).

Extensive research efforts have been directed toward developing suitable resin materials in response to the various desired properties before and after curing. Important considerations for choosing the resin are chemistry, photospeed, vat life, viscosity, glass transition temperature, etc. These can directly influence the achievable properties of the prototype such as accuracy, strength, toughness, humidity resistance, temperature resistance, optical properties, environmental resistance, etc. Most of the SL resins are commercially manufactured by Ciba (e.g., SL 5195, SL 5510, SL 5520, SL 5530HT, etc.), Du Pont (e.g., SOMOS 2110, SOMOS 7110, SOMOS 7120, SOMOS 8100, etc.), and various other companies.

9.4.1.3 Geometric Aspects of Laser-Cured Resin during Stereolithography

The important parameters during the SL which determine the quality (accuracy, distortion, shrinkage) of the prototype are depth and width of the cured line. These geometric features can be estimated from the extent of interaction of the laser beam with the resin. During SL operation, the energy of the incident laser beam is absorbed by the resin. The absorption of laser energy in the resin is given by the Beer–Lambert exponential law (Fuh et al. 1995):

$$E(z) = E_0 \exp\left(-\frac{z}{D_\mathrm{p}}\right), \tag{9.1}$$

where, z is depth from the surface of resin; $E(0)$ is laser exposure at the surface of resin; $E(z)$ is laser exposure at z and D_p is penetration depth defined as the depth from the surface at which laser irradiance is $1/e$ times the irradiance at the surface. As the depth below the surface of the resin increases the laser exposure decreases. Polymerization takes place when the laser exposure is greater than certain critical value, E_c. This defines the maximum cured depth (C_d) beyond which no curing takes place since the laser exposure is insufficient to cause the polymerization (Fig. 9.8). For a Gaussian laser beam with maximum energy corresponding to E_max, this critical value is given by (Fuh et al. 1995):

$$E_\mathrm{c} = E_\mathrm{max} \exp\left(-\frac{C_\mathrm{d}}{D_\mathrm{p}}\right). \tag{9.2}$$

Thus,

$$C_\mathrm{d} = D_\mathrm{p} \ln\left(-\frac{E_\mathrm{max}}{E_\mathrm{c}}\right). \tag{9.3}$$

The width of the cured line is given by:

$$L_\mathrm{w} = \sqrt{2}\, W_0 \sqrt{\ln\left(\frac{E_\mathrm{max}}{E_\mathrm{c}}\right)}, \tag{9.4}$$

where W_0 is the Gaussian half width. Figure 9.9 shows a typical cross section of the single cured line for SCR-300 resin. The depth and width of the single cured line increase with increasing the maximum laser exposure (Fig. 9.10).

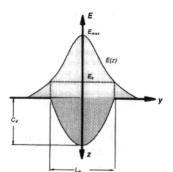

Fig. 9.8 Schematic of cross-sectional geometry of single cured line in stereolithography (Reprinted from Fuh et al. 1995. With permission. Copyright Elsevier.)

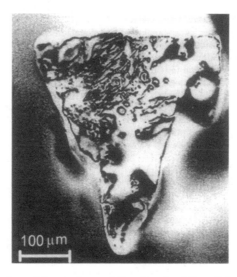

Fig. 9.9 Typical cross section of the single cured line (Reprinted from Fuh et al. 1995. With permission. Copyright Elsevier.)

Each layer in SL is built by scanning the laser beam in two distinct patterns: contour and hatching patterns (Fig. 9.11). Contour pattern is used for creating (by solidifying the resin) the borders of the slice area by joining a number of linear segments, whereas the hatching pattern is used for creating the interior of the slice area by solidifying the number of closely spaced adjacent line segments (Giannatsis et al. 2001). The adjacent lines are generally overlapped to avoid the entrapment of uncured resin between the lines. Figure 9.12 shows the effect of the percentage of overlap on the cross-sectional microstructure of single cured layer (Fuh et al. 1995). The part is built by solidifying multiple layers of resin according to the corresponding horizontal cross-sectional geometry of the part. Schematic of the multilayer architecture obtained by point-by-point laser scanning is presented in Fig. 9.13 (Wang et al. 1996). The important geometric parameters here are the scan pitch (hatch spacing) and layer pitch (layer thickness). Due to Gaussian distribution of the energy in the laser beam, the cross section of each cured scan is nearly semielliptical resulting in nonuniform thickness of the cured layer. Between two adjacent cured lines, the areas of uncured resin (often referred as undercure) exist, which depend on the degree of overlap between cured lines. During part building, subsequent layers of resin are cured such that the depth of curing is longer than the layer pitch resulting in partial overcuring of the previous layer (often referred as overcure).

Build style refers to a set of predefined laser scanning parameters selected for creating the part in LM. Various build styles which are commonly used in SL are STAR-WEAVE (staggered hatch alternate sequencing retracted hatch weave), ACES (accurate clear epoxy solid), and QuickCast. The first two are solid build styles, whereas the third one is hollow build style. STAR-WEAVE build style was developed for building solid acrylate parts and important laser scanning parameters are nonoverlapping crosshatch vectors within a single laser scan, alternating

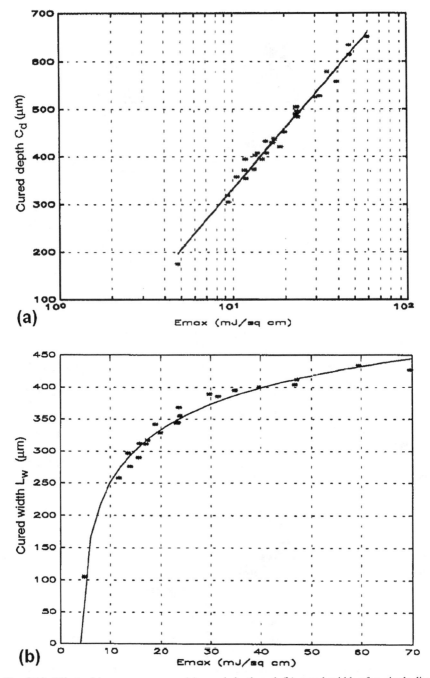

Fig. 9.10 Effect of laser exposure on (**a**) cured depth and (**b**) cured width of a single line (Reprinted from Fuh et al. 1995. With permission. Copyright Elsevier.)

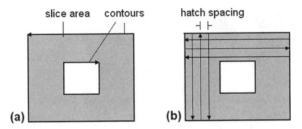

Fig. 9.11 Laser scanning patterns in stereolithography (**a**) contour and (**b**) hatching patterns

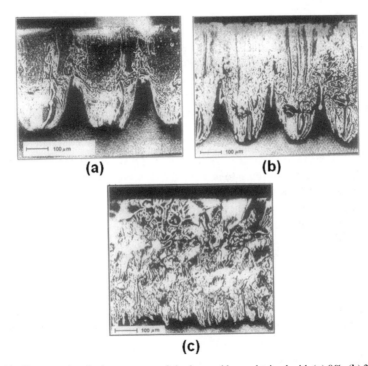

Fig. 9.12 Cross-sectional microstructure of single cured layer obtained with (**a**) 0%, (**b**) 25%, and (**c**) 50% overlap between the adjacent cured lines (Reprinted from Fuh et al. 1995. With permission. Copyright Elsevier.)

crosshatch draw order sequencing, retracted crosshatch from layer border vectors, and staggered crosshatch vector layout from layer to layer. ACES style was developed for solid epoxy parts and involves the overlapping crosshatch vectors. This laser scanning procedure ensures the complete and uniform curing of resin such that parts are virtually free from postcuring distortion and internal stresses. QuickCast build style is used for creating hollow patterns particularly useful in rapid tooling applications. The hollow patterns made by this build style offers significant advantages during shell investment casting such as easy burnout of the pattern and better casting quality (Williams et al. 1996).

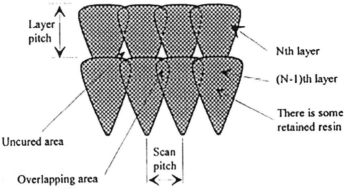

Fig. 9.13 Schematic of multilayered architecture of the stereolithography part (Reprinted from Wang et al. 1996. With permission. Copyright Elsevier.)

9.4.1.4 PostProcessing of Stereolithography Parts

During part building by SL process, resins do not undergo complete curing after completion of the process. In general, the extent of resin curing after the part is built ranges up to 65–90%. In order to achieve the desired mechanical properties complete curing of the resin is required. This necessitates the additional postcuring operation for the built parts. The most common postcuring operation involves the exposure of the built parts to UV radiation; however, the parts can also be cured in microwave or conventional ovens.

The volume of the uncured resin in the built part can be found by the equation:

$$\text{Degree of cure } (\%) = (1 - H_{spc} / H_{liq}) \times 100, \tag{9.5}$$

where H_{spc} and H_{liq} are the thermal energy released from laser-cured and liquid resin respectively. DSC (differential scanning calorimetry), (differential scanning photocalorimetry) can be used to determine the thermal energy released and thus the percentage of curing (Fuh et al. 1999; Colten and Blair 1999). Since DSC gives only the average thermal energy released during UV scanning, Raman spectroscopy can also be used to determine the percentage of curing at different locations of the cross-sectional area. Figure 9.14 presents the Raman spectra of uncured (liquid) and fully UV-cured resin (SCR-310). The peak on Raman spectra at delta wavenumber $1{,}663\,\text{cm}^{-1}$ is a result of carbon double bond (C = C) and intensity of this peak is a measure of amount of monomer present in the resin. The percentage of cure in the resin can be quantified by taking the ratio of intensities of peaks at wavenumbers $1{,}663\,\text{cm}^{-1}$ (which corresponds to carbon bond) and $1{,}444\,\text{cm}^{-1}$ (which corresponds to methylene deformation) and comparing the ratios with calibration curves as shown in Fig. 9.15. For this analysis, area under the peaks can also be used in a similar way as intensities (Fuh et al. 1999).

As mentioned earlier, the mechanical properties of the finished SL part depend on the degree of curing achieved during postcuring operations. This in turn depends

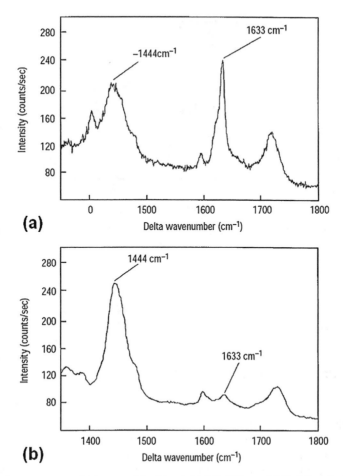

Fig. 9.14 Raman spectra of (**a**) uncured liquid and (**b**) fully UV-cured SCR-310 resin (Reprinted from Fuh et al. 1999. With permission. Copyright Emerald.)

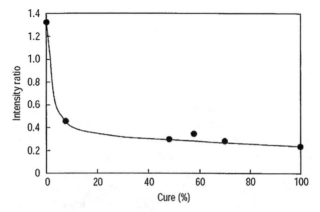

Fig. 9.15 Variation of intensity ratio with percentage of cure (Reprinted from Fuh et al. 1999. With permission. Copyright Emerald.)

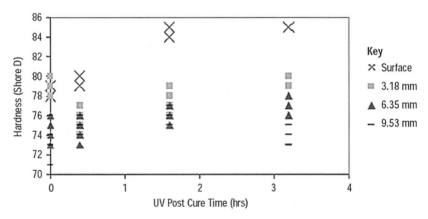

Fig. 9.16 Variation of hardness at various depths of samples (made of SOMOS 7110 resin) with UV postcure time (Reprinted from Colten and Blair 1999. With permission. Copyright Emerald.)

on the time (postcure time) for which the resin is exposed to UV radiation. Figure 9.16 illustrates that hardness at various depths of the UV-cured part (made of SOMOS 7110 resin) increases with postcure time (Colten and Blair 1999).

9.4.1.5 Process Planning for Stereolithography

SL is a complex process involving a number of variables. Hence, selection of appropriate process parameters to achieve the desired characteristics of the fabricated part becomes important. This constitutes the process planning for SL. The various goals of process planning for SL can be specified as build time, surface finish, and accuracy. Build time goal is directed toward the minimization of time required to build the prototype, whereas surface finish and accuracy goals are directed toward minimization of geometric deviation of the prototype from the user specifications (West et al. 2001). Two distinct approaches for process planning are algorithmic and decision support methods (Dutta et al. 2001). The first is based on geometric methods that operate on CAD model. Decision support methods seek to balance the competing and sometimes conflicting process goals during process planning. The process is based on setting up the preferences for the specific goals to meet the desired characteristics of the prototype.

Process planning tasks for SL have been identified as orientation determination, support generation, slicing, and path planning (Kulkarni et al. 2000). These four tasks can be categorized into two distinct domains: model domain and layer domain (Fig. 9.17). Orientation determination and support generation are the model domain tasks and its process planning requires geometric information from the 3D model. The process of slicing divides the CAD model into a number of horizontal layers corresponding to the cross sections of the prototype. Slicing and path planning are the layer domain tasks and related with the generation of data required for building each layer of the model. Due to distinct similarities of the basic methodology in the

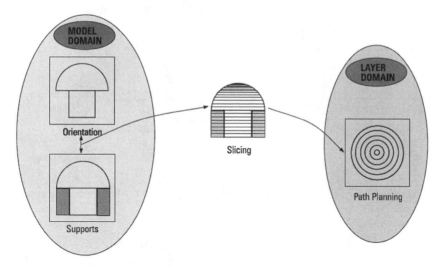

Fig. 9.17 Model and layer domain tasks for stereolithography process planning (Reprinted from Kulkarni et al. 2000. With permission. Copyright Emerald.)

various LM processes, these process planning tasks, in general, can also be applied to other LM processes.

Orientation

Process planning starts with the determination of optimum orientation among the various orientations in which the part can be built. Determination of orientation is done by analysis of 3D model either on its STL file or original CAD file. Determination of optimum orientation involves the consideration of widely different factors affecting the cost, quality, and ease of manufacturing. Most important factors are build height, quality of the surfaces, area of the base, mechanical properties, distortion/shrinkage, and external support structure (Kulkarni et al. 2000).

In most of the LM processes, minimizing the height of the part in build direction directly translates into reduction of build time by minimizing the number of layers for the complete fabrication of the part (exception is the adaptive slicing methods). This is based on the assumption that fabrication time for each layer is generally same irrespective of the complexity of the laser scanning path in each layer. Based on this criterion, the preferred build orientation corresponds to the minimum height in the build direction.

In SL, due to 2.5 dimensions of each layer (i.e., extrusion of the 2D surface), the sloping surfaces of the parts are inevitably represented by the stepped surface features (staircase effect). Better surface quality of the built part can be obtained by selecting the build orientation such that staircase effects are minimized. One of the ways to minimize the staircase effect is to maximize the areas of perfectly vertical and horizontal surfaces which do not cause stepped textures. Figure 9.18 indicates

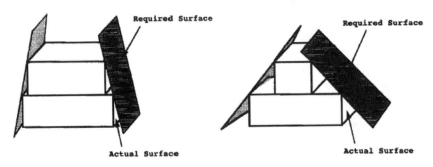

Fig. 9.18 Schematic showing the effect of steepness of the surfaces on the achievable surface finish of the part in stereolithography (Reprinted from Lan et al. 1997. With permission. Copyright Elsevier.)

that near vertical (or horizontal) surfaces better approximate the desired surface geometry of the part. Also, surface quality of the built part can be maximized by minimizing the contacts of the surfaces with the support structures. Area of the base is important from the consideration of stability of the part during part building and initial part curl. However, this must also be considered in relation with the desired surface finish on the base surface. It is important to determine the build orientation which fabricates the functional and aesthetic surfaces when oriented vertically, horizontally, or upward-facing (Lan et al. 1997).

Determination of orientation is also important from the view point of the desired mechanical properties of the built part. It has been observed that the mechanical properties in the built part differ in different part direction. Hence, build direction must be determined such that best mechanical properties are achieved in the most critical direction of the part. Also, build orientation should ensure the minimum distortion/shrinkage of the part. Finally, the optimum orientation determination should take into consideration the area of support structure in contact with the part and the total volume of the support structures. These support parameters need to be minimized to ensure reduction in build time and finishing time of the final part and thus the reduction in the fabrication cost. Based on this discussion optimum build orientations of a random part for achieving specific preferred objective are schematically shown in Fig. 9.19 (Kulkarni et al. 2000).

Support Structures

Support structures are used in SL to hold the overhanging surfaces of the prototype while it is built. Supports can either be external structures (for overhangs) or internal structures (for top surfaces of hollow parts). Support structures can be determined from the STL file or the original CAD model. While using the STL file, the generation of the support structures involves the determination of the surfaces of the model which need the support structures while part building. Generally, these are the down-facing surfaces with vertical normal greater than 45° from the

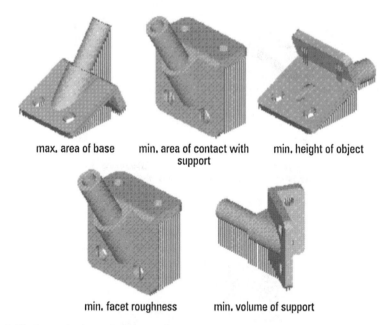

max. area of base min. area of contact with min. height of object
         support

min. facet roughness min. volume of support

Fig. 9.19 Determination of optimum build orientation based on preferred build objective (Reprinted from Kulkarni et al. 2000. With permission. Copyright Emerald.)

horizontal (West et al. 2001). When CAD model is used for the determination of support structure, the Gaussian map of the surface indicates the regions that need support. After the complete fabrication of the part support structures are removed in finishing operations. The area affected by the support structures can be calculated by tessellation of the CAD model in the given orientation. Support structures have detrimental effect on the surface quality (especially surface roughness) of the affected surfaces. Hence, process planning is directed toward minimizing the support structures required for the fabrication of the part.

Support structures are generally added at the end-edge of the cantilevered features of the part instead of the whole down-facing surfaces of the part. Also, the supports can be added in various ways: right-angled or diagonal supports (Fig. 9.20). Right-angled supports provide the part more mechanical stability, whereas the diagonal supports minimize the support material. This makes use of the number of supports as a decision factor instead of volume or area of the supports because it accurately indicates the relative quantity of supports required by an object in various orientations (Lan et al. 1997).

Slicing

During LM like SL, one of the important tasks is the slicing of the model into a number of layers. The important objective of the slicing procedures is to arrive at

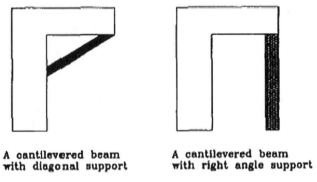

**A cantilevered beam
with diagonal support**

**A cantilevered beam
with right angle support**

Fig. 9.20 Two different ways of support structures for cantilever features of part (Reprinted from Lan et al. 1997. With permission. Copyright Elsevier.)

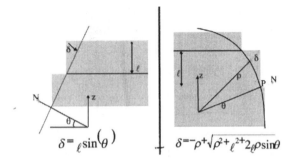

$$\delta = \ell \sin(\theta)$$

$$\delta = -\rho + \sqrt{\rho^2 + \ell^2 + 2\ell\rho \sin\theta}$$

δ – Cusp height; ℓ – layer thickness; N – Surface normal;

z – Build direction; ρ – Radius of curvature

Fig. 9.21 Caculation of cusp height and layer thickness for planar and curved surfaces (Reprinted from West et al. 2001. With permission. Copyright Elsevier.)

the optimum thickness and the contour profile of each layer. These parameters significantly affect the surface finish and the build time of the prototype. Slicing can be done by two approaches: uniform and adaptive. In uniform slicing approach, the model is divided into layers of constant thicknesses and is useful for the building of planes and cones/cylinders with vertical feature axis. Adaptive slicing approach allows the layer thickness to vary across the extent of the surface and is useful for the curved surfaces such as spheres, B-splines, and cones/cylinder which do not have a vertical feature axis. Adaptive slicing offers significant advantages over uniform slicing approaches: better surface quality due to less staircase effect and reduction in the build time.

The important parameter while determining the layer thickness is the cusp height, δ (Fig. 9.21). The cusp height may be taken as a measure of dimensional tolerance. The figure shows the geometry of planar and curved surfaces for the calculation of cusp height (West et al. 2001). For curved surfaces, the surface normal changes from point to point and hence the cusp height (Xu et al. 1997). Based on a maximum specified cusp height, the slicing scheme needs to be implemented for the calculation of layer thicknesses and number of layers. For a specified maximum cusp height, uniform slicing scheme requires large number of slices compared to the adaptive slicing scheme, thus allowing faster manufacturing by adaptive slicing (Kulkarni and Dutta 1996).

Path Planning

Path planning refers to the trajectory followed by the curing medium (photopolymer) during SL and involves the determination of the geometry of the paths and the appropriate process parameters. The important parameters for SL process planning are fill overcure, hatch overcure, sweep period, and z-level wait period. The first two are the layer parameters and the other two are the recoating parameters. Layer parameters determine the way each layer is cured in the vat, whereas recoating parameters determine the way a new layer of resin is deposited on the previous layer. These parameters again significantly affect the surface finish, accuracy, and build time.

In the context of SL, the path planning involves the determination of exterior and interior path for the solidification of resin. As discussed in the previous section, the two distinct scanning patterns for SL are the contour (exterior path) and the hatching patterns (interior path). Hatch spacing is the important parameter while scanning the interior path to fill the inside of the contour with solidified resin. For longer hatch spacing, the laser scans less number of hatch vectors per unit cross-sectional area which results in shorter build time. However, this may result in excessive part shrinkage and poor dimensional accuracy due to greater extent of uncured resin between the adjacent scans (McClurkin and Rosen 1998).

Various commonly used path planning methods are crosshatch, zigzag, and offset contour (Huang et al. 2003). Figure 9.22 shows the two distinct scan raster patterns for SL of suspended beam: short raster pattern and long raster pattern (Huang and Jiang 2003). Both simulation and experimental results have shown that the short scan raster pattern causes less curl distortion compared to the long scan raster pattern (Fig. 9.23).

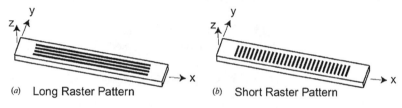

(a) Long Raster Pattern   (b) Short Raster Pattern

Fig. 9.22 (**a**) short raster and (**b**) long raster patterns for stereolithography of suspended beam (Reprinted from Huang and Jiang 2003. With permission. Copyright Springer.)

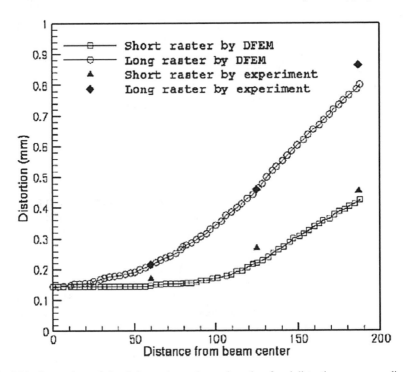

Fig. 9.23 Comparison of simulation and experimental results of curl distortions corresponding to short and long raster patterns for stereolithography of suspended beam (Reprinted from Huang and Jiang 2003. With permission. Copyright Springer.)

9.4.1.6 New Topics in Stereolithography

Stereolithography for Ceramic, Metallic, and Composite Parts

SL has been conventionally used for fabrication of plastic prototypes particularly based on acrylic and epoxy resins. Recently, increased research interests have been directed toward extending the capabilities of SL for the rapid fabrication of parts based on various other material systems such as metals and ceramics. The inspiration for such research is primarily based on increased utilization of advanced metallic and ceramic materials in the biomedical applications. For example, metallic materials (such as titanium alloys) and bioceramics (such as alumina, zirconia, and hydroxyapatite) can be used in medical implant such as bones and tooth prostheses (Licciulli et al. 2004). RP for such applications, accompanied by continuous advances in the related converging technologies such as medical imaging and image processing, present a great commercial potential. Also, SL of metallic materials is of particular importance in applications where the final product needs to be conductive (Lee et al. 2006).

The main challenge in the rapid fabrication of ceramics and metallic materials by SL is the preparation of photoreactive suspension containing ceramic or metallic

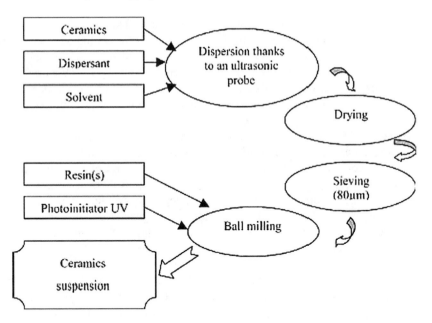

Fig. 9.24 General procedure for the preparation of UV curable ceramic suspensions for stereolithography (Reprinted from Dufaud et al. 2002. With permission. Copyright Elsevier.)

powder and resin with properties (such as viscosity and UV reactivity) similar to those of commercial resins. The general procedure for preparing such suspensions is the dispersion of ceramic particle into appropriate photopolymerizable resin in which a photoinitiator is dissolved. The schematic of the procedure for preparing the UV-curable suspension is shown in Fig. 9.24 (Dufaud et al. 2002). During SL, the resin gets polymerized and forms a matrix around the ceramic particles. Subsequently, the organic phase is removed by appropriate thermal treatments and the green ceramic body is sintered to achieve the final desired properties. An important consideration for green sintering of the ceramic and the consequent final properties is the fraction of particles in the suspension. In general, the volume fraction of ceramic powders should be greater than 0.5 (Hinczewski et al. 1998). However, this decreases the depth of cure causing unfavorable effects on the object cohesion. Hence, the understanding of the rheological behavior of ceramic suspension is very important for SL process planning.

Microstereolithography

Microstereolithography is a new technique evolved from the conventional SL for manufacturing high-resolution, small, and complex 3D objects particularly for microengineering applications. The improvement in the resolution of microstereolithography is based on two broad approaches: vector-by-vector and integral.

In conventional vector-by-vector processes, a focused laser beam is deflected by scanning mirrors on the surface of the resin. This causes the continuous changes in the dimensions of the beam along the laser scan which is detrimental to the resolution of the built part. In the first approach of microstereolithography, focused laser beam is not deflected but kept static and precisely focused on the surface of the resin. The layer building is carried out by moving the object to be built along with its photoreactor (Bertsch et al. 2003).

In integral methods, a complete layer during part building is obtained by single light irradiation. The build time for these processes are drastically reduced compared to the conventional vector-by-vector process irrespective of the complexity of the parts. Also, polymerization of unwanted areas due to thermal effects can be minimized. The important component of the integral microlithography systems is the dynamic pattern generator. This dynamic pattern generator allows the shaping of the light beam such that it carries the image of the layer to be solidified which is subsequently reduced and focused on the surface of the resin with suitable optical system. Dynamic pattern generators can be based on LCD (liquid crystal display) or digital micrometer device (DMD). LCD panel consists of a large array of pixels (cells) and each pixel can be set into its transparent or opaque state by changing the orientation of the molecules it contains. When such LCD screens are placed in the optical path of light, the transmitted path can be modulated such that complex images corresponding to the cross sections of the part can be transferred to the surface of the resin for microstereolithography. The key issue is that the dynamic pattern generators are not compatible with UV light and commercial resins. Hence, most of the integral microstereolithography experiments were carried out on the photopolymerizable resins that can be sensitized by visible light (Bertsch et al. 2000, 2003, 2004). Figure 9.25 shows the schematic of the integral microstereolithography system developed at Swiss Federal Institute of Technology in Lausanne (EPFL). One of the examples representing the capabilities of the integral microstereolithography is presented in Fig. 9.26. The scale model of a car shown in the figure is manufactured in 673 layers of 5 μm thickness each and is built in about 3 h

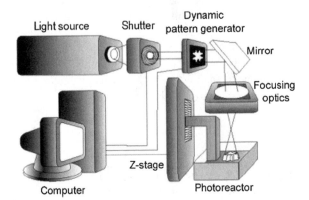

Fig. 9.25 Schematic of the integral microstereolithography system (Reprinted from Bertsch et al. 2000. With permission. Copyright Emerald.)

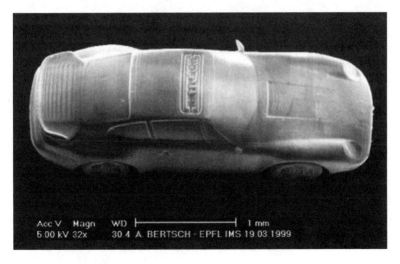

Fig. 9.26 Scale model of a small car fabricated by integral microstereolithography (Reprinted from Bertsch et al. 2000. With permission. Copyright Emerald.)

(Bertsch et al. 2000). Recently, the interests are further extended toward direct creation of the objects inside the resin without superimposing the layers. In such efforts, two-photon process and single-photon processes for undersurface polymerization can be cited (Bertsch et al. 2003)

9.4.2 Selective Laser Sintering

SLS is an RP process used for layered fabrication of 3D physical objects by controlled and selective fusing of powder material using a high-power laser. SLS was developed and patented at the University of Texas at Austin by Dr. Carl Deckard in the mid-1980s. The process was subsequently commercialized by DTM. In 2001, DTM was acquired by 3D Systems, which now leads the activities in the manufacture and sales of SLS systems and powder materials. Electro optical systems (EOS GmbH) of Munich, Germany, also manufactures and supplies SLS systems.

9.4.2.1 Selective Laser Sintering Process

As with most of the RP processes, the SLS process begins with the creation of a suitable CAD model of the object. The CAD data are tessellated into STL format which is the standard for the SLS systems. Various softwares are available which can import one or more STL files and determine the optimum build orientations such that overall build time of the objects is reduced. SLS systems allow the simultaneous fabrication of a number of parts.

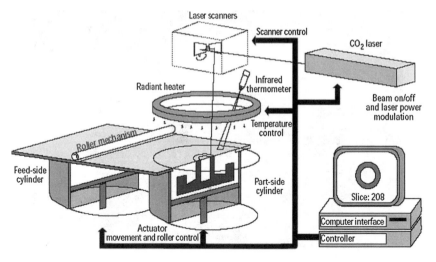

Fig. 9.27 Schematic of the SLS process (Reprinted from Subramanian et al. 1995. With permission. Copyright Emerald.)

The schematic of the apparatus for the SLS is shown in Fig. 9.27 (Subramanian et al. 1995). The important units of the machine are:

1. A build piston (build cylinder) where the part is built
2. One or more feed pistons (build cylinders) which supplies a controlled amount of powder for layer building
3. A roller/sweeper for uniform spreading of powder on the table
4. Laser system with its optics

At the start of the process, the build piston will be at the top position. During part building, the build piston lowers in increments corresponding to the thickness of each cross-sectional layer of the part. For the building of each layer, one of the feed pistons moves up and delivers a controlled amount of powder which is subsequently spread on the build area by a roller/sweeper. The powder is then preheated to the temperature just below the melting point of the material to minimize the thermal distortion and shrinkage after subsequent laser sintering. The laser is scanned on the surface of the powder bed corresponding to the cross section of the part to be built. The laser energy is sufficient to cause the sintering of the preheated powder by partial melting, or softening of the powder or its coatings. Once a complete layer of the powder is sintered and cooled, the build piston is lowered and the processes of powder spreading and laser scanning are repeated to build the subsequent layers of the part until the complete part is built. After the process is complete, the part is in the form of a sintered cake surrounded by the unsintered loose powder. The unsintered powder is removed from the part at the break-out station (BOS) by using brushes and the excess powder is recycled. Various finishing operations such as cleaning, polishing, grit blasting, and painting can be done to improve the surface finish.

The significant advantages associated with the SLS process can be summarized as:

Table 9.2 Specifications of some of the selective laser sintering systems available from 3D Systems (Courtesy of 3D Systems)

Model	Sinterstation Pro	Sinterstation HiQ
Maximum build volume, XYZ (mm³)	Pro 140: 550 × 550 × 460	HiQ: 381 × 330 × 457
	Pro 230: 550 × 550 × 750	HiQ+HS: 381 × 330 × 457
	70 Watt CO_2	HiQ: 30 Watt CO_2
Laser system		HiQ+HS: 50 Watt CO_2
	10	HiQ: 5
Scan speed (m/s)		HiQ+HS: 10
	STL	STL
Input data file		

1. No need of the support structures: The powder bed in the build area acts as a support for the overhanging parts. Thus, the powder bed in SLS setup is self-supporting without any need of building extra supports. This is also associated with the elimination of some of the postprocessing steps resulting in faster and cleaner process.
2. Possibility of building a number of parts: A number of STL files can be imported and correspondingly a number of parts can be built with individual optimum orientation.
3. Wide range of build materials: Finished parts can be produced by SLS using any type of build material which is produced in the powder form and sintered by thermal energy.

The specifications of various commercial SLS systems available from 3D Systems are given in Table 9.2.

9.4.2.2 Materials and Mechanisms of SLS

As mentioned in the previous section, SLS can be used for building parts with a wide range of materials. Perhaps, any powder material which is susceptible to be sintered by thermal energy can be used as build material for SLS. Various approaches of SLS for broad classes of build materials are explained below.

SLS of Metals and Ceramics

Metals and ceramics are the choice of materials for the prototypes when mechanical properties better than the polymers are desired. Approaches of SLS for these materials include various direct and indirect methods.

In direct SLS methods, a laser beam is used to directly sinter the metal and ceramic powders without addition of any binding material such as polymers. One of the direct SLS methods is based on the solid-state sintering (SSS) of powder. During SSS, the powder is heated to a temperature at which the fusion between the

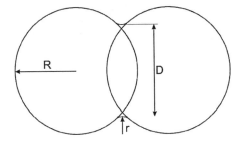

Fig. 9.28 Schematic of solid-state sintering mechanism (Reprinted from Kruth et al. 2005. With permission. Copyright Emerald.)

particles occurs without actual melting. This temperature generally lies between $T_m/2$ and T_m, where T_m is the melting point. The sintering process proceeds with the formation of the neck between the adjacent powder particles and subsequent fusion by reducing the surface energy between the particles (Fig. 9.28). The fusion process is driven by the diffusion of vacancies. Higher vacancy concentration at the curved neck and lower vacancy concentration at the flat surface establishes the vacancy gradient resulting in a flux of vacancies from the neck (and flux of atoms to the neck) thus increasing the size of neck. The role of SSS in the consolidation of powders has been demonstrated for metallic and ceramic materials. For Titanium powders, the degree of sintering is found to be dependant on the processing temperature employed for SSS. It was found that the degree of sintering increased significantly above α/β-transition temperature possibly due to high self-diffusion coefficient of β-phase compared to the α-phase (Kruth et al. 2005). The direct SLS process based on SSS of single component powders is generally a slow process and almost always associated with high porosity in the final built part (Agarwala et al. 1995).

Strong and dense metallic parts can also be fabricated by irradiating the powder with high-power laser such that selective melting of powder takes place. The process is also termed as selective laser melting (SLM). SLM characteristics of the various metallic materials such as aluminum, copper, iron, stainless steel, chromium, and nickel-based alloys have been studied using pulsed Nd:YAG laser with mean power of 50 W and maximum peak power of 3 kW. It was found that nickel-based powders exhibited the best densified condition after melting (Abe et al. 2001). Due to significant shrinkage associated with the solidification, the process is generally accompanied with formation of various defects such as cracks. The process can be improved by employing dual laser scanning system where additional CO_2 laser is used for reheating. Also, additional effects such as formation of spherical ball of material occur during SLM. When the temperature is close to the melting point, the viscosity and surface-tension effects cause the molten particles to aggregate and form larger spherical ball-type droplets. This effect is also aggravated due to larger beam size of the laser such that a number of particles melt simultaneously and form the larger droplets. These larger droplets can be fused with other droplets only at certain points resulting in high porosity in the built part (Kathuria 1999). Hence, the process parameters need to be optimized to minimize the effect of balling to achieve high-quality dense metallic parts. Figure 9.29 illustrates the effect of scanning speed on the morphology of single track obtained by SLM of M2 steel powder.

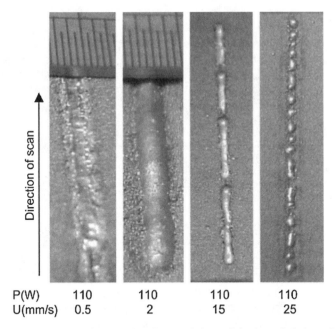

| P(W) | 110 | 110 | 110 | 110 |
| U(mm/s) | 0.5 | 2 | 15 | 25 |

Fig. 9.29 Effect of laser scanning speed on the morphology of single track during selective laser melting of M2 steel with laser beam diameter of 1.1 mm (Reprinted from Childs et al. 2005. With permission. Copyright Institution of Mechanical Engineers.)

The figure indicates that balling effects become predominant at higher scanning speeds (Childs et al. 2005).

Alternatively, the direct method of SLS can be used with two-component powders. The material consists of a mixture of two powders: a high melting point structural powder and a low melting point binder powder. During SLS process, the laser parameters are selected such that the temperature on the surface of powder bed reaches between the melting points of the binder and the structural powder. This causes the binder material to melt and flow through the pores between the structural powder particles assisted by various forces such as liquid pressure, viscosity, and capillary forces. In the case of direct SLS method with two-component powder mixtures, the densification of the powder layer is achieved by liquid phase sintering (LPS). Figure 9.30 presents the important stages in the LPS process. The LPS process goes through melting of low melting components, wetting, rearrangement, and densification (Kathuria 1999). In some of the powder systems, the laser beam irradiated may cause the SSS prior to the melting of low melting binder particles. LPS begins with the melting of binder powder particles and subsequent wetting of the solid structural powder particle. The capillary forces exerted by the liquid causes the rearrangement of the solid particles resulting in the densification of the compact. The rearrangement of the solid particles is further improved by the redistribution of liquid into the small pores. The rate of densification due to rearrangement of the solid particles is faster in the initial stage due to low viscosity of the liquid.

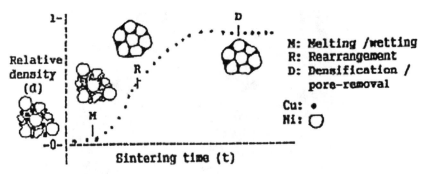

Fig. 9.30 Various stages in the liquid phase sintering of powder (Reprinted from Kathuria 1999. With permission. Copyright Elsevier.)

Eventually the viscosity increases resulting in slow densification by rearrangement of solid particles. The final stage of densification of the compacts is dominated by the effects dependent on the solid solubility in the liquid and diffusivity in the liquid. This final stage is also associated with the changes in the shape of the grains (Rahaman 2003). The important considerations in the two-component direct SLS method are powder particle sizes and the ratio of binding powder to the structural powder. Enhanced densification can be obtained with the binder particles smaller than the structural powder particles due to better wetting ability of smaller binder particles to the larger structural powder particles.

Extensive investigations of the SLS of metallic components based on LPS have been carried out. The typical combination of metals include low melting point metals such as tin (T_m ~232 °C) and copper (T_m ~1,083 °C) with high melting point metals such as iron (T_m ~1,540 °C) and nickel (T_m ~1,455 °C). The detailed studies on the SLS of Ni alloy and Fe structural powders using Cu as binder metals have been reported. In the case of Fe–Cu system, the LPS proceeds with the melting of low melting point Cu. Figure 9.31 shows the micrographs of the Fe–Cu system before and after SLS (Tolochko et al. 2003). However, when the melting points of the binder and the structural powders are comparable, the LPS proceeds by the melting of the material with high absorptance for the laser radiation. This is evident in the case of Ni alloy–Cu system where, LPS starts with the partial surface melting of Ni particles followed by dissolution of Cu and thus forming a molten layer around Ni particles (Fig. 9.32). Cu can also be coated on the surface of Ni alloy and Fe particle prior to SLS for uniform densification of the compacts (Tolochko et al. 2003). SLS by LPS mechanism has also been extended for the direct fabrication of hard metal parts which are used as inserts for stamping dies, deep drawing dies, and cutting tools. Various combinations of the powders investigated are WC–Co, TiC–Ni/Co/Mo, TiCN–Ni, TiB_2–Ni, ZrB_2–Cu, and Fe_3C–Fe (Wang et al. 2002).

In the indirect SLS methods, organic materials such as polymers or inorganic material which exhibit low viscosity upon heating are used as binder materials. The structural metallic or ceramics powders are mixed with or coated with thin layer of binder prior to laser sintering. During sintering, the laser energy is selected such

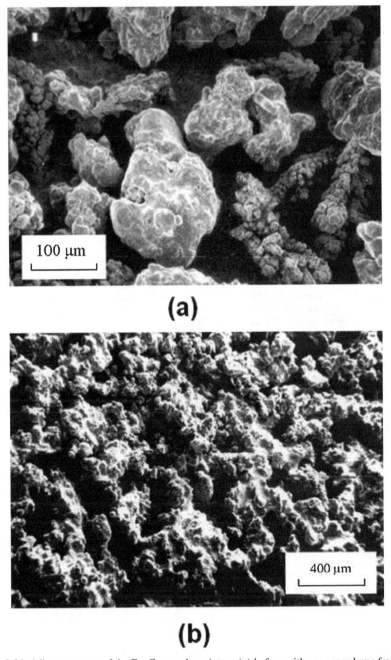

Fig. 9.31 Microstructure of the Fe–Cu powder mixture (**a**) before with copper volume fraction of 0.25 and (**b**) after selective laser sintering with laser energy density of 6.9 J/mm² and copper volume fraction of 0.5 (Reprinted from Tolochko et al. 2003. With permission. Copyright Emerald.)

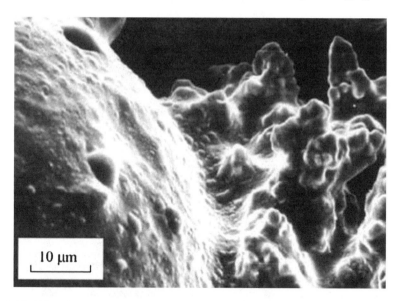

Fig. 9.32 SEM micrograph showing the surface melting of nickel particle and subsequent inter-action with copper particle during SLS of Ni alloy–Cu powder system (Reprinted from Tolochko et al. 2003. With permission. Copyright Emerald.)

that the binder material melts and forms a bond between the structural particles without actually melting them. The melting of the binder is exhibited by its high infrared absorption and low melting point (Kathuria 1999). The sintered parts obtained by the indirect SLS are highly porous (~45% porosity) and hence needs to be further densified to achieve the desired mechanical properties. Various post-processing approaches can be used for converting the green structures into highly densified finished parts. One of the approaches is to chemically convert the binder into a higher melting point component (Lee and Barlow 1993). For example, dense alumina ceramic compacts were successfully fabricated by using inorganic binders such as ammonium dihydrogen phosphate. During sintering, the binder melts and forms dense green part. Subsequent firing converts the binder to aluminum phos-phate (Lakshminarayana and Marcus 1991). Aluminum can also be used as binder for the alumina ceramic powders. The firing of such green parts converts the aluminum into alumina (Subramanian et al. 1995). The organic polymer binder can be thermally decomposed and eliminated from the system such that subsequent high temperature sintering or infiltration with suitable material provides highly dense structures (Vail et al. 1996). For example, Cu melts at 1,083 °C and can be used for infiltration of porous metallic structures after burning out the binder component of the green parts. Figure 9.33 shows the optical micrographs of the sintered SinterSteel metal part before and after Cu infiltration. The Cu infiltrated steel parts are characterized by very good tensile strength (475 MPa) and conductivity (185 W/m K) and can be easily finished by surface grinding, milling, drilling, polishing, surface plating, etc. (Pham et al. 1999). Recently, considerable interest has been directed toward infiltration of water suspended colloids such as colloids of

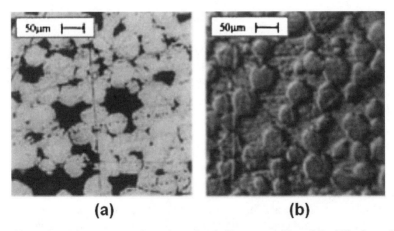

Fig. 9.33 Surface microstructure of (**a**) sintered and (**b**) copper infiltrated RapidSteel metal part (Reprinted from Pham et al. 1999. With permission. Copyright Institution of Mechanical Engineers.)

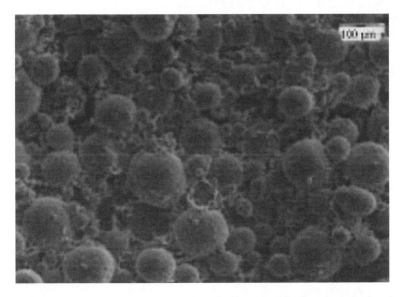

Fig. 9.34 Micrograph showing the fracture surface of infiltrated and fired part made from agglomerated 2 μm alumina (Reprinted from Subramanian et al. 1995. With permission. Copyright Emerald.)

alumina or silica in the polymer-bound ceramic compacts. After drying of the colloids, the ceramic compacts are subjected to firing to eliminate the polymer binder. Additional sintering step is generally required for rapid sintering of colloid. Figure 9.34 presents a fracture surface of a SLS part made from alumina agglomerated (from 2 μm alumina) particles infiltrated with alumina colloid and fired. Infiltration of alumina colloid into alumina ceramic parts tends to substantially improve the part strength during debinding and sintering (Subramanian et al. 1995).

Some ceramic parts can also be fabricated directly by SLS without the use of binders. This has been reported for SiC ceramics with chemically induced binding. When the SiC particles are heated to a very high temperature, it partially disintegrates into Si and C. The free Si thus formed tends to form SiO_2, which can act as a binder between the SiC particles. Thus the final part consists of a mixture of SiO_2 and SiC. The full densification of the parts can then be obtained by infiltrating it with Si (Klocke and Wirtz 1997).

SLS of Polymers

Polymeric materials are the most widely used material for SLS. Among the various types of polymeric materials, thermoplastic materials are the most suitable for SLS. Various types of thermoplastic polymeric materials for SLS include amorphous, semicrystalline and reinforced, or filled polymers. The thermal behavior of the amorphous, semicrystalline, and crystalline polymers can be represented by specific volume versus temperature plots (Fig. 9.35).

Amorphous polymers such as polycarbonates consist of long chain molecules which are arranged in a random manner. The important characteristic of the amorphous polymers is the glass transition temperature, T_g. During SLS, the amorphous polymers are preheated to the temperature slightly less than T_g such that

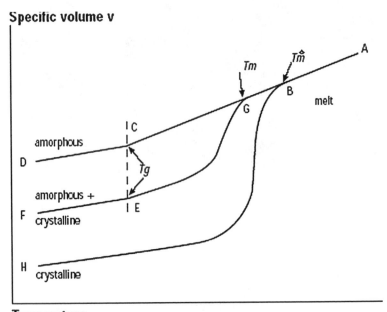

Fig. 9.35 Specific volume versus temperature behavior of amorphous, semicrystalline and crystalline polymers (Reprinted from Gibson and Shi 1997. With permission. Copyright Emerald.)

subsequent irradiation with the laser transforms the polymer from rigid glassy state to highly viscous liquid state by increasing the temperature above glass transition temperature (Gibson and Shi 1997; Childs et al. 1999). The parts fabricated from amorphous polymeric powders are generally characterized by good dimensional accuracy, feature resolution, and surface finish. However, the difficulties related with incomplete sintering may arise, thus limiting the application of SLS parts (Kruth et al. 2003).

Crystalline polymers consist of long chain molecules which are arranged in an orderly structure like nylon. The perfectly or fully crystalline polymers are not generally encountered in practice and most of the polymers in this class exist in semicrystalline state characterized by distinct glass transition and melting temperatures. During SLS, these semicrystalline polymers are preheated to the temperature slightly less than the melting point, such that powders can be rapidly melted and sintered by the application of relatively low thermal input from the laser beam (Nelson et al. 1993). SLS of semicrystalline powders is associated with highly dense parts with better mechanical properties, thus making them suitable for high-strength functional prototypes. However, semicrystalline polymers also exhibit large specific volume changes close to melting point resulting in significant shrinkage and distortion (Gibson and Shi 1997). The total shrinkage associated with the SLS of semicrystalline polymer powders is of the order of about 3–4% (Kruth et al. 2003).

Recently, significant interests have been directed toward the SLS of reinforced polymers. Polymer powder can be readily mixed with other materials to rapidly fabricate the composite parts with improved thermal and mechanical properties. Most of the work was concentrated on the glass reinforced polyamide and metal (Cu) reinforced polyamide. The Cu reinforced polyamide (Cu-PA) is used as high-strength material for inserts for injection molds (Kruth et al. 2003). Also, studies on fabrication of nanocomposites based on clay nanoparticles reinforced in nylon 6 have been reported (Kim and Creasy 2004). The utilization of SLS of reinforced polymers for advanced medical applications has been attempted. SLS can allow the direct fabrication of customized implants for solid and porous bone replacement and implants of controlled porosity for use as tissue scaffolds. For example, SLS offers a great potential for the fabrication of hydroxyapatite (HA) particulate reinforced high density polyethylene (HDPE) composites (HA-HDPE) which is a bioactive ceramic polymer composite material (Savalani et al. 2006). Figure 9.36 indicates the uniform distribution of HA particles in the rage of 3–10 µm in HDPE matrix after SLS.

9.4.2.3 Process Parameters in SLS

Interaction of the laser beam with the powder bed during SLS is a complex process and is influenced by a number of factors such as processing parameters, material properties, and geometry of the object. All these factors determine the amount of energy delivered into the surface of the powder bed. The final

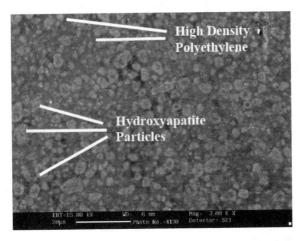

Fig. 9.36 Microstructure of hydroxyapatite (HA) reinforced high density polyethylene (HDPE) fabricated by selective laser sintering (Reprinted from Savalani et al. 2006. With permission. Copyright Institution of Mechanical Engineers.)

geometric properties (accuracy, extent of curling, distortion edge quality, etc.) and mechanical properties (density, strength, etc.) of the built part depend on this energy input to the surface of the powder bed. The schematic in Fig. 9.37 shows the various parameters which needs consideration during the design of SLS process. A typical scan pattern in the SLS process produced by partial overlapping of linear single scans is presented in Fig. 9.38 (Williams and Deckard 1998). The overlap, O, between the successive scan can be calculated from hatch spacing, h_s, and spot size, D, using equation:

$$O = 1 - \frac{h_s}{D}. \tag{9.6}$$

Due to overlap between the successive adjacent scans, the hatch spacing is always less than the laser beam radius and some points on the powder bed are exposed to multiple scanning. The total number of effective exposures is given by:

$$N_{eff} = \frac{D}{h_s} - 1. \tag{9.7}$$

Also, the delay time between successive irradiation exposures is given by:

$$t_d = \frac{L}{V}. \tag{9.8}$$

These geometric parameters combined with material properties and laser parameters determine the energy stored at the surface during SLS processing.

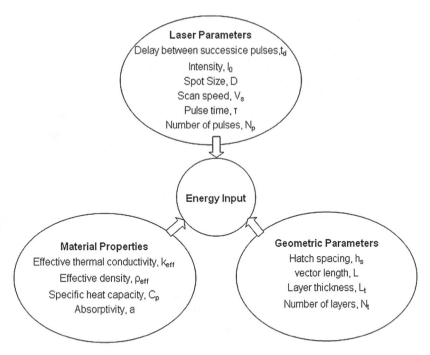

Fig. 9.37 Various geometric, laser, and material parameters influencing the energy input on surface of powder bed during SLS

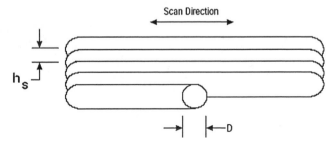

Fig. 9.38 Typical laser scan pattern in SLS (Reprinted from Williams and Deckard 1998. With permission. Copyright Emerald.)

9.4.2.4   Modeling of Selective Laser Sintering Process

SLS is based on sintering the preheated powder by irradiating it with laser beam. The process is carried out in a layered fashion such that each layer is built by scanning the laser according to the cross-sectional geometry of the part. The important physical processes relevant with the SLS are heat transfer and sintering of powder. Thus, the modeling of SLS primarily involves the submodels corresponding to heat transfer and sintering (Bugeda et al. 1999).

Thermal Submodel

Properties of the powder bed relevant for the thermal modeling of the SLS process differ significantly from the properties of the solid material. Hence, the first step in the thermal modeling of the SLS process involves the determination of the effective properties of the powder bed. The most important property in this context is the thermal conductivity. The thermal conductivity of the powder bed can vary from point to point across the powder bed depending on local porosity, condition of contacts between the particles, and local temperature. One of the approaches for calculating the effective thermal conductivity of the powder bed is based on a model which takes into account the conduction, convection, and radiation effects in the powder bed (Yagui and Kunni 1989). According to this model, the effective conductivity, k_{eff}, of the packed powder bed is given by:

$$k_{eff} = \frac{\mu_R \, k_s}{1 + \phi \dfrac{k_s}{k_g}}, \tag{9.9}$$

where μ_R is the solid fraction, k_s is the conductivity of solid material, k_g is the conductivity of air, and Φ is the empirical coefficient generally taken equal to $0.02 \times 10^{2(0.7-\mu R)}$. According to simplified Nelson's model, the effective conductivity of the packed powder bed is given by:

$$k_{eff} = k_s \left(\frac{\rho}{\rho_s} \right), \tag{9.10}$$

where ρ and ρ_s are the local powder density and solid density respectively (Berzins et al. 1996). In addition to the effective properties of the packed powder bed, the temperature dependence of the thermal properties such as conductivity and specific heat also needs to be established for improving the accuracy of the results obtained by thermal modeling of SLS process. For example, temperature dependence of specific heat, C_p, and solid conductivity, k_s, of Bisphenol-A polycarbonate is given by:

$$C_p(T) = 932 + 2.28T, \tag{9.11}$$

$$k_s(T) = 0.0251 + 0.005T, \tag{9.12}$$

where T is the temperature (Nelson 1993).

During SLS, the laser beam is scanned over the surface of the powder bed. The laser energy absorbed by the surface of the powder is either conducted into the material or lost through convection and radiation at the surface. The formulation of the heat transfer problem can be expressed in terms of classical transfer equation (Carslaw and Jeager 1959):

$$\rho C_p \frac{\partial T}{\partial t} = \nabla(k \nabla T) + g(x, y, x, t), \tag{9.13}$$

where the term on the left represents the energy storage and the terms on the right represent the sum of the heat due to 3D conduction and volumetric heat generated by the internal source.

Two boundary conditions at the top surface ($z = 0$) and bottom surface ($z = zp$) of the powder bed can be expressed as (Williams and Deckard 1998):

$$1. \text{At } z = 0 : -k \left(\frac{\partial T}{\partial z} \right)_{z=0} = \varepsilon \sigma \left(T_{z=0}^4 - T_{suf}^4 \right) + h \left(T_{z=0} - T_{env} \right), \qquad (9.14)$$

$$2. \text{At } z = z_p : -k \left(\frac{\partial T}{\partial z} \right)_{z=z_p} = 0, \qquad (9.15)$$

where ε is emissivity, σ is Stefan–Boltzmann constant, h is convective heat transfer coefficient, and T_{sur} and T_{env} are the temperatures of the surroundings and the environment respectively. The first boundary condition represents the heat loss at the surface through radiation and convection, whereas the second boundary condition represents that no heat is lost from the bottom surface of the powder bed.

The initial condition of uniform temperature distribution throughout the powder bed prior to sintering can also be applied as:

$$T(x, y, x, 0) = T_0 \qquad (9.16)$$

The solution of this problem gives the spatial and temporal temperature distribution throughout the powder bed. This temperature data can be further used for calculating the depth of sintering and the cooling rate during sintering process.

Sintering Submodel

Early models of sintering are based on the assumption of Frenkel that the energy dissipated in viscous flow is equal to the energy gained by the decrease in surface area during densification (Frenkel 1945). Based on this assumption, Mackenzie and Shuttleworth proposed the sintering model for viscous body containing closed spherical pores of identical sizes (Fig. 9.39). The rate of densification is given by:

$$\frac{d\rho}{dt} = \frac{3\gamma}{2a\eta} (1 - \rho), \qquad (9.17)$$

where γ is the particle surface energy, a is the radius of spherical pore, and η is the viscosity (Mackenzie and Shuttleworth 1949). Since most of the sintering beds have very high initial open porosity, the Mackenzie and Shuttleworth's model assuming closed spherical pores is generally applied for the last stages of sintering where the relative density is greater than 0.94.

Recent models of sintering for low density materials have been proposed (Scherer 1977a, b, c). Scherer's sintering model is based on the cubic array of intersecting cylinders assuming that the cylinder radius corresponds to the average particle radius in the preform (Fig. 9.40). During sintering, the cylinder heights

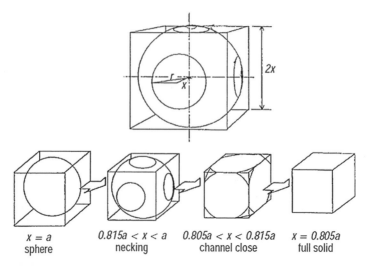

Fig. 9.39 Schematic of closed spherical pore model of sintering proposed by Mackenzie and Shuttleworth (Reprinted from Bugeda et al. 1999. With permission. Copyright Emerald.)

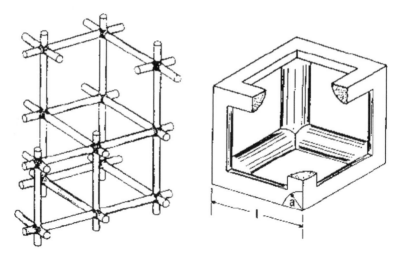

Fig. 9.40 Scherer model of sintering based on an array of intersecting cylinders with radius equal to the radius of particle (Reprinted from Scherer 1977a. With permission. Copyright American Ceramic Society.)

collapse until the cylinder walls touch each other and the cell contains a closed pore. The relative density at this final stage is 0.94. Beyond this density, the structure cannot be explained by an array of intersecting cylinders. Hence, Scherer's model is best suited for the initial stages of densification, whereas the final stages of densification are best explained by the Mackenzie and Shuttleworth model.

In the Scherer's geometric model, the amount of energy dissipated in the viscous flow as the height of cylinders decreases can be expressed as (Scherer 1977a):

$$\dot{E}_f = \left(\frac{3\pi\eta r^2}{h}\right)\left(\frac{dh}{dt}\right)^2, \tag{9.18}$$

where r and h are radius and height of the cylinder respectively. The energy gained by the decrease in surface area can be expressed as:

$$\dot{E}_s = \gamma\left(\frac{dS_c}{dt}\right), \tag{9.19}$$

where S_c is the area of single full cylinder. Equating the energy dissipated in the viscous flow and the energy gained by the decrease in surface area based on Frenkel's assumption leads to:

$$\frac{dx}{dt} = \left(\frac{\gamma}{2\eta}\right)\left(\frac{1}{l}\right), \tag{9.20}$$

where l is the length of the side of the cell and x is equal to a/l. Introducing the density of the cell, ρ, in the above equation gives

$$l(t) = \frac{l_0(\rho_0/\rho_s)^{1/3}}{(3\pi x^2 - 8\sqrt{2}x^3)^{1/3}}, \tag{9.21}$$

where ρ_0 and l_0 are initial values of ρ and l respectively. ρ_s is the theoretical density of solid phase. Substituting Eq. (9.21) in Eq. (9.20) and integrating leads to:

$$K(t - t_0) = \int_0^x \frac{2dx}{(3\pi - 8\sqrt{2}x)^{1/3} x^{2/3}}, \tag{9.22}$$

where, t_0 is the fictitious time at $x = 0$. K is given by

$$K = \left(\frac{\gamma}{\eta l_0}\right)\left(\frac{\rho_s}{\rho_0}\right)^{1/3}. \tag{9.23}$$

The Eq. (9.22) gives the variation of distance, x with time, t from which the density of the cell as a function of time can be calculated. These equations were success-fully applied to the sintering of glass (Scherer 1977a).

9.4.2.5 Advances in Selective Laser Sintering Process

Selective Laser Sintering/Hot Isostatic Pressing (SLS/HIP)

SLS/HIP is a hybrid direct free-form fabrication technique being developed at the University of Texas for high performance materials (Das et al 1998a, b). The SLS/HIP process is associated with the advantages of both the techniques. SLS facilitates the

SLS/HIP

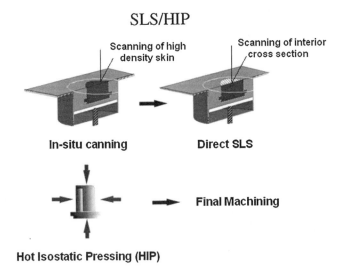

Fig 9.41 Schematic of SLS/HIP process (Reprinted from Das et al. 1998b. With permission. Copyright Emerald.)

rapid free-form fabrication of the part which can subsequently be fully densified by HIP. The schematic of the processing steps in SLS/HIP are presented in Fig. 9.41. The initial steps of the fabrication involve the SLS of the metal powder by scanning a laser beam on the surface of the powder bed. The laser scanning is carried out in such a way that boundary of the part is sintered to a very high density (>98% of theoretical density) to a thickness of around 1–3 mm, whereas the interior of the part is sintered to a density typically exceeding 80% of the theoretical density. The high-density integral boundary of the part acts as an impermeable "can." Finally the encapsulated part is evacuated, sealed, and postprocessed by HIP (Das et al. 1998a).

Various advantages of the SLS/HIP can be listed as:

1. There is no need of secondary canning process since the integral can is produced from the same material as the part during SLS.
2. Full densification of the part can be achieved by the HIP of the SLS part.
3. Preprocessing steps in the fabrication of container and postprocessing steps for container removal are eliminated.
4. Adverse interactions between the container and powder are eliminated.

The process has been successfully demonstrated for the fabrication of parts using two high performance materials: 625 superalloy and Ti-6Al-4V. The SLS/HIP technique with Ti-6Al-4V was used for the fabrication of guidance section housing for the AIM-9 Sidewinder missile (Fig. 9.42). It was found that the microstructure and mechanical properties of the SLS/HIP process were comparable with the conventionally processed material along with additional advantages associated with SLS/HIP (Das et al. 1998a).

Fig. 9.42 A titanium guidance section housing for a Sidewinder missile (Reprinted from Das et al. 1998a. With permission. Copyright The Minerals, Metals, and Materials Society.)

9.4.3 Laminated Object Manufacturing (LOM)

LOM is an RP technology in which the models are built by unique additive and subtractive processes. The process involves layered fabrication in which an individual layer is created by laser cutting of adhesive bonded sheets to the dimensions corresponding to each cross section of the part to be built (Feygin and Hsieh 1991). The process is used for building 3D parts from high-quality paper, plastic, and composite materials. LOM was developed by Michael Feygin of Helisys and the technology was commercialized in 1991. Recently, Helisys was succeeded by Cubic Technology and is leading the manufacturing and marketing of LOM machines.

9.4.3.1 Laminated Object Manufacturing Process

The LOM process begins with the creation of standard files such as STL files for input to the LOM machine. Subsequently, the STL file is sliced into a number of layers corresponding to the cross sections of the part to be built. The important consideration is the determination of optimum orientation for minimizing the build time and better accuracy of the curved surfaces. The schematic of the LOM process is given in Fig. 9.43. Various important units of the LOM machine are:

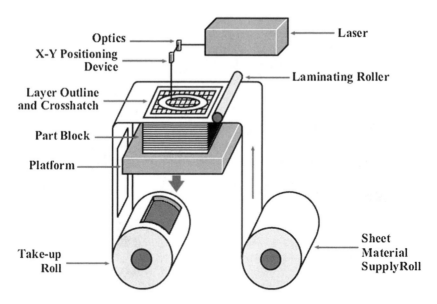

Fig. 9.43 Schematic of the Laminated Object Manufacturing process (Courtesy Cubic Technologies.)

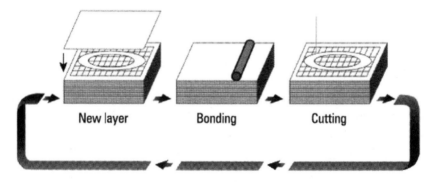

Fig. 9.44 Sequence of part building in LOM (Courtesy Cubic Technologies.)

1. Sheet material supply roll
2. Take-up roll
3. Elevator platform
4. Hot laminating roller
5. Laser system and associated optics

During LOM process, the sheet material supply roll feeds a necessary amount of paper across the platform. The hot laminating roller is then rolled along the surface of the paper. The roller activates the thermally sensitive adhesive on the surface of the paper and also applies downward pressure on the paper. This facilitates the bonding of the paper to the previous layer. A laser is used to cut the outline of each cross section of the part in the paper (Fig. 9.44). To prevent the adhesion of the excess paper to the parts, uniform crosshatches are scribed in the excess paper such that its easy removal can be facilitated after the completion of part building. At the end of the layer building, a rectangular cut is made in the paper which encompasses the excess material. After building each layer, the elevator platform is lowered corresponding to the thickness of the layer and the process of part building is repeated. As the process of building layers continues, the crosshatched excess material acts as a support structure for the subsequent layers having overhanging parts. The laser parameters are selected such that only the thickness of each subsequent layer is cut without damaging the previously built layers (Cooper 2001).

Various processes based on LOM have been developed for specific purposes. One such process is the paper lamination technology (PLT) developed by Kira, Japan. The major differences between the two techniques lie in the material being used and the method of cutting the outline in the sheet of material for building each layer. In PLT, a computerized knife is used instead of laser beam to cut the cross-sectional outline of object in each layer. The process consists of printing the cross sections of the part on the paper. The process begins with the placement of a paper on the build platform which is subsequently bonded using a hot roller. Next, a computer-driven knife cuts the outline of the cross section of the part in the layer and also creates the crosshatches in the excess material. The process is repeated till the complete building of the object. Finally, the excess material is removed (decubing) using a wooden chisel (Pham and Dimov 2001; Cooper 2001).

Various specific advantages of the LOM technique can be listed as (Mueller and Kochan 1999):

1. Large parts (500 × 800 × 500 mm) can be built by LOM with relative ease.
2. LOM parts are characterized by low internal tension and consequent low distortion, shrinkage, and deformation.
3. Highly durable parts with low brittleness and fragility can be used.
4. Variety of organic and inorganic materials with widely differing properties can be used for part building.
5. Materials used for LOM are generally nontoxic and nonreactive and hence easy to handle and dispose.
6. LOM systems are generally lower in cost that other RP systems.

However, the process is also associated with some inherent disadvantages such as delamination of layers, dimensional errors in z-direction and labor-intensive 'decubing' (waste-removal) process.

Specifications of various commercially available LOM systems from Cubic Technologies are presented in Table 9.3.

9.4.3.2 Process Parameters in Laminated Object Manufacturing

LOM process needs to be optimized for achieving minimum build time and waste material, and good quality parts without any debonding or delamination between the layers. Various important process parameters for achieving such objectives are

Table 9.3 Specifications of some of the laminated object manufacturing systems available from Cubic Technologies (Courtesy of Cubic Technologies)

Model	LOM-1015Plus	LOM-2030H
Maximum part envelop, LWH (mm^3)	381 × 254 × 356	813 × 559 × 508
	32	204
Maximum part weight (kg)	25 W CO_2 laser	50 W CO_2 laser
Laser system	0.20–0.25	0.203–0.254
Laser beam diameter (mm)	0.08–0.25	0.076–0.254
Material thickness (mm)	Roll width: 256	Roll width: 711
Material size (mm)	Roll Diameter: 256	Roll Diameter: 711
	Windows-based	Windows-based
System software	LOMSlice	LOMSlice
	Temperature: 20–27 °C	Temperature: 20–27 °C
Facility requirements	Humidity: <50%	Humidity: <50%

crosshatch size, left and right heater (roller) margins, heater speed and temperature, and laser power and speed (Park et al. 2000).

Size of the crosshatches primarily influence the build time and ease of decubing, i.e., removal of extra materials at the end of the LOM process. If the size of the crosshatches is very small (typically less than 6 mm), then the extra material can be removed easily. However, this will increase the build time substantially due to the longer path that the laser needs to travel for complete crosshatching of the excess material. The distances traveled by the hot roller past the left and right support boundaries of the part are referred to as left and right roller margins. Longer boundaries result in better bonding at the part edge, however, build time may substantially increase. The heater speed and temperature influence the lamination of sheets during LOM. For obtaining better interlaminar strength, the heater speed must be slow enough to form a good bond between the layers. Heater speed also influences the surface temperature of the sheets. Hence, for higher heater speed, the heater temperature should be sufficiently high to cause better bonding. Laser power and speed are selected such that the outline of the part is cut in a single sheet of build material. Thus, the power and speed depends on the type of build materials and its thickness. For instance, single layer LPH 042 paper is generally cut using laser powers of 5–11% (1.4–3.1 W) of maximum power and laser cutting speed of 10.16–20.32 cm/s using LOM 1015 machine (Park et al. 2000). Additional parameters such as material advance margin, support wall thickness, and compression also need to be considered for fast building and better part quality (Cooper 2001).

Due to the complexity of the process parameters involved in the LOM process, comprehensive understanding of the process and the effects of the process parameters is required to achieve desired quality of the part and overall productivity of the process. In this context, thermomechanical modeling of the process instead of trial-and-error approaches produces more comprehensible results. One of such modeling efforts analyzed the states of stress and temperature during LOM using finite element method. The numerical solutions of the model resulted in the influence of process parameters such as roller temperature, velocity, and indentation on the temperature and stress distribution in the laminates. Some of the results of this analysis are (Sonmez and Hahn 1998):

1. Normal stresses at the interface induced by the roller decays to zero while the adhesive is still above the glass transition temperature. In order to facilitate the adhesion, the pressure needs to be maintained at the interface until the glass transition is reached. This may be done by the application of the secondary roller.
2. Better adhesion can be obtained by using larger roller diameters. Small diameter rollers tend to cause concentrated stress distribution at the interface, thus allowing shorter times for bonding.
3. Thin laminates show higher stresses. Uniform distribution of stresses at the interfaces can be obtained by controlling the roller indentation.
4. There is an upper limit of the roller speed above which bonding is not possible.

9.4.3.3 Laminated Object Manufacturing of Ceramics and Composites

Traditionally, the LOM process has been extensively used for the rapid fabrication of parts by sequential cutting, stacking, and bonding of papers. Due to increasingly stringent requirements of properties for the prototype, increased interests have been directed toward extending the capabilities of LOM for fabricating the prototypes from advanced materials such as ceramics and composites. The basis of using LOM for composite fabrication is in the capability of this process for handling flat-sheet materials such as tapes and prepregs, for producing geometrically complex objects and for operating with a high degree of automation.

In this context, the fabrication of both particulate and fiber reinforced polymer matrix composites (PMCs) and ceramic matrix composites (CMCs) have been demonstrated using LOM. For example, the starting material for fabricating SiC particulate ceramic composites using LOM consisted of the SiC tapes (~250 µm thick) fabricated by standard tape-casting processes. The typical composition of the SiC tape was made of bimodal SiC powder (60 and 2–3 µm sizes), carbon black and graphite powder, and polymer binder. During LOM, the ceramic tapes are laminated using standard roller which causes the melting of the binder resulting in adhesion between the layers. Subsequently, a CO_2 laser cuts the ceramic tapes corresponding to the cross section of the object. Various postprocessing steps such as binder burn-out and reaction sintering are required. During reaction sintering, liquid Si is infiltrated into the porous SiC parts. Si reacts with C to form SiC which along with the free Si forms the matrix for the original SiC particles. Such SiC components were found to be uniform without evidences of layered fabrication and also characterized by very low porosity and good silicon infiltration (Klosterman et al. 1998, 1999). Also, a modified LOM process was developed for the fabrication of a curved surface of fiber reinforced PMCs such that continuity of the fibers is retained. The process consists of fabricating a matched mandrel by conventional LOM.

Fig. 9.45 Monolithic armor panel placed on the curved mandrel after curved layer LOM and decubing (Reprinted from Klosterman et al. 1999. With permission. Copyright Emerald.)

The sheets of the desired material generally, CMC prepreg are transported and laminated on the curved mandrel using flexible thermoforming mechanism. The cutting of the outline of the part is done using a focused laser beam. Figure 9.45 shows a monolithic armor panel placed on the curved mandrel after curved layer LOM and decubing (Klosterman et al. 1999).

Recently, a CAM-LEM process has been developed for the fabrication of geometrically complex shapes from the green ceramic tapes. The process involves laser cutting of outlines of the part cross sections. These cross-sectional slices are extracted from the sheets and then stacked and bonded separately. Finally, the parts are processed using conventional binder-removal and sintering procedures (Cawley et al. 1996).

9.4.4   Laser Engineered Net Shaping (LENS)

LENS is an RP method being developed at Sandia National Laboratory and is based on powder forming technology. In addition to the fabrication of near net-shape prototypes, the process can also be used for direct fabrication of fully dense metal parts directly from the CAD models.

9.4.4.1   LENS Process

The LENS system is shown in Fig. 9.46 (Ye et al. 2006). The important units are a high-power Nd:YAG laser, a three-axis computer-controlled positioning system, a powder feed unit (deposition head), and a controlled atmosphere glove. During LENS process, a substrate is used for building the metallic part. Part building starts with the creation of a melt pool on the substrate by focusing the laser beam and simultaneously feeding the powder particles in the melt pool. A thin cross section of the part is then built on the substrate by moving the substrate in X-plane and Y-plane beneath the laser. Each layer is created by scanning the border of the cross-sectional area and the appropriate fill pattern inside the area. The fabrication of successive layers is facilitated by incrementing the powder delivery nozzle and focusing lens assembly in the vertical direction corresponding to the thickness of the previous layer such that 3D objects are created. After part building is completed, the part is mechanically removed from the substrate plate. The important consideration during deposition process is the uniformity of the distribution of powder fed around the laser beam. Powder is generally delivered under gravity or pressurized carrier gas through the deposition head coaxial to the focus of the laser beam. However, designs with off-axis powder delivery nozzles are also possible. Inert gas such as argon is often used to prevent the melt pool from contamination during part building and thus ensures better adhesion between the layers.

Another similar technique for direct metal deposition is directed light fabrication (DLF) being developed at Los Alamos National Laboratory. The process involves the fusing of metal powders delivered by gas into the focal zone of high-power laser

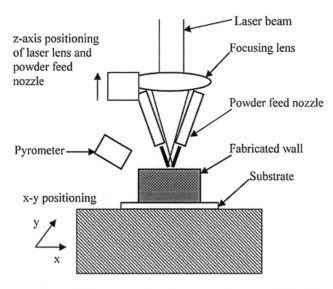

Fig. 9.46 Schematic of LENS process (Reprinted from Ye et al. 2006. With permission. Copyright Elsevier.)

Table 9.4 Specifications of some of the commercially available laser engineered net shaping systems available from Optimec (Courtesy of Optomec)

Model	LENS 750	LENS 850-R
Work envelop (mm³)	300 × 300 × 300	900 × 1500 × 900
Motion control	3-axes:	5-axes:
	XY linear table motion	XYZ linear gantry motion
	Z–gantry motion	Tilt–Rotate worktable
Laser system	500 W Nd:YAG laser	1 kW or 2 kW IPG fiber laser
Positional accuracy (mm)	±0.25	±0.25
Linear resolution (mm)	±0.025	±0.025
Motion velocity (mm/s)	60	60
Deposition rate (kg/hr)	Up to 0.5	Up to 0.5
Powder feeder	One feeder: 14 kg powder holding capacity	Two feeder: 14 kg powder holding capacity each

beams. The process has successfully demonstrated the fabrication of geometrically complex fully dense metal parts such as hemispherical shapes using five-axis DLF machines (Lewis 1994; Milewski et al. 1998).

Specifications of some of the commercially available LENS systems available for Optomec are presented in Table 9.4.

9.4.4.2 Materials for LENS

As mentioned in the introduction of LENS, the process can be applied to the fabrication of parts from a variety of metals and alloys. This includes stainless steels (304 and 316), tool steels, iron–nickel alloy, nickel aluminide, titanium alloys, Inconel, etc. Rapid solidification during the LENS process has significant influence on the microstructure and properties of the component. Hence during fabrication of metallic components by LENS, various thermal parameters such as cooling rates and temperature gradients need to be evaluated. Extensive thermal modeling of the LENS process for AISI 316 stainless steel has indicated that the temperature gradient at the molten pool is of the order of 5×10^2 K/mm. Also, the temperature shows sharp changes around the molten pool and decreases away from the molten region (Ye et al. 2006). It has been observed that Ni–Mo alloys deposited from the blend of 75% elemental nickel and 25% elemental molybdenum powders exhibited rapidly solidified microstructures characterized by dendritic morphology and single FCC phase with short-range chemical ordering at ¼ < 4 2 0 > type. The absence of equilibrium and metastable phases in the microstructures suggest the very high cooling rates associated with LEN process (Banerjee et al. 2003).

Also, the metallic parts can be fabricated in graded composition by using elemental powder blends in the system with multiple hoppers. Figure 9.47 shows the x-ray patterns of the TiC/Ti composite fabricated by LENS with compositions changing from pure α-Ti in the bottom layer to approximately 95% TiC in the top

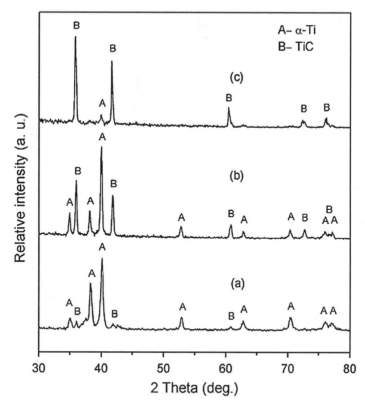

Fig. 9.47 XRD patterns from (**a**) bottom, (**b**) intermediate, and (**c**) top layer of Ti-TiC composite fabricated by LENS in graded composition (Reprinted from Liu and DuPont 2003. With permission. Copyright Elsevier.)

layer. This graded composition was fabricated by delivering the constituent materials from different powder feeders in a controlled manner. Such functionally graded materials also exhibit the distribution of microstructure and properties along the length of the specimen. In TiC/Ti graded composites with TiC ceramic at one end and titanium metal at the other end offers unique advantages such as smooth transition of thermal stresses along the thickness direction and minimized stress concentration at the other end (Liu and DuPont 2003).

9.5   Applications of Rapid Prototyping Processes

RP technologies have found applications in many fields for automated fabrication of geometrically complex parts using a broad variety of material. The techniques have evolved from the traditional fabrication of prototypes to the direct fabrication of

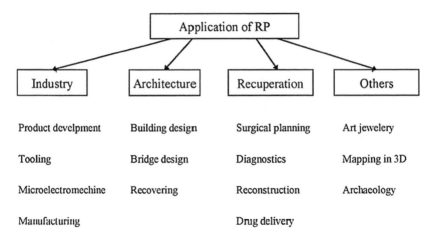

Fig. 9.48 Various applications of rapid prototyping processes (Reprinted from Kochan et al. 1999. With permission. Copyright Elsevier.)

parts. Various applications of the RP technologies are shown in Fig. 9.48 (Kochan et al. 1999). Some of these applications are discussed in the following sections.

9.5.1   Applications in Rapid Prototyping

RP processes have been extensively used in the fabrication of prototypes. Various types of prototypes fabricated by RP processes can be classified as: visual, functional, material, and production prototypes. Visual prototypes are simple solid models primarily used for revealing design defects; functional prototypes are form-and-fit-type models used for detecting assembly flaws; material prototypes are designed to have the same geometric and mechanical properties as the final part; and production prototypes are fabricated by the same processes as the final part (Karapatis et al. 1998). SLS is widely used for fabricating prototypes which can be subjected to various functional testing. These prototypes are essentially made of Nylon-based materials. Prototypes can also be subjected to the various engineering analysis such as photoelastic stress analysis, thermoelastic tension analysis, flow pattern analysis. Such analyses of the RP models facilitate the testing of physical models in a much early stage of development (Pham and Dimov 2001). Also, RP processes such as SLA and SLS are extensively used in the fabrications of patterns for investment and vacuum casting. During investment casting, RP pattern is immersed in ceramic slurry to form the shell around the pattern. The lost pattern is burntout such that a hollow cavity is formed which can be subsequently filled with molten material for creating high sound casting. Porous patterns are better suited for investment casting and it minimizes the chances of shell cracking during burnout of the pattern. Dense patterns can be efficiently used for vacuum casting (Ferreira and Mateus 2003).

9.5.2 Applications in Rapid Tooling

RP processes play an important role in the direct fabrication of manufacturing tools such as dies and molds. RP in combination with rapid tooling allows the faster development of the product. Various approaches for rapid tooling are soft, hard, and bridge tooling. SLS and LENS are capable of direct fabrication of the metallic tooling which can be used for injection molding of plastic parts. Important considerations for use of rapid tooling in these applications are the strength of the tools and the achievable surface finish of final parts. SLS tools need to be postprocessed prior to its use as tools. One of the benefits of rapid tooling is in the fabrication of molding inserts for the design of cooling lines, ejector pin guides, gates, and runners (Pham and Dimov 2001).

9.5.3 Applications in Rapid Manufacturing

Apart from the direct fabrication of prototypes and toolings, RP processes can also be used in the direct fabrication of metal, cermet, and ceramic components especially for low-volume production. For example, SLS and LENS have been demonstrated for the direct free-form fabrication of components from metal (bronze–nickel) and metal–ceramic (tungsten carbide–cobalt–nickel) material systems. The density of SLS parts is typically around 80% of the theoretical densities. Further improvements (up to 95% of theoretical density) in the densities can be achieved by high-temperature postprocessing operations sintering (Das et al. 1998).

9.5.4 Medical Applications

RP is increasingly finding applications in medical fields. This is derived from the successful collaborations between the engineers and the medical practitioners. The major step in such efforts was the interfacing of the data from CT and MR images to the RP machines. Research is still ongoing for improving resolution of the CT data. Major areas of applications in the medical sector can be listed as:

1. Implantable prostheses: RP techniques can be used to fabricate the geometrically accurate models of the human organs directly from CT-scan data after appropriate preprocessing. The models are then used as patterns for manufacturing the custom implants from biocompatible materials. For example, SL techniques have been demonstrated for the fabrication of skull-defect model. Such models were then used as negative surfaces for final cranial prosthesis (Webb 2000).
2. Surgical planning: RP models of the deformities allow the surgeons to plan the corrective surgery and predict the appearance of the outcome for the first time. This also enables the surgeon to design the accurate corrective devices.

Some of the studies have indicated that SLA model reveals better details of the defects than the CT image, thus helping in planning the corrective surgery (Webb 2000).

3. Educational models: RP models of the defects can also be used for medical training of the students.

References

Abe F, Osakada K, Shiomi M, Uematsu K, Matsumoto M (2001) The manufacturing of hard tools from metallic powders by selective laser melting. Journal of Materials Processing Technology 111:210–213.

Agarwala M, Bourell D, Beaman J, Marcus H, Barlow J (1995) Direct selective laser sintering of metals. Rapid Prototyping Journal 1:26–36.

Banerjee R, Brice CA, Banerjee S, Fraser HL (2003) Microstructural evolution in laser deposited Ni-25at. % Mo alloy. Materials Science and Engineering A 347:1–4.

Bertsch A, Bernhard P, Vogt C, Renaud P (2000) Rapid prototyping of small size objects. Rapid Prototyping Journal 6:259–266.

Bertsch A, Jiguet S, Bernhard P, Renaud P (2003) Microstereolithography: a review. Materials Research Society Symposium Proceedings 758:LL1.1.1–LL1.1.13.

Bertsch A, Jiguet S, Renaud P (2004) Microfabrication of ceramic components by microstereolithography. Journal of Micromechanics and Microengineering 14:197–203.

Berzins M, Childs TH, Dalgarno KW, Ryder GR, Stein G (1996) The selective laser sintering of polycarbonate. Annals of CIRP 45:187–190.

Bugeda G, Cervera M, Lombera G (1999) Numerical prediction of temperature and density distributions in selective laser sintering processes. Rapid Prototyping Journal 5:21–26.

Carslaw HS, Jaeger JC (1959) Conduction of heat in solids. Oxford University Press, Oxford.

Cawley JD, Heuer AH, Newman WS, Mathewson BB (1996) Computer-aided manufacturing of laminated engineering materials. Amercan Ceramic Society Bulletin 75:75–79.

Chandru V, Manohar S, Prakash CE (1995) Voxel-based modeling for layered manufacturing. IEEE Computer Graphics and Applications 15:42–47.

Childs TH, Berzins M, Ryder GR, Tontowi A (1999) Selective laser sintering of an amorphous polymer-simulations and experiments. Proceedings of the Institution of Mechanical Engineers B: 213:333–349.

Childs TH, Hauser C, Badrossamay M (2005) Selective laser sintering (melting) of stainless and tool steel powders: experiments and modeling. Proceedings of the Institution of Mechanical Engineers B: Journal of Engineering Manufacture 219:339–357.

Colton J, Blair B (1999) Experimental study of post-build cure of stereolithography polymers for injection molds. Rapid Prototyping Journal 5:72–81.

Cooper KG (2001) Rapid prototyping: selection and application. Marcel Dekker, New York.

Corcione CE, Greco A, Maffezzoli A (2003) Photopolymerization kinetics of an epoxy-based resin for stereolithography. Journal of Thermal Analysis and Calorimetry 72:687–693.

Das S, Wohlert M, Beaman JJ, Bourell DL (1998a) Producing metal parts with selective laser sintering/hot isostatic pressing. JOM 50:17–20.

Das S, Beaman JJ, Wohlert M, Bourell DL (1998b) Direct laser freeform fabrication of high performance metal components. Rapid Prototyping Journal 4:112–117.

Decker C (1999) High-speed curing by laser irradiation. Nuclear Instruments and Methods in Physics Research B 151:22–28.

Dufaud O, Marchal P, Corbel S (2002) Rheological properties of PZT suspensions for stereolithography. Journal of the European Ceramic Society 22:2081–2092.

Dutta D, Prinz FB, Rosen D, Weiss D (2001) Layered manufacturing: current status and future trends. Journal of Computing and Information Science in Engineering, Transactions of the ASME 1:61–71.

Ferreira JC, Mateus A (2003) A numerical and experimental study of fracture in RP stereolithography patterns and ceramic shells for investment casting. Journal of Materials Processing Technology 134:135–144.

Feygin M, Hsieh B (1991) Laminated object manufacturing: A simpler process. Proceedings of Solid Freeform Fabrication Symposium, Austin, Texas.

Frenkel J (1945) Viscous flow of crystalline bodies under the action of surface tension. Journal of Physics 9:385–391.

Fuh JY, Choo YS, Nee AY, Lu L, Lee KC (1995) Improvement of the UV curing process for the laser lithography technique. Materials and Design 16:23–32.

Fuh JY, Lu L, Tan CC, Shen ZX, Chew S (1999) Curing characteristics of acrylic photopolymer used in stereolithography process. Rapid Prototyping Journal 5:27–34.

Gibson I, Shi D (1997) Material properties and fabrication parameters in selective laser sintering process. Rapid Prototyping Journal 3:129–136.

Giannatsis J, Dedoussis V, Laios L (2001) A study of the build-time estimation problem for Stereolithography systems. Robotics and Computer Integrated Manufacturing 17:295–304.

Hinczewski C, Corbel S, Chartie T (1998) Ceramic suspensions suitable for stereolithogiaphy. Journal of the European Ceramic Society 18:583–590.

Huang YM, Jiang CP (2003) Curl distortion analysis during photopolymerisation of stereolithography using dynamic finite element method. International Journal of Advanced Manufacturing Technology 21:586–595.

Huang YM, Jeng JY, Jiang CP (2003) Increased accuracy by using dynamic finite element method in the constrain-surface stereolithography system. Journal of Materials Processing Technology 140:191–196.

Karapatis NP, van Griethuysen JP, Glardon R (1998) Direct rapid tooling: a review of current research. Rapid Prototyping Journal 4:77–89.

Kathuria YP (1999) Microstructuring by selective laser sintering of metallic powder. Surface and Coatings Technology 116–119:643–647.

Kim J, Creasy TS (2004) Selective laser sintering characteristics of nylon 6/clay-reinforced nanocomposite. Polymer Testing 23:629–636.

Klocke F, Wirtz H (1997) Selective laser sintering of ceramics. Proceedings of LANE'97 (Laser Assisted Net Shape Engineering) Conference, 2:589–96.

Klosterman D, Chartoff R, Graves G, Osborne N, Priore B (1998) Interfacial characteristics of composites fabricated by laminated object manufacturing. Composites Part A 29:1165–1174.

Klosterman DA, Chartoff RP, Osborne NR, Graves GA, Lightman A, Han G, Bezeredi A, Rodrigues S (1999) Development of a curved layer LOM process for monolithic ceramics and ceramic matrix composites. Rapid Prototyping Journal 5:61–71.

Kochan D, Kai CC, Zhaohui D (1999) Rapid prototyping issues in the 21st century. Computers in Industry 39:3–10.

Kruth JP (1991) Material incress manufacturing by rapid prototyping techniques. Annals of the CIRP 40:603–614.

Kruth JP, Wang X, Laoui T, Froyen L (2003) Lasers and materials in selective laser sintering. Assembly Automation 23:357–371.

Kruth JP, Mercelis P, Vaerenbergh JV (2005) Binding mechanisms in selective laser sintering and selective laser melting. Rapid Prototyping Journal 11:26–36.

Kulkarni PK, Dutta D (1996) An accurate slicing procedure for layered manufacturing. Computer-Aided Design 28:683–697.

Kulkarni PK, Marsan A, Dutta D (2000) A review of process planning techniques in layered manufacturing. Rapid Prototyping Journal 6:18–35.

Lakshiminarayan U, Marcus HL (1991) Microstructural and mechanical properties of Al_2O_3/P_2O_5 and Al_2O_3/B_2O_3 composites fabricated by selective laser sintering. In: Marcus HL, Beaman JJ,

Barlow JW, Bourell DL, Crawford RH (eds) Proceedings of the Solid Freeform Fabrication Symposium, pp 205–212.

Lan PT, Chou SY, Chen LL, Gemmill D (1997) Determining fabrication orientations for rapid prototyping with stereolithography apparatus. Computer-Aided Design 29:53–62.

Lee GH, Barlow JW (1993) Selective laser sintering of bioceramic materials for implants. In: Marcus HL, Beaman JJ, Barlow JW, Bourell DL, Crawford RH (eds) Proceedings of the Solid Freeform Fabrication Symposium, pp 376–380.

Lee JW, Lee IH, Cho DW (2006) Development of micro-stereolithography technology using metal powder Microelectronic Engineering 83:1253–1256.

Lewis GK, Nemec RB, Milewski JO, Thoma DJ, Cremers D, Barbe MR (1994) Directed light fabrication. Proceedings of the ICALEO'94. Laser Institute of America, Orlando, Florida, p 17.

Licciulli A, Corcione CE, Greco A, Amicarelli V, Maffezzoli A (2004) Laser stereolithography of ZrO$_2$ toughened Al$_2$O$_3$. Journal of the European Ceramic Society 24:3769–3777.

Liu W, DuPont JN (2003) Fabrication of functionally graded TiC/Ti composites by laser engineered net shaping. Scripta Materialia 48:1337–1342.

Mackenzie JK, Shuttleworth RA (1949) Phenomenological theory of sintering. Proceedings of the Physical Society 62:833–852.

McClurkin JE, Rosen DW (1998) Computer-aided build style decision support for stereolithography. Rapid Prototyping Journal 4:4–13.

Milewski JO, Lewis GK, Thoma DJ, Keel GI, Nemec RB, Reinert RA (1998) Directed light fabrication of a solid metal hemisphere using 5-axis powder deposition. Journal of Materials Processing Technology 75:165–172.

Mueller B, Kochan D (1999) Laminated object manufacturing for rapid tooling and patternmaking in foundry industry. Computers in Industry 39:47–53.

Nelson JC, Xue S, Barlow JW, Beaman JJ, Marcus HL (1993) Model of the selective laser sintering of bisphenol-A polycarbonate. Industrial and Engineering Chemistry Research 32:2305–2317.

Park J, Tari MJ, Hahn HT (2000) Characterization of the laminated object manufacturing (LOM) process. Rapid Prototyping Journal 6:36–49.

Pham DT, Dimov S (2001) Rapid manufacturing: the technologies and applications of rapid prototyping and rapid tooling. Springer, London.

Pham DT, Dimov S, Lacan F (1999) Selective laser sintering: applications and technological capabilities. Proceedings of Institution of Mechanical Engineers B 213:435–449.

Rahaman MN (2003) Ceramic processing and sintering. Marcel Dekker, New York.

Savalani MM, Hao L, Harris RA (2006) Evaluation of CO$_2$ and Nd:YAG lasers for the selective laser sintering of HAPEX. Proceedings of Institution of Mechanical Engineers B: Journal of Engineering Manufacture 220:171–182.

Scherer G (1977a) Sintering of low density glasses: I, theory. Journal of American Ceramic Society 60:236–239.

Scherer G (1977b) Sintering of low density glasses: II, experimental study, Journal of American Ceramic Society 60:239–245.

Scherer G (1977c) Sintering of low density glasses: III, effect of a distribution of pore size. Journal of American Ceramic Society 60:245–248.

Sonmez FO, Hahn HT (1998) Thermomechanical analysis of the laminated object manufacturing (LOM) process. Rapid Prototyping Journal 4:26–36.

Subramanian K, Vail N, Barlow J, Marcus H (1995) Selective laser sintering of alumina with polymer binders. Rapid Prototyping Journal 1:24–35.

Tolochko N, Mozzharov S, Laoui T, Froyen L (2003) Selective laser sintering of single- and two-component metal powders. Rapid Prototyping Journal 9:68–78.

Vail NK, Balasubramanian B, Barlow JW, Marcus HL (1996) A thermal model of polymer degradation during selective laser sintering of polymer coated ceramic powders. Rapid Prototyping Journal 2:24–40.

Wang WL, Cheah CM, Fuh JY, Lu L (1996) Influence of process parameters on stereolithography part shrinkage. Materials and Design 17:205–213.

Wang XC, Laoui T, Bonse J, Kruth JP, Lauwers B, Froyen L (2002) Direct selective laser sintering of hard metal powders: experimental study and simulation. International Journal of Advanced Manufacturing Technology 19:351–357.

Webb PA (2000) A review of rapid prototyping (RP) techniques in the medical and biomedical sector. Journal of Medical Engineering and Technology 24:149–153.

West AP, Sambu SP, Rosen DW (2001) A process planning method for improving build performance in stereolithography. Computer-Aided Design 33:65–79.

Williams JD, Deckard CR (1998) Advances in modeling the effects of selected parameters on the SLS process. Rapid Prototyping Journal 4:90–100.

Williams RE, Komaragiri SN, Melton VL, Bishu RR (1996) Investigation of the effect of various build methods on the performance of rapid prototyping (stereolithography). Journal of Materials Processing Technology 61:173–178.

Xu F, Wong YS, Loh HT, Fuh JY, Miyazawa T (1997) Optimal orientation with variable slicing in stereolithography. Rapid Prototyping Journal 3:76–88.

Xu, F, Loh HT, Wong YS (1999) Considerations and selection of optimal orientation for different rapid prototyping systems. Rapid Prototyping Journal 5:54–60.

Yagui S, Kunni D (1989) Studies on effective thermal conductivities in packed beds. Journal of American Institute of Chemical Engineers (AIChe) 3:373–381.

Ye R, Smugeresky JE, Zheng B, Zhou Y, Laverni EJ (2006) Numerical modeling of the thermal behavior during the LENS process. Materials Science and Engineering A 428:47–53.

Chapter 10
Laser Welding

10.1 Introduction

Laser welding has evolved as an important industrial manufacturing process for joining a variety of metallic and nonmetallic materials. With the developments in the high-power laser technology over the past few decades, laser welding is now capable of joining thicker sections with higher processing speed and better weld quality. Due to the noncontact nature of laser processing, high degree of automation is possible providing economic advantages in the typical industrial environment. Even though, laser welding seems to be a simpler process, it presents significant challenges to produce defect-free welds at high speed and under reproducible conditions. This requires a strong understanding of the underlying concepts of laser–material interactions in the regime of laser welding conditions. Since extensive literature is published on the topic of laser welding in dedicated books, book chapters, and journals, this chapter is intended to present only brief discussions on important aspects of laser welding process.

10.2 Laser Welding Process

When a laser beam is irradiated on the surface of a material, the absorbed energy causes the heating, melting, and/or evaporation of the material depending on the absorbed laser power density. The general condition of laser welding process is to create a pool of molten material (weld pool) at the overlapping workpiece surfaces. There are two general approaches for laser welding processes. In the first approach sometimes referred as conduction welding, the laser processing conditions are such that the surface of the weld pool remains unbroken. In this approach the energy transfer into the depth of the material takes place by conduction (Fig. 10.1a). The second and the most important approach referred as deep penetration welding corresponds to the laser processing conditions which create a "keyhole" in the weld pool. Generally, the transition from the conduction mode to the deep penetration welding is associated with the increase in laser power intensity or irradiation time

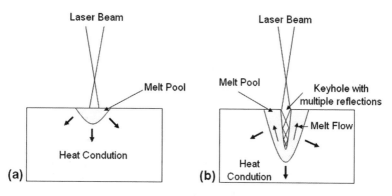

Fig. 10.1 Schematic of the cross sections of (**a**) conduction and (**b**) deep penetration laser welding showing various effects

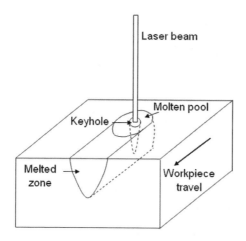

Fig. 10.2 Schematic of the deep penetration welding process

such that surface vaporization at the molten weld pool begins (Duley 1999). The resulting evaporation-induced recoil pressure forms a small depression in the weld pool which subsequently develops into a keyhole by the upward displacement of molten material sideways along the keyhole walls (Fig. 10.1b). The subsequent ionization of the vapor results in the formation of the plasma plume. The laser energy entering the keyhole wall is determined by the attenuation due to absorption of laser energy in the plasma plume. Within a keyhole, the laser energy is reflected repeatedly (multiple reflection) with efficient Fresnel absorption of energy at the keyhole walls. Thus, the keyhole plays an important role in transferring and distributing the laser energy deep into the material. The surface vaporization continues at the keyhole wall during laser welding to maintain the cavity. The vaporized materials act against the surface tension to keep the keyhole open. The melt accelerated upward flows continuously out of the cavity. Figure 10.2 presents the schematic geometry of deep penetration laser welding. Various mechanisms of energy absorption

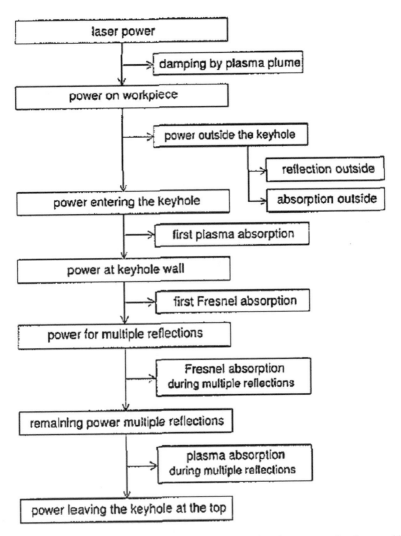

Fig. 10.3 Flowchart of various absorption mechanisms during deep penetration laser welding. (Reprinted from Kaplan 1994. With permission. Copyright Institute of Physics.)

during laser welding process are presented in Fig. 10.3 (Kaplan 1994). Under dynamic condition such as a moving laser beam, the welding speed is determined by the stability of the keyhole. Stable welding conditions correspond to the speed at which the keyhole speed achieves steady state (Sickman and Morijn 1968). The symmetry of the keyhole is also influenced by the welding speed. At low welding speed, the keyhole may be approximated by the rotational symmetry, whereas at high welding speed, the keyhole profile differs significantly at the front and back wall (Kaplan 1994).

10.3   Analysis of Laser Welding Process

Extensive literature is available on the modeling of the laser welding process. Most of these models are based on the theory of heat flow due to a moving source of heat (Rosenthal 1946). Excellent reviews of laser welding process modeling have been published (Mackwood and Crafer 2005; Duley 1999). This section briefly discusses various important laser welding models and the current understanding on the subject.

The general equation for three-dimensional heat transfer is given by (Carslaw and Jaeger 1959):

$$\alpha\left[\frac{\partial^2 T(x,y,z,t)}{\partial x^2} + \frac{\partial^2 T(x,y,z,t)}{\partial y^2} + \frac{\partial^2 T(x,y,z,t)}{\partial z^2}\right] = \frac{\partial T(x,y,z,t)}{\partial t} \quad (10.1)$$

where, T is the temperature, t is the time, α is the thermal diffusivity, and, x, y, and z are the coordinates. The earliest analytical solution applicable to the laser welding process was derived by Rosenthal (1946). The model considered the heat flow by moving point, linear, or plane source of heat in solids of infinite size or bounded by planes (one-dimensional, two-dimensional, and three-dimensional heat flow). The schematic of the model geometry for heating with moving point source is presented in Fig. 10.4.

For a point heating source moving with constant velocity, v, in the x-direction, the three-dimensional heat transfer equation (10.1) can be rewritten with point source as origin as (Rosenthal 1946):

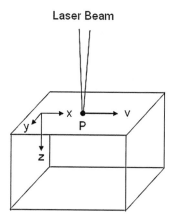

Fig. 10.4 Schematic of the model geometry for laser welding using moving point source of heat

$$\alpha \left[\frac{\partial^2 T(\xi, y, z, t)}{\partial \xi^2} + \frac{\partial^2 T(\xi, y, z, t)}{\partial y^2} + \frac{\partial^2 T(\xi, y, z, t)}{\partial z^2} \right]$$
$$= v \frac{\partial T(\xi, y, z, t)}{\partial \xi} + \frac{\partial T(\xi, y, z, t)}{\partial t} \qquad (10.2)$$

where ξ is the distance of a considered point from the point source. This distance can be expressed as:

$$\xi = x - vt. \qquad (10.3)$$

For a solid much longer than the extent of heat dissipation, the temperature distribution around the point heat source becomes constant such that an observer located at the moving point source fails to notice the temperature changes around the point source as it moves on. This type of heat flow is generally referred as quasistationary heat flow. This state of the heat flow is defined as:

$$\frac{\partial T(\xi, y, z, t)}{\partial t} = 0 \qquad (10.4)$$

Substituting in Eq. (10.2) yields:

$$\alpha \left[\frac{\partial^2 T(\xi, y, z, t)}{\partial \xi^2} + \frac{\partial^2 T(\xi, y, z, t)}{\partial y^2} + \frac{\partial^2 T(\xi, y, z, t)}{\partial z^2} \right] = -v \frac{\partial T(\xi, y, z, t)}{\partial \xi}. \qquad (10.5)$$

The Eq. (10.5) can be further simplified by using:

$$T = T_0 + e^{-\lambda v \xi} \varphi(\xi, y, z), \qquad (10.6)$$

where T_0 is the initial temperature and φ is the function which needs to be determined to calculate the temperature distributions. The Eq. (10.5) becomes:

$$\left[\frac{\partial^2 \varphi(\xi, y, z)}{\partial \xi^2} + \frac{\partial^2 \varphi(\xi, y, z)}{\partial y^2} + \frac{\partial^2 \varphi(\xi, y, z)}{\partial z^2} \right] - \left(\frac{v}{2\alpha} \right)^2 \varphi(\xi, y, z) = 0. \qquad (10.7)$$

These equations can be solved with appropriate boundary conditions to obtain the temperature profiles around the moving point source of heat for an infinite and semi-infinite solid.

For the two-dimensional situation, the solution of heat transfer equation is given by:

$$T - T_0 = \frac{Q'}{2\pi k} e^{-\frac{v\xi}{2\alpha}} K_0 \left(\frac{vR}{2\alpha} \right).$$
(10.8)

For the three-dimensional situation, the solution is given by:

$$T - T_0 = \frac{Q}{4\pi k} e^{-\frac{v\xi}{2\alpha}} \frac{e^{-\frac{vR}{2\alpha}}}{R},$$
(10.9)

where Q' is the rate of heat per unit length, Q is the rate of heat, k is the thermal conductivity, and R is the distance to the heat source (Rosenthal 1946). Swift-Hook and Gick (1973) simplified the standard moving source heat solution. Their analytical model of continuous laser welding considered a moving line heat source and the shape of the fusion zone was given by the location of isotherm corresponding to the melting temperature. The model provides the simple relationship between the normalized laser power per unit depth (M) and the normalized width of the weld (Y):

$$Y \sim 0.484 \, M, \text{for } vr/2\alpha > 1$$
(10.10)

and

$$Y \sim \exp\left[1.50 - 2\pi/M\right], \text{for } vr/2\alpha < 1,$$
(10.11)

where $Y = vW/\alpha$ and $M = P/akT_m$. In these expressions, v is the velocity, W is the width of weldment, α is the thermal diffusivity, P is the laser power, a is the penetration depth, k is the thermal conductivity, T_m is the melting point, and r is the beam radius (Swift-Hook and Gick 1973). Most of these earliest models neglected the formation of keyhole which plays an important role in the laser welding process.

Earliest attempts to incorporate the effects of the formation of the keyhole in the modeling of laser welding were done by Klemens (1976). The model assumed a moving cylindrical vapor-filled cavity (keyhole) formed by continuous vaporization of material due to a laser beam. The cavity shape was determined by the balance of the liquid flow and surface tension which tends to close the cavity and the continuous vaporization which tends to maintain the cavity. However, the model only gives the qualitative analysis of the variation of cavity radius with the depth and the shape of the molten zone based on heat flow, vapor flow, surface tension, and gravity (Klemens 1976). Further analysis of the keyhole during laser welding was done by Andrews and Atthey (1976). The model considered a stationary beam and semi-infinite molten pool and took into account the gravity and surface tension effects for the prediction of the depth of the hole and profile. However, the model was based on many simplified assumptions. The model neglected the vapor pressure distribution inside the hole and assumed that vapor pressure inside the cavity was equal to the atmospheric pressure. Furthermore, the model assumed that all the incident radiation was absorbed by the surface without any attenuation by the plasma and also without any reflection from the surface. Also, the entire absorbed energy was assumed to be used for evaporation without any conduction. With these simplified assumptions, the laser welding process is controlled by two-dimensionless parameters:

$$Q = q /(g\rho\rho_g h^2 a)^{1/2}, \tag{10.12}$$

$$\tau = T / \rho g a^2, \tag{10.13}$$

where q is the laser power density, a is the beam radius, h is the heat per unit mass to boil the material, T is the surface tension coefficient, and, ρ and ρ_g are the densities of the liquid and vapor respectively. Two distinct physical regimes governing the depth of penetration were given. For shallow holes, the depth is determined by power density (W/a^2, where W is the laser power), whereas for deep holes, the depth is determined by the power per unit radius (W/a) when gravity is the only restoring force. Inclusion of surface tension tends to reduce the depth of penetration (Andrews and Atthey 1976). Due to neglecting of heat conduction, this model is not particularly suitable for predicting the temperature distribution during laser welding.

A more realistic analytical model of laser welding considering the conditions of keyhole formation and conduction of heat was presented by Cline and Anthony (1977). They analyzed the temperature distribution, cooling rate distribution, and depth of melting as a function of various laser parameters (beam spot size, velocity, laser power, etc.) due to a moving Gaussian beam (heating and melting regimes). The model extended to deep penetration welding and assumed that the onset of penetration welding occurs when the surface temperature reaches the boiling point. The keyhole of depth, Z_v, is then formed by evaporation-induced pressure which pushes the liquid–vapor interface deeper. The keyhole depth is then determined by the balance between the vapor pressure at the tip and, the buoyancy and capillary forces. This balance can be expressed as:

$$\text{Pressure} = \rho_L Z_v g + (\gamma_{LV} / r_v), \tag{10.14}$$

where ρ_L is the liquid density, g is the acceleration due to gravity, γ_{LV} is the surface tension of the liquid–vapor interface, and r_v is the radius of the tip of vapor protrusion (keyhole). The temperature distribution at the keyhole can be determined by the finite, moving line source (of length Z_v) with spherical heat flow conditions at the tip (Fig. 10.5). The solution was based on the method of images which requires

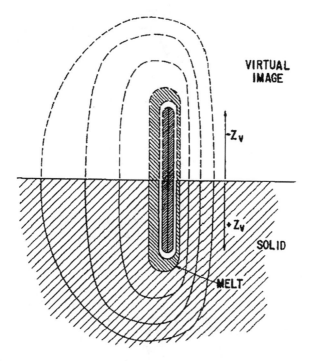

Fig. 10.5 Schematic of the deep penetration welding showing the depths of melting and keyhole. The temperature distribution at the keyhole was determined using a moving finite heat source of depth Z_v. (Reprinted from Cline and Anthony 1977. With permission. Copyright American Institute of Physics.)

an image source from 0 to $-Z_v$ for a distributed source from 0 to Z_v. The power (P) is related with the keyhole depth as:

$$\frac{P}{P_V} = \frac{2(Z_v / Z_0)}{1 - \exp\left[-2(Z_v / Z_0)\right]},$$ (10.15)

where P_V is the power level at which the temperature reaches the boiling point (T_v), and Z_0 is the penetration depth. The fixed depth of the melt, Z_m, below the keyhole depth at high laser power levels is given by:

$$Z_m = Z_v + Z_0 \ln\left(T_v / T_m\right),$$ (10.16)

where T_m is the melting point. Figure 10.6 presents the variation of calculated normalized melt and keyhole depths as a function of normalized power for laser welding of 304-stainless steel. The figure indicates that keyhole depth (Z_v) is zero

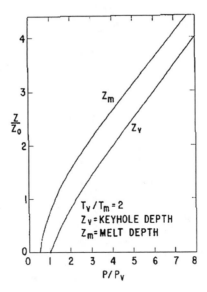

Fig. 10.6 Calculated normalized depths of melting and keyhole as a function of normalized power for laser welding of 304 stainless steel. (Reprinted from Cline and Anthony 1977. With permission. Copyright American Institute of Physics.)

corresponding to the power (P_v) required to initiate the vaporization at the surface and then increases with the power (Cline and Anthony 1977).

Deep penetration welding process is also modeled by calculating the keyhole profile using a point-by-point determination of energy balance at the keyhole wall (Fig. 10.7) (Kaplan 1994). The heat conduction from the wall of the keyhole inside the material was modeled based on the theory of heat flow by moving line source of heat. The local energy balance at the keyhole wall gives the local wall angle (θ_w):

$$\tan(\theta_w) = \frac{q_v(x)}{I_a(x,z)}, \tag{10.17}$$

where q_v is the heat flow at a point at the evaporation temperature and I_a is the absorbed laser power. If the absorbed power is considered to be equal to the intensity absorbed by Fresnel absorption, $I_{a,Fr}$, the equation becomes:

$$\tan(\theta_w) = \frac{q_v}{I_{a,Fr}} = \frac{q_v}{\alpha_{Fr}(\theta_w)I}, \tag{10.18}$$

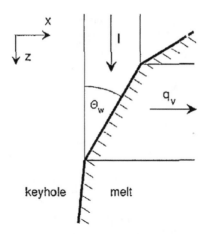

Fig. 10.7 Calculation of keyhole geometry from local energy balance at the wall. (Reprinted from Kaplan 1994. With permission. Copyright Institute of Physics.)

where α_{Fr} is the Fresnel absorption coefficient. Including other absorption mechanisms such as plasma absorption due to inverse Bremsstrahlung, $I_{a,iB}$, and Fresnel absorption during multiple reflection, $I_{a,mr}$, the energy balance becomes:

$$\tan(\theta_w) = \frac{q_v - I_{a,iB} - I_{a,mr}}{I_{a,Fr}(\theta_w)}.$$
(10.19)

The plasma absorption can be described by Beer–Lambert law. The absorbed fraction of the beam is given by:

$$\alpha_{pl} = 1 - \exp(-\alpha_{iB} h_{pl}),$$
(10.20)

where α_{iB} is the plasma absorption coefficient and h_{pl} is the mean height of the metal vapor plume over the workpiece. The laser beam intensity after n_{mr} reflections is given by:

$$I_n = I_i (\rho_{Fr})^{n_{mr}},$$
(10.21)

where ρ_{Fr} is the angle-dependent Fresnel reflection coefficient. The number of multiple reflections is given by:

$$n_{mr} = \frac{\pi}{4\bar{\theta}_w}.$$
(10.22)

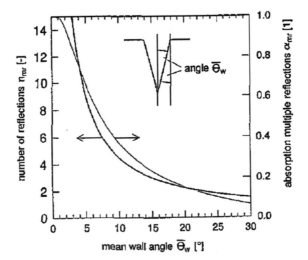

Fig. 10.8 Dependence of number of reflections, n_{mr} (thick line) and corresponding absorption coefficient, α_{mr} (thin line) on mean wall angle, $\bar{\theta}_w$. (Reprinted from Kaplan 1994. With permission. Copyright Institute of Physics.)

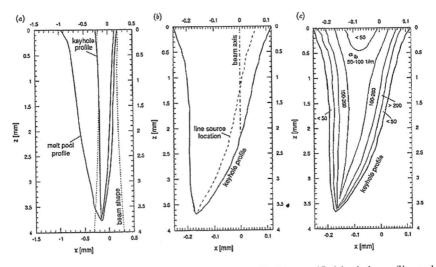

Fig. 10.9 (a) Profile of beam, keyhole, and molten pool; (b) magnified keyhole profile; and (c) distribution of plasma absorption coefficient in keyhole. (Reprinted from Kaplan 1994. With permission. Copyright Institute of Physics.)

For a conical structure of the keyhole, the dependence of the number of reflections, n_{mr}, on the mean wall angle, $\bar{\theta}_w$, is presented in Fig. 10.8. Figure 10.9 presents the calculated profile of the keyhole and the molten pool relative to the beam axis during laser welding of steel with laser power of 4 kW and welding speed of 50 mm/s. The figure clearly indicates that the front wall of the keyhole is closer to the beam axis compared to rear keyhole wall resulting in overall curvature in the keyhole profile. This is due to the efficient heat conduction at the front wall due to the convective effect of the welding speed. The figure also presents the distribution of plasma absorption coefficient, α_{iB}, in the keyhole (Kaplan 1994).

Most of the models of laser welding dealing with the analysis of the keyhole assume that the keyhole is symmetric and either coaxial with the laser beam or slightly displaced in the direction opposite to the direction of laser travel. These models also assumed that the balance between the recoil pressure and surface tension pressure determines the stability of the keyhole. These assumptions necessitate that the entire keyhole (i.e., both front and rear walls of the keyhole) remains within the area of the laser beam (Klemens 1976; Dowden et al. 1987; Kroos et al. 1993; Klein et al. 1994; Postacioglu et al. 1991; Dowden and Kapadia 1995; Kaplan 1994). Recent experimental investigations have shown that for higher welding speeds (>20–50 mm/s), only the front wall of the keyhole is exposed to the laser beam and the rear wall remains well outside the area of significant laser intensity. Matsunawa and Semak (1997) developed a simplified numerical model of laser welding assuming that only the front wall of the keyhole is exposed to the high-intensity laser beam and that the propagation of keyhole inside the workpiece is due to melt expulsion (when the recoil pressure exceeds the surface tension and hydrostatic pressure as in laser drilling). The simulation results indicated that when the component of keyhole wall propagation velocity along the sample translational velocity becomes negative (i.e., keyhole wall moving away from the laser beam axis), the formation of humps takes place on the keyhole wall surface. The formation of such humps on the front wall of the keyhole is termed as "runaway" instability and has been experimentally observed during laser cutting. The evolution of "runaway" instability is presented in Fig. 10.10 for the

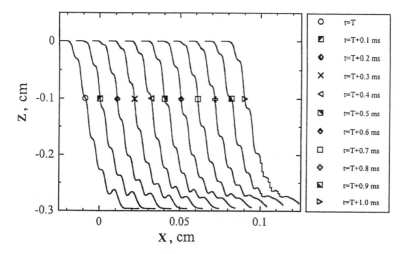

Fig. 10.10 Evolution of "runaway" instability on the front wall of the keyhole. Each keyhole front wall profile is shifted by 0.01 cm on x-axis relative to previous profile (laser parameters: power 3 kW; beam radius 140 mm; welding speed 10 cm/s; drilling velocity delay time 50 μs). (Reprinted from Matsunawa and Semak 1997. With permission. Copyright Institute of Physics.)

laser power of 3 kW, beam radius of 140 mm, welding speed of 10 cm/s, and drilling velocity delay time (T) of 50 μs. The hump velocity is found to be increased from zero to around 180 cm/s during first 0.5 ms and then decreased to zero when the hump approaches the bottom (Matsunawa and Semak 1997).

One of the major objectives of the modeling of laser welding process is the prediction of the overall dimensions of the weld cross section for the given processing parameters and material properties. Mazumder and Steen (1980) developed one of the earliest numerical models of continuous laser welding process using finite difference technique. The model assumed quasisteady-state heat transfer and Gaussian power distribution in the incident laser beam. The model also assumed that the reflectivity is zero when the temperature exceeds the boiling point. This is due to the proposition that a keyhole is formed by vaporization and acts as black body. The radiation penetrating the substrate is considered to be absorbed according to Beer–Lambert law. Matrix points considered to have evaporated remain in solid conduction network at fictitiously high temperature to simulate the high convection and radiation effects within the plasma (Mazumder and Steen 1980). However, the model neglected the temperature dependence of the thermophysical properties. The model could be used for the prediction of the temperature profile, maximum welding speed, heat-affected zone, thermal cycle at any location or speed, the effect of thickness or other parameters (such as reflectivity and conductivity), and the effect of supplementary heating or cooling. Some of the results of the model are presented in Figs. 10.11 and 10.12. Figure 10.11 presents the comparison of experimental and predicted fusion zones for partial penetration bead-on-plate weld in mild steel. The figure also includes the effect of varying absorption coefficient on the fusion zone profile. The model can also be used to calculate maximum speed at which melting point is just reached at the lower surface. Figure 10.12 presents the comparison of variation in predicted and experimental welding speed with laser power for titanium

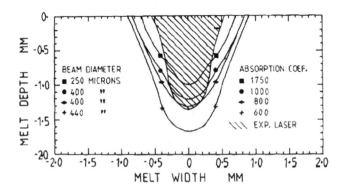

Fig. 10.11 Variation of fusion zone profiles with beam diameter and absorption coefficient during laser welding of mild steel (parameters: laser power 1,570 W; reflectivity 0.8; welding velocity 33.5 mm/s). (Reprinted from Mazumder and Steen 1980. With permission. Copyright American Institute of Physics.)

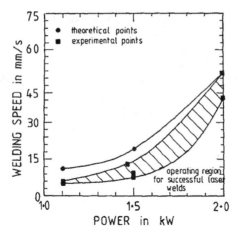

Fig. 10.12 Variation of predicted and experimental welding speeds with laser power during laser welding of 2 mm thick titanium alloy. (Reprinted from Mazumder and Steen 1980. With permission. Copyright American Institute of Physics.)

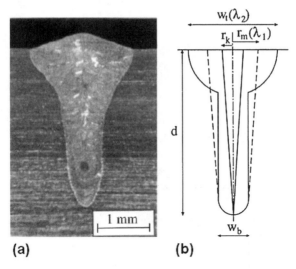

Fig. 10.13 (a) Actual cross section of the laser weld, and (b) corresponding simplified geometry for calculating various weld dimensions. (Reprinted from Lampa et al. 1997. With permission. Copyright Institute of Physics.)

alloy weld. The predicted results are in good agreement with the experimental data.

A typical cross-section of the fusion zone profile in a laser weld is presented in Fig. 10.13a. As indicated in the figure, the width of the weld at the top differs significantly from that at the bottom. Recently significant efforts have been made by Lampa et al. (1997) to predict the overall dimensions of the weld cross section for a given set of processing parameters. They simplified the weld geometry such that the complete description of the weld geometry is given by the width of the top, w_t, the width of the bottom, w_b, and the penetration depth, d (Fig. 10.13b). At the top of the weld, the thermal transport is accelerated by thermocapillary (Marangoni) flow resulting in the widening of the weld. At the bottom of the weld thermal transport is primarily due to conduction mechanism. To account for these differences in the thermal transport mechanisms at the top and the bottom of the weld, the model artificially assigned high thermal conductivity at the top (around 2.5 times the thermal conductivity at the bottom). Figure 10.14 presents the calculated weld pool profiles for laser welding of austenitic stainless steel at various welding speeds. The figure indicates that an increase in welding speed is associated with a proportional decrease in the amount of material melted. However, the decrease in penetration depth is much slower with increasing welding speed. This model incorporating artificial high conductivity at the top of the weld (to account for thermocapillary flow) yielded very good agreement (accuracy of predictions >90%) with the experimental values of weld widths (top and bottom) and penetration depths (Lampa et al. 1997).

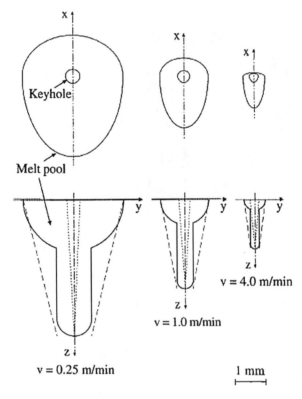

Fig. 10.14 Calculated weld profiles in austenitic stainless steel at various welding speeds (laser parameters: power 1,400 W; wavelength 10.6 μm; raw-beam diameter 18 mm). (Reprinted from Lampa et al. 1997. With permission. Copyright Institute of Physics.)

10.4   Quality Aspects

Laser welding is attracting increased interests for welding of various materials in automotive and aerospace industries. These applications necessitate very high quality of the welds to ensure the desired performance of the laser-welded components. Due to the complexity of the effects involved in the laser welding process, it is often associated with various geometric and metallurgical defects. This section discusses various common laser welding defects.

10.4.1   Porosity

Porosity is one of the major defects associated with the laser welding process. There are various sources of porosity in the laser welds. Porosities appearing in various locations of the weld may be characterized by distinct morphologies and distributions. For example, large voids are commonly formed at the bottom of the weld, whereas small distributed voids are formed towards the surface of the weld. Figure 10.15 presents the microstructure of the weld cross section in tantalum showing the formation of large empty voids and small bubble-like voids. The formation of small bubbles is often related with the trapping of gas bubble (formed either due to degasification or from surrounding gas) during solidification. The origin of large voids is believed to be the solidification of molten material before the complete filling of the keyhole (Fig. 10.16) (Girard et al. 2000).

The formation of porosity is greatly influenced by a number of laser processing parameters such as laser power, focusing conditions, welding speed, shielding gas, etc.

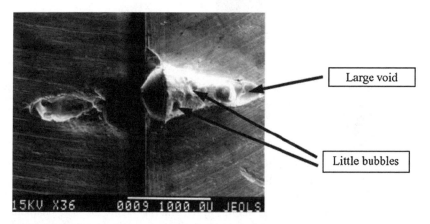

Fig. 10.15 Formation of large and small voids (bubbles) in the weld microstructure of tantalum (laser parameters: Nd:YAG laser; power 3 kW; spot size 300 μm). (Reprinted from Girard et al. 2000. With permission. Copyright Institute of Physics.)

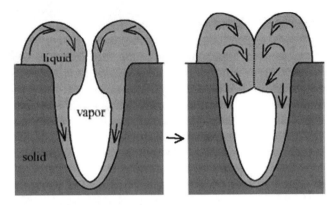

Fig. 10.16 Schematic showing the formation of large void due to incomplete filling of keyhole during laser welding. (Reprinted from Girard et al. 2000. With permission. Copyright Institute of Physics.)

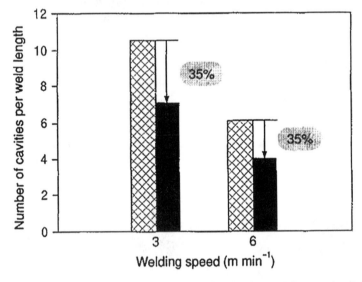

Fig. 10.17 Number of cavities per weld length as a function of the welding speed and the beam quality (K) in laser welding of Al–Mg–Si alloy with CO_2 lasers (hatched: $K = 0.21$; Solid: $K = 0.3$). (Reprinted from Rapp et al. 1995. With permission. Copyright Springer.)

The formation of porosity can be minimized by optimizing the laser processing parameters such that the stability of the keyhole is improved and the turbulence in the fluid flow is minimized. In general, the number of pores per unit weld length can be reduced by increasing the welding speed and improving the laser beam quality (Fig. 10.17). Also, increased power density is expected to stabilize the keyhole in deep penetration welding resulting in reduction in porosity (Rapp et al. 1995). Proper selection of focusing condition can also result in reduction in porosity

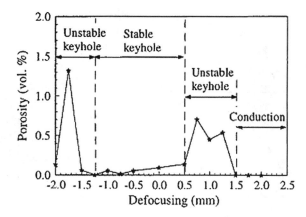

Fig. 10.18 Porosity as a function of laser focus conditions for welding of Nd:YAG laser welding of 5754 aluminum alloy (processing parameters: laser power 3 kW; welding speed 150 in/min; helium shielding gas) (Courtesy of Professor DebRoy, Pennsylvania State University)

during laser welding. Figure 10.18 shows that, there exists some optimum focus condition for which the keyhole is stable resulting in minimum porosity (Pastor 1998). Various ways have been suggested to reduce the porosity during penetration welding. The application of two laser beams (tandem-beam laser welding) is reported to produce cavity-free welds due to the stabilization of the keyhole (Glumann et al. 1993). Also, use of filler wire during welding is found to improve the weld quality in some type of materials (Rapp et al. 1995).

Porosity in the weld may also result due to the rejection of the dissolved gases during solidification, from the entrapment of gases due to melt turbulence or solidification shrinkage (Cao et al. 2003).

10.4.2 Cracking

Cracking in the weld structures is one of the most serious laser welding defects. Most of the cracks in the welds originate from the restrictions to the free contractions of the material during the cooling cycle. Such restrictions result in the setting up of high tensile stresses causing cracking. Hot cracking in the weld can either be solidification cracking or liquation cracking. Cracking in the fusion zone during solidification is called solidification cracking. Cracking may be caused due to the liquation of low melting components in the partially melted zone (liquation cracking). Liquation cracking is susceptible in the heat-treatable alloys which may form eutectic phases with low melting point during laser welding (Zhao et al. 1999; Cao et al. 2003).

Extensive studies have been conducted to understand the mechanism of solidification cracking during laser welding and the effects of the various parameters on the susceptibility to cracking. Various theories have been put forward explaining

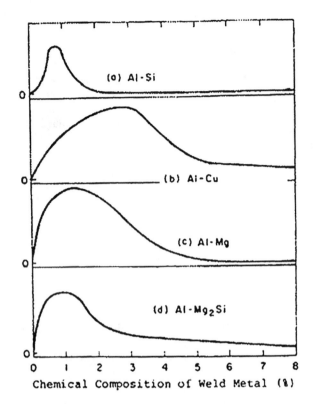

Fig. 10.19 Effect of chemical composition on relation solidification crack susceptibility (ordinate) for various aluminum alloys (Reprinted from Dudas and Collins 1966. With permission. Copyright American Welding Society.)

the mechanisms of solidification cracking. Among these, "strain theory" (Pellini 1952), "brittleness temperature range theory" (Pumphrey and Jennings 1948), and "generalized theory" (Borland 1960a, 1961b) are popular. According to strain theory, cracking occurs due to the local strain concentration in the thin liquid film formed during the final stages of solidification. According to brittle temperature range (BTR) theory, alloys pass through BTR during solidification and cracking occurs when the thermal tensile strains exceed the ductility of the weld in BTR. During laser welding, solidification primarily proceeds through the formation and growth of dendritic network. If the surrounding liquid is able to flow freely and fill the interdenditic space, effective crack healing may take place. Based on the solidification behavior, the generalized theory of Borland suggests the critical range of composition (difference between nominal liquidus and solidus temperature) corresponding to the highest solidification crack susceptibility. The crack susceptibility is greatly influenced by the composition of the alloy. In the case of aluminum alloys, compositions corresponding to 0.8% Si in Al–Si, 1–3% Cu in Al–Cu, 1–1.5% Mg in Al–Mg, and 1% Mg_2Si in Al–Mg–Si seem to have the highest solidification crack susceptibility (Fig. 10.19) (Dudas and Collins 1966).

The tendency to crack during laser welding can be minimized by various ways such as careful design of the material, optimization of laser processing parameters, additional designs (such as welding fixtures and joint design). Other defects associated with laser welding may include incomplete penetration, spatter, etc.

10.4.3 Heat-Affected Zone

The regions adjacent and below the fusion zone of laser weld may undergo some transformations due to the heating and cooling cycles (maximum temperature less than melting point). This area is referred as heat-affected zone (HAZ). HAZs often show the altered properties and microstructure and mark the transition region between the fusion zone and the bulk material. Figure 10.20 presents the microstructure of a single-pass laser weld in 1.2343 grade steel. Figure 10.21 presents the hardness profile in the laser weld suggesting the presence of HAZ as indicated by a drop in hardness below the fusion region (Vedani 2004). The extent of HAZ can be characterized by the width of the altered region between the fusion zone and the bulk material. Also, the extent of HAZ is expected to be influenced by various laser processing parameters such as laser power, welding speed, etc.

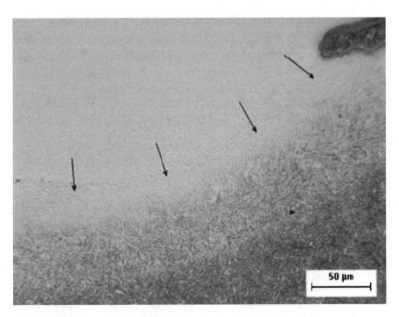

Fig. 10.20 Microstructure of single-pass laser weld in 1.2343 steel showing heat-affected zone (HAZ). Arrows indicate the location of fusion line. (Reprinted from Vedani 2004. With permission. Copyright Springer.)

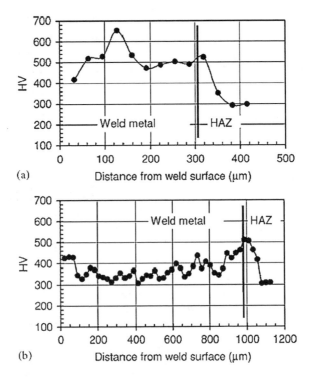

Fig. 10.21 Microhardness profile in (**a**) single-pass and (**b**) double-pass laser weld in 1.2311 steel grade. (Reprinted from Vedani 2004. With permission. Copyright Springer.)

10.4.4 Mechanical Properties of Laser Welds

Various mechanical properties such as hardness, tensile strength, fatigue strength, formability of the laser-welded joints have been extensively studied. The mechanical properties of the weld joints are significantly influenced by various welding defects such as cracking and porosity. Also, other parameters such as laser joint configuration, laser welding parameters, material composition play an important role in determining the mechanical strength of the joint. Due to the complexity of the parameters involved in laser welding, the reported mechanical properties should not be regarded as general trends.

Hardness of the weld zone may be higher, lower, or equivalent to the base material depending on the type of material. In general, the hardness of the laser weld zone in nonheat-treatable alloys is generally higher than the base material with an intermediate transition zone corresponding to HAZ. The increase in the hardness of the weld zone may be due to refinement of the microstructure. Some of the heat treatable alloys show reduction in hardness in the weld joints due to the loss of precipitates in the weld and overaging in the HAZ (Cao et al. 2003). Figure 10.22

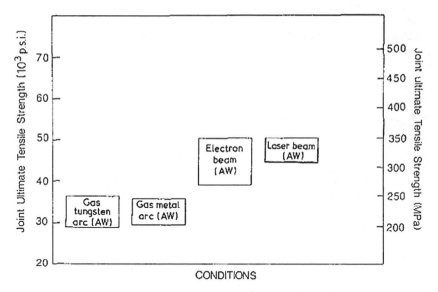

Fig. 10.22 Tensile strength of aluminium–lithium welds produced by various welding processes. (Reprinted from Molian and Srivatsan 1990. With permission. Copyright Springer.)

compares the tensile properties of an aluminum–lithium alloy joint welded by various processes. The better tensile strength of the laser-welded joint is primarily due to the small HAZ and the finer structure in the weld zone (Molian and Srivatsan 1990). Similarly, good fatigue strength is reported for laser-welded joints of 6 mm thick HSLA (high-strength low-alloy) steels with fatigue limit in the range of 260 MPa (Onoro and Ranninger 1997).

10.5   Practical Considerations

This section briefly explains the various laser welding parameters which influence the geometric and metallurgical aspects of the welds, speed of welding, etc.

10.5.1   Effect of Laser Parameters

The various important laser parameters in laser welding include laser type, laser power, and beam dimensions.

   Laser welding has been reported of using a variety of different lasers for various specific materials and applications. By far, the most widely used lasers in welding applications are CO_2 (Wavelength: 10.6 μm) and Nd:YAG (Wavelength: 1.06 μm) lasers. CO_2 lasers are popular particularly due to higher power outputs (>3 kW). Recently, Nd:YAG lasers are attracting significant attention in the laser welding applications. Nd:YAG lasers are now capable of delivering continuous power in the range of 4 kW. Nd:YAG lasers also offer significant advantages over CO_2 lasers. Due to short wavelength of the Nd:YAG lasers, the radiation is much more strongly absorbed in the material. This results in lower threshold power for laser welding with Nd:YAG laser. Once the keyhole is formed, the absorptivity of the material is relatively less important because the keyhole causes the multiple reflection of radiation resulting in enhanced absorption. Also, Nd:YAG laser welding is associated with less severe plasma effects. In general, for a given thickness of workpiece, Nd:YAG laser welding gives higher penetration depths and welding speeds compared to CO_2 laser welding. The most important advantage of the Nd:YAG laser welding is the flexibility offered by fiber optic delivery (Duley 1999).

Laser power directly determines the penetration depth and welding speed for a given material. There is a minimum power for a given thickness of the workpiece which gives efficient welding. Figure 10.23 presents such data for minimum power requirements for the various thicknesses of the materials. As indicated in the figure, for a given thickness of the material, higher power is required for welding copper than steel. This may be due to the high initial reflectivity of copper (Locke et al. 1972). In general, the weld penetration depth increases with increasing laser power. Also, higher welding speed can be achieved by increasing laser power for a given thickness of the workpiece. Figure 10.24 presents the variation of welding speed with laser power for CO_2 laser welding of 4 mm thick Inconel 718 alloy. The figure also presents the corresponding changes in the weld pool morphology with increasing laser speed (Gobbi et al. 1996).

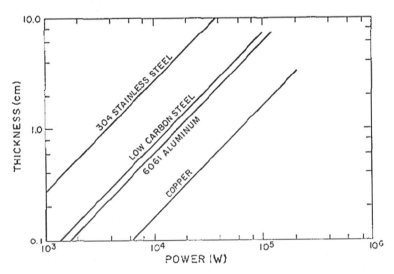

Fig. 10.23 Minimum power for efficient welding of various materials. (Reprinted from Locke et al. 1972. With permission. Copyright Institute of Electrical and Electronics Engineers.)

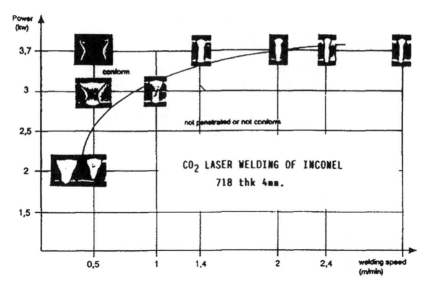

Fig. 10.24 Variation of welding speed with laser power for CO_2 laser welding of 4 mm thick Inconel 718 alloy. (Reprinted from Gobbi et al. 1996. With permission. Copyright Elsevier.)

10.5.2 Effect of Focusing Conditions

Focusing conditions of the laser beam with respect to the specimen surface needs to be optimized to obtain the welds with desired shape and minimum defects. For similar laser processing conditions, the highest depth of the keyhole and the weld penetration corresponds to the beam focus conditions close to the sample surface (Jin and Li 2004). In general, if the position of the focus is away from the sample surface, significantly higher incident laser energy is required to obtain the same depth of penetration. Also, the focus position determines the shape of the molten pool. Figure 10.25 presents the schematic of the shapes of the molten pool obtained at various focusing conditions for aluminum. The figure indicates that defect-free and uniform weld zones are obtained when the focal point is away from the surface (+3 and −3 mm). Negative values indicate the focal point above the sample surface, whereas positive values indicate the focal point inside the sample. When the beam is focused close to the surface (focal point positions: + 1, 0, −1 mm), the weld zone consists of several irregular porosities. This may be due to the unstable keyhole formed when the beam is focused close to the surface (Kim et al. 1995).

10.5.3 Effect of Shielding Gas

During laser welding with sufficiently higher powers, the surface of the workpiece undergoes rapid vaporization. The subsequent ionization of the vapors creates a standing ionization cloud (plasma) above the workpiece surface. This standing

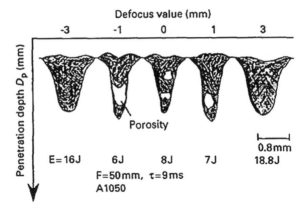

Fig. 10.25 Schematic of the influence of laser beam focus conditions on the shape of weld zones in aluminum. (Reprinted from Kim et al. 1995. With permission. Copyright Springer.)

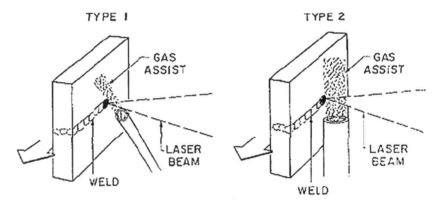

Fig. 10.26 Various configurations of assist gas during laser welding. (Reprinted from Locke et al. 1972. With permission. Copyright Institute of Electrical and Electronics Engineers.)

ionization cloud absorbs and scatters the laser radiation. For effective welding, it is necessary to remove the cloud. Various shielding gases such as argon, helium, CO_2 can be used for this purpose. Two common configurations of assist gases can be used (Fig. 10.26). In one, the assist gas is directed at the laser–material interaction point; while, in the other, the assist gas is directed slightly offset from the workpiece. It has been reported that the assist gas directed slightly offset from the interaction zone results in uniform, even, and smooth welds. Direct impingement of the assist gas causes the nonuniform penetration in the weld. Also, the top surface of the weld is blown off due to dynamic pressure exerted on the molten pool during welding (Locke et al. 1972).

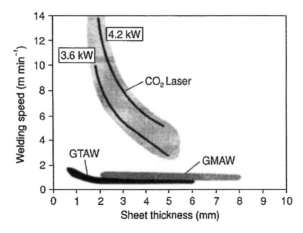

Fig. 10.27 Variation of welding speed with sheet thickness for welding of aluminum with various welding processes. (Reprinted from Rapp et al. 1995. With permission. Copyright Springer.)

10.5.4 Effect of Welding Speed

Thickness of the workpiece determines the welding speed for producing deep welds. In general, as the thickness of the workpiece increases, the welding speed for producing deep welds decreases. Similarly, the depth of penetration decreases with increasing welding speed. Figure 10.27 presents the variation of welding speed as a function of sheet thickness for welding of aluminum with various welding processes. The figure also shows that for a given thickness of the workpiece, laser welding gives higher welding speeds compared to some conventional welding processes (Rapp et al. 1995).

10.5.5 Joint Configurations

Laser welding can be used with a variety of joint configurations such as butt weld, lap weld, edge weld, T-weld, fillet weld, etc. (Fig. 10.28). The welding may be carried out in a continuous seam or through spot welding methods in these joints. The important consideration here is the achievement of tight tolerances by minimizing air gaps between the overlapping workpieces.

Various other practical parameters which influence the weld geometry and quality may include the workpiece surface conditions, use of filler material, etc.

Joint types

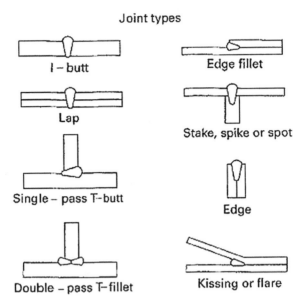

Fig. 10.28 Commonly used joint configurations in laser welding. (Reprinted from Sun and Ion 1995. With permission. Copyright Springer.)

10.6   Laser Welding of Various Materials

Laser welding is demonstrated for a variety of different materials. However, all materials cannot be welded with equal ease due to widely differing thermophysical properties. This section briefly discusses the results of laser welding studies conducted on broad classes of materials.

10.6.1   Metallic Materials

Metallic materials are probably the most extensively studied material systems for laser welding. Laser welding is extensively studied for the joining of steel, aluminum, magnesium, and titanium metals and alloys.

Comprehensive studies on the laser welding of steels were conducted by David and coworkers. They reported that the weld pool cooling rate and the postsolidification solid-state cooling rate during laser welding greatly influence the microstructure in the weld. Depending on the rapid solidification characteristics and the composition of the steel, the microstructure may be fully austenitic, ferritic, or duplex austenitic (David et al. 1987). The microstructure of the weld zone may also be influenced by postwelding tempering. Defocused beam can be used to temper the weld zone such that a finer microstructure is obtained. Laser welding of stainless steel may be associated with the changes in the composition of the weld metal due to the pronounced vaporization of the alloying elements from the weld pool. The pronounced effect of laser welding on the decrease in manganese content in the weld pool has been reported. Furthermore, the effect of manganese vaporization was observed to be more pronounced at low powers. Welding speed and shielding gas flow rates did not significantly influence the composition changes (Mundra and DebRoy 1993; Khan et al. 1988).

Laser welding of aluminum alloys for applications in aerospace, automotive, and other industries is very attractive. Aluminum alloys are also susceptible to the loss of alloying elements in the weld pool. Excellent review of the research progress in the laser welding of aluminum alloys (5000, 6000 series) is published by Cao et al. (2003). This review provides a detailed discussion on the influence of laser type (CO_2, Nd:YAG), laser parameters, process parameters, and material properties on the geometric and metallurgical aspects of the weld. Discussions of the welding defects and the mechanical properties of laser welds are also presented in this review.

Titanium alloys, because of their excellent properties, find wide applications in aerospace, ship, and chemical industries. Due to high chemical activity, conventional welding of titanium is often associated with serious weld defects. Recently significant interests have been directed towards welding of titanium alloys. Titanium alloys show a fine acicular α martensitic phase in the weld zone. Various studies have reported the development of microstructure in the weld pool, the mechanical properties of the joint, and the influence of the processing parameters during laser welding of titanium alloys (Li et al. 1997; Yunlian et al. 2000). Recently, laser welding of titanium alloys is reported to be advantageous particularly for dental applications (Liu et al. 2002). The important considerations during the laser welding of titanium are the attainment of desired tolerances without welding defects and the minimization of welding defects.

10.6.2   Ceramics

Ceramic materials strongly absorb the radiation from the common industrial laser sources (Nd:YAG and CO_2). Hence, laser welding of ceramics seems attractive. However, laser welding of ceramics is limited due to serious quality issues. Most importantly, the rapid solidification during laser processing is associated with the development of extended cracks in the weld zone. Furthermore, other serious defects, such as porosity, HAZ, and grain growth makes welding ineffective. Very few studies have been published on the subject of laser welding of ceramics. The most comprehensive report on laser welding of sintered zirconia was published by Maruo et al. (1986). They studied the fusion welding and surface melting characteristics of sintered zirconia ceramics fully stabilized with CaO, MgO, or Y_2O_3 using CO_2 laser. It was reported that preheating at temperatures exceeding $1,200\,°C$ prevents the macroscopic cracks in the base material. However, microscopic intergranular cracks were observed in the fusion zone in most of the compositions of zirconia ceramic. In one of the compositions of CaO stabilized zirconia containing 1.7% SiO_2, 1% MgO, and 0.7% Al_2O_3, no microcracking was observed. This may be due to the formation of low melting compounds at the grain boundaries which plays an important role in releasing the shrinkage stress in the fusion zone. Furthermore, it was observed that the strength of the weld joint is reduced due to the porosity formed at the HAZ at lower welding speeds and the porosity formed at the bead center at higher welding speeds. Increasing laser welding speed is reported to suppress the formation of porosities, thus increasing the weld joint efficiency (Maruo et al. 1986).

10.6.3 Polymers

The laser welding of polymers was first reported in 1972 (Ruffler and Gürs 1972). These early reports dealt with the welding of low density polyethylene sheets (thickness up to 1.5 mm) using CO_2 laser. Recently there are renewed interests in the welding of polymers primarily directed towards increasing polymer joining speeds in packaging applications (Duley and Mueller 1992). Various lasers such as CO_2, Nd:YAG, and diode lasers can be used for polymer welding. Major challenge in the welding of polymers is to optimize the laser processing parameters such that thermal decomposition of the polymer is avoided. This limits the temperature range for laser welding between the melting point and decomposition temperature. The welding characteristics of the polymer may be altered by various additives and colorants in the plastic material (Coelho et al. 2000). Extensive research efforts are still needed to develop the laser welding of polymers in industrial practices.

10.6.4 Composites

Composite materials are increasingly used in various aerospace and automotive applications due to the unique properties derived from the combination of reinforcing and the matrix phases. However, welding and joining of composite materials is often challenging due to the differing interaction of thermal energy with the constituent phases of the composite and the corresponding effect of the matrix reinforced material compatibility. Extensive efforts have been directed towards laser welding of SiC–Al metal matrix composites (Dahotre et al. 1989, 1991, 1992, 1993). The studies were mostly conducted on the composites consisting of 10–20 vol. % SiC (10 μm average size) particles distributed in A356 aluminum matrices. The microstructure of the laser-welded composites showed significant variation across the weld zone. It was reported that the laser-welded microstructure consisted of platelike and blocky precipitates distributed in the dendritic matrix in the fusion zone, whereas HAZ showed platelike precipitates along with SiC particulates and small blocks of Si (Fig. 10.29). The average length and width of the platelike

Fig. 10.29 Microstructure of SiC$_p$ reinforced aluminum (A356) matrix composites after laser welding with specific energy of 5.3×10^4 J/cm^2: (**a**) heat-affected zone, (**b**) laser-melted zone, and (**c**) magnified view of laser-melted zone showing platelike structure. (Reprinted from Dahotre et al. 1989. With permission. Copyright American Institute of Physics.)

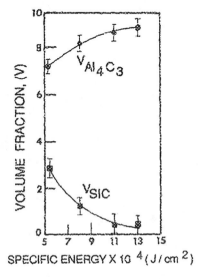

Fig. 10.30 Variation of volume of SiC particulate and Al$_4$C$_3$ platelike precipitates in the laser-melted region as a function of specific energy. (Reprinted from Dahotre et al. 1989. With permission. Copyright American Institute of Physics.)

precipitates is influenced by the laser energy density. The platelike precipitates (Al$_4$C$_3$) in the weld zone were formed according to the reaction:

$$4Al_{(l)} + 3SiC_{(s)} = Al_4C_{3(s)} + 3Si_{(s)}.$$

The above reaction suggests that the platelike precipitates are formed at the expense of SiC particulates. This is supported by the observation that with increasing specific energy, the volume fraction of the Al$_4$C$_3$ precipitates increases and the volume fraction of SiC particulate decreases (Fig. 10.30). Figure 10.31 presents the variation of microhardness in the various planes of the laser-welded region of SiC/Al356 MMC. The high hardness in the laser-melted zone seems to be due to the formation of precipitates and associated effects (Dahotre et al. 1989).

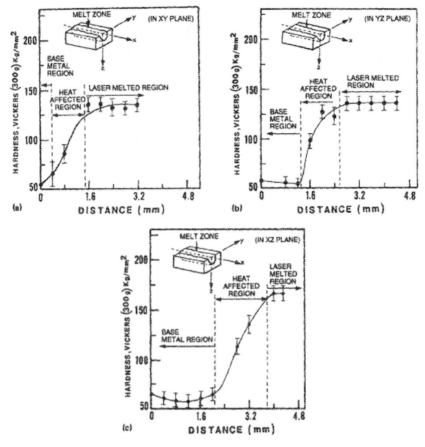

Fig. 10.31 Microhardness profiles in the laser weld of SiC/A356 aluminum matrix composite in various planes: (a) the $x–y$ plane, (b) the $y–z$ plane, and (c) the $x–z$ plane (specific energy: 8.0×10^4 J/cm^2). (Reprinted from Dahotre et al. 1989. With permission. Copyright American Institute of Physics.)

10.6.5 Dissimilar Materials

Welding of dissimilar metals presents significant challenges due to differing physical properties such as absorptivity, melting point, thermal conductivity, and thermal expansion coefficient. Such differing properties of the materials involved in the welding joint may result in the formation of intermediate phases detrimental to the mechanical properties of the joint. Laser welding offers significant advantages over conventional fusion welding processes for dissimilar welding applications. This includes better weld quality, higher productivity, and better flexibility. Most importantly, laser welding parameters can be optimized to cover a range of dissimilar metal combinations. Excellent review of the laser welding of dissimilar metals is published by Sun and Ion (1995). This review discusses in detail the underlying principles, research trends, and applications of dissimilar metal laser welding. The discussions on the laser welding of stainless steel–carbon steel, steel–copper, steel–aluminum, steel–nickel, aluminum–lead dissimilar metal combinations were presented (Sun and Ion 1995).

10.7   Advances in Laser Welding

Significant progress has been made to improve the quality of the laser welds, welding efficiency, and productivity. Some of the major areas which are attracting considerable interests are briefly explained in this section.

10.7.1   Arc-Augmented Laser Welding

This process involves rooting of an electric arc into the keyhole or the laser–material interaction zone produced during laser welding. Arc-augmented laser welding was first demonstrated by Eboo et al. (1978) using a TIG welding arc coupled with a laser beam for the welding of tin plates. It was observed that such an arc augmentation gives a welding speed which is almost four times the speed obtained with laser welding alone. The arc augmented laser welding shows many characteristics similar to the laser welding process. The arc current plays an important role in determining whether full penetration in the weld would occur (Eboo et al. 1978; Steen and Eboo 1979; Steen 1980). Significant studies have been conducted since then to determine the influence of arc current on the welding speed and the susceptibility of forming welding defects. It was reported that maximum welding speed increases with increasing arc current (Alexander and Steen 1980). Also, coupling of the arc increases the dissipation of energy within the keyhole resulting in increased liquid volume around the keyhole and increased weld width. Increased volume of liquid around the keyhole is expected to reduce the susceptibility of forming welding defects. Arc-augmented laser welding with coupled energy sources may present an alternative to the high-power welding process (Duley 1999).

10.7.2   Multibeam/Dual-Beam Laser Welding

The use of two laser beams for welding was first demonstrated by Glumann et al. (1993). The process is reported to offer significant advantages in terms of quality of the weld. It was observed that dual-beam laser welding of aluminum sheets resulted in significant reduction in porosity when the second beam was defocused. This was attributed to the better stabilization of the keyhole by optimization of the laser processing parameters using independently controlled laser beams. The dual-beam laser welding offers flexible processing since the power and other laser parameters can be independently altered in two laser beams (Glumann et al. 1993). Since the average power output of the Nd:YAG lasers (4 kW or greater) is less than the CO_2 lasers, dual-beam processing with Nd:YAG lasers is particularly attractive. Due to high absorptivity of the materials to Nd:YAG laser radiation, coupling of the laser powers is expected to offer significant advantages over single-beam laser welding (Punkari et al. 2003).

Laser welding with two split beams from a single laser source is also expected to offer significant control over the weld microstructure and productivity. One of the beams can be used for melting of the workpiece during welding, while, the other beam can be used either for preheating of workpiece ahead of welding beam or postheating (heat treatment) of the welds past the welding beam. The distance between the two beams on the workpiece can be adjusted to control the thermal effects. The split-beam laser welding is expected to offer significant quality advantages over single beam (high intensity) laser welding in case of high hardenability materials. These quality advantages are derived from lower cooling rates associated with split-beam laser welding compared to those obtained with single-beam welding (Kannatey-Asibu 1991).

10.7.3   Laser Welding of Tailor-Welded Blanks

One of the biggest challenges the present day automotive industry is faced with is to produce the high-performance cars at minimum costs. An approach to this is to reduce the overall weight of the vehicle which also improves the fuel economy and reduces emission. Significant reduction in the weight of the vehicle can be achieved by welding the blanks (of various materials, thicknesses, properties, surface treatments, etc.) into a single "tailor-welded blank". This is followed by forming of the tailor-welded blank (TWB). Laser welding has demonstrated great success in the welding of TWBs, which result in substantial cost benefits due to reduction in the number of dies, minimization of scrap, reduction in the number of process steps, etc. Also, TWB offers significant improvements in the dimensional accuracy and properties of the formed components (Kinsey et al. 2000). Laser welding of the TWBs is challenging due to differing properties of the starting blanks. However, laser processing parameters can be effectively optimized to achieve the desired results.

References

Alexander J, Steen WM (1980) The effect of process variables on arc augmented laser welding, Symposium OPTIKA'80, Budapest.
Andrews JG, Atthey DR (1976) Hydrodynamic limit to penetration of a material by a high-power beam. Journal of Physics D: Applied Physics 9:2181–2194.

Borland JC (1960a) Generalized theory of super-solidus cracking in welds and (castings). British Welding Journal 7:508–512.

Borland JC (1961b) Suggested explanation of hot cracking in mild and low alloy steel welds. British Welding Journal 8:526–540.

Cao X, Wallace W, Immarigeon JP, Poon C (2003) Research and progress in laser welding of wrought aluminum alloys. II. Metallurgical microstructures, defects, and mechanical properties. Materials and Manufacturing Processes. 18:23–49.

Carslaw HS, Jaeger JC (1959) Conduction of heat in solids. Oxford University Press, Oxford.

Cline HE, Anthony TR (1977) Heat treating and melting material with a scanning laser or electron beam. Journal of Applied Physics 48:3895–3900.

Coelho JP, Abreu MA, Pires MC (2000) High-speed laser welding of plastic films. Optics and Lasers in Engineering 34:385–395.

Dahotre NB, McCay TD, McCay MH (1989) Laser processing of a SiC/Al–alloy metal matrix composites. Journal of Applied Physics 65:5072–5077.

Dahotre NB, McCay MH, McCay TD, Gopinathan S, Allard LF (1991) Pulse laser processing of a SiC/Al–alloy metal matrix composite. Journal of Materials Research 6:514–529.

Dahotre NB, McCay MH, McCay TD, Gopinathan S, Sharp CM (1992) Laser joining of metal matrix composites. In: Srivatsan TS, Bowden D (eds) Machining of composite materials, ASM International, Metals Park, Ohio.

Dahotre NB, McCay TD, McCay MH (1993) Laser-induced reaction joining of metal matrix composite. Proceedings of ICALEO'93, Laser Institute of America, Orlando, FL, vol. 73.

David SA, Vitek JM, Reed RW, Hebble TL (1987) Effect of rapid solidification on stainless steel weld metal microstructures and its implication of Schaeffler diagram. Technical Report, Oak Ridge National Laboratory.

Dowden J, Kapadia P (1995) A mathematical investigation of the penetration depth in keyhole welding with continuous CO_2 lasers. Journal of Physics D: Applied Physics 28: 2252–2261.

Dowden J, Postacioglu N, Davis M, Kapadia P (1987) A keyhole model in penetration welding with a laser. Journal of Physics D: Applied Physics 20:36–44.

Dudas JH, Collins FR (1966) Preventing weld cracks in high-strength aluminum alloys. Welding Journal 45:241S–249S.

Duley WW (1999) Laser welding. Wiley, New York.

Duley WW, Mueller RE (1992) CO_2 laser welding of polymers. Polymer Engineering and Science 32:582–585.

Eboo M, Clarke J, Steen WM (1978) Arc-augmented laser welding. Advances in Welding Processes 1:257–265.

Girard K, Jouvard JM, Naudy P (2000) Study of volumetric defects observed in laser spot welding of tantalum. Journal of Physics D: Applied Physics 33:2815–2824.

Glumann C, Rapp J, Dausinger F, Hügel H (1993) Welding with a combination of two CO_2-lasers – advantages in processing and quality. Proceedings of the SPIE vol. 230, p. 672.

Gobbi S, Zhang L, Norris J, Richter KH, Loreau JH (1996) High power CO_2 and Nd:YAG laser welding of wrought Inconel 718. Journal of Materials Processing Technology 56:333–345.

Jin X, Li L (2004) An experimental study on the keyhole shapes in laser deep penetration welding. Optics and Lasers in Engineering 41:779–790.

Kannatey-Asibu E (1991) Thermal aspects of the split-beam laser welding concept. Journal of Engineering Materials and Technology 113:215–221.

Kaplan A (1994) A model of deep penetration laser welding based on calculation of the keyhole profile. Journal of Physics D: Applied Physics 27:1805–1814.

Khan PA, DebRoy T, David S (1988) Laser beam welding of high-manganese stainless steels – examination of alloying element loss and microstructural changes. Welding Research Supplement 67:1S–7S.

Kim JS, Watanabe T, Yoshida Y (1995) Effect of the beam-defocusing characteristics on porosity formation in laser welding. Journal of Materials Science Letters 14:1624–1626.

Kinsey B, Liu Z, Cao J (2000) A novel forming technology for tailor welded blanks. Journal of Materials Processing Technology 99:145–153.

Klein T, Vicanek M, Kroos J, Decker I, Simon G (1994) Oscillations of the keyhole in penetration laser beam welding. Journal of Physics D: Applied Physics 27:2023–2030.

Klemens PG (1976) Heat balance and flow conditions for electron beam and laser welding. Journal of Applied Physics 47:2165–2174.

Kroos J, Gratzke U , Simon G (1993) Towards a self-consistent model of the keyhole in penetration laser beam welding. Journal of Physics D: Applied Physics 26:474–480.

Lampa C, Kaplan A, Powell J, Magnusson C (1997) An analytical thermodynamic model of laser welding. Journal of Physics D: Applied Physics 30:1293–1299.

Li Z, Gobbi SL, Norris I, Zolotovsky S, Richter KH (1997) Laser welding techniques for titanium alloy sheet. Journal of Materials Processing Technology 65:203–208.

Liu J, Watanabe I, Yoshida K, Atsuta M (2002) Joint strength of laser-welded titanium. Dental Materials 18:143–148.

Locke EV, Hoag ED, Hella RA (1972) Deep penetration welding with high-power CO_2 lasers. IEEE Journal of Quantum Electronics 8:132–135.

Mackwood AP, Crafer RC (2005) Thermal modelling of laser welding and related processes: a literature review. Optics & Laser Technology 37:99–115.

Maruo H, Miyamoto I, Arata Y (1986) Laser welding of sintered zirconia: CO_2 laser welding of ceramics III. Quarterly Journal of the Japan Welding Society 4:192–198.

Matsunawa A, Semak V (1997) The simulation of front keyhole wall dynamics during laser welding. Journal of Physics D: Applied Physics 30:798–809.

Mazumder J, Steen WM (1980) Heat transfer model for cw laser material processing. Journal of Applied Physics 51:941–947.

Molian PA, Srivatsan TS (1990) Weldability of aluminium–lithium alloy 2090 using laser welding. Journal of Materials Science 25:3347–3358.

Mundra K, DebRoy T (1993) Towards understanding alloying element vaporization during laser beam welding of stainless steel. Welding Journal 72:1S–9S.

Onoro J, Ranninger C (1997) Fatigue behavior of laser welds of high-strength low-alloy steels. Journal of Materials Processing Technology 68:68–70.

Pastor M (1998) M.S. Thesis. Pennsylvania State University, PA, USA.

Pellini WS (1952) Strain theory of hot tearing. Foundry 80:124–133.

Postacioglu N, Kapadia P, Dowden J (1991) Theory of the oscillations of an ellipsoidal weld pool in laser welding. Journal of Physics D: Applied Physics 24:1288 –1292.

Pumphrey WI, Jennings PH (1948) A consideration of the nature of brittleness at temperature above the solidus in castings and welds in aluminum alloys. Journal of Institute of Metals 75:235–256.

Punkari A, Weckman DC, Kerr HW (2003) Effects of magnesium content on dual beam Nd: YAG laser welding of Al–Mg alloy. Science and Technology of Welding and Joining 8:269–281.

Rapp J, Glumann C, Dausinger F, Hügel H (1995) Laser welding of aluminium lightweight materials: problems, solutions, readiness for application. Optical and Quantum Electronics 27:1203–1211.

Rosenthal D (1946) The theory of moving sources of heat and its application to metal treatment. Transactions of ASME 48:849–866.

Ruffler C, Gürs K (1972) Cutting and welding using CO_2 laser. Optics and Laser Technology 4:265–269.

Sickman JG, Morijn R (1968) Philips Research Reports.

Steen WM, Eboo M (1979) Arc-augmented laser beam welding. Metal Construction 11:332–335.

Steen WM (1980) Arc-augmented laser processing of materials. Journal of Applied Physics 51:5636–5641.

Sun Z, Ion JC (1995) Review – Laser welding of dissimilar metal combinations. Journal of Materials Science 30:4205–4214.

Swift-Hook DT, Gick AE (1973) Penetration welding with lasers. Welding Journal 52:492S–499S.

Vedani M (2004) Microstructural evolution of tool steels after Nd:YAG laser repair welding. Journal of Materials Science 39:241–249.

Yunlian Q, Ju D, Quan H, Liying Z (2000) Electron beam welding, laser beam welding and gas tungsten arc welding of titanium sheet. Materials Science and Engineering A 280:177–181.

Zhao H, Whire DR, DebRoy T (1999) Current issues and problems in laser welding of automotive aluminum alloys. International Materials Review 44:238–266.

Part IV
Special Topics in Laser Processing

Chapter 11
Laser Interference Processing

11.1 Introduction

Previous chapters deal mostly with the processing of materials with a single beam of laser (as in drilling, cutting) or with multiple beams from different laser sources (as in laser shaping). Novel material processing methods can be developed based on the interference patterns produced by the superposition of two or more laser beams. Historically, interference phenomena have been the means of establishing the wave nature of light and have found significant practical applications in spectroscopy and metrology (Born and Wolf 1980). Recently, the interference phenomena have been utilized for the surface processing of materials in a wide range of applications such as micromachining and biomedical applications. Laser interference processing is a relatively new technique finding increased utilization in the areas of extended-area surface processing of materials. This chapter briefly discusses the theory and the applications of laser interference phenomena in surface processing of materials.

11.2 Theory of Interference

When a beam of light is divided by a suitable apparatus into two or more beams which are subsequently superposed, the intensity in the region of superposition shows the unique variation. The intensity in the region of superposition varies from point to point between the maxima (exceeding the sum of intensities in the beams) and the minima (may be zero). This superposition of two or more beams is referred as interference. Interference patterns are generally obtained by the superposition of beams which are coherent with each other. The beams coming from different sources are mutually incoherent and no interference is generally observed under common experimental condition. However, if the two beams originate from the same source, the fluctuations in the beam are generally correlated and the beams are said to be completely or partially coherent. The superposition of such coherent beams originating from the same source gives rise to interference patterns. There are two methods for obtaining beams from a single source: division of wave front and division of amplitude. In the first method, a beam is divided by passage through

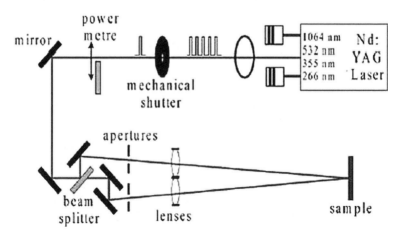

Fig. 11.1 The basic experimental arrangement for two-beam laser interference processing. (Reprinted from Daniel et al. 2003. With permission. Copyright Elsevier.)

adjacent apertures. In the other method, a beam is divided by partially reflecting surfaces where a part of light is reflected and the other part is transmitted (Born and Wolf 1980). In most of the laser interference processing methods, optical devices such as beam splitters are used which split a beam of light into two by partially reflecting and transmitting a beam. In a simplified arrangement, a beam splitter consists of two triangular glass prisms which are joined together at the base using suitable resin. In other arrangements, thin films deposited on glass surfaces which enhance the reflectivity can be used as beam splitters. A beam splitter in an interferometer divides an incident beam into two beams.

Typical experimental arrangement for materials processing using two-beam interference technique is shown in Fig. 11.1. The various elements of this arrangement are the laser source, interferometer, and the imaging surface (Daniel et al. 2003).

The geometry of the interference patterns formed by the superposition of two or more coherent and linearly polarized beams depends on the wavelength and the angle between the beams. The intensity distribution resulting from the superposition of two linearly polarized beams with their E vectors in the x-direction can be expressed as (Daniel et al. 2003):

$$I(x) = 2I_0 \left[\cos\left(\frac{2\pi x}{l}\right) + 1 \right], \tag{11.1}$$

where I_0 is the intensity of a laser beam, λ is the wavelength, θ is the angle between the beams, and l is the period. The superposition of two beams produces an interference pattern with spatially modulated light field with intensity distribution oscillating between zero and $4I_0$. The interference of the two beams generates a one-dimensional periodic pattern. The periodicity of the two-beam interference pattern is expressed as:

$$l = \frac{\lambda}{2\sin(\theta/2)}. \tag{11.2}$$

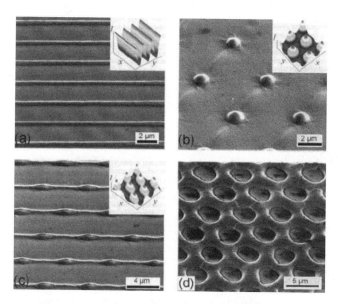

Fig. 11.2 Various possible one-dimensional and two-dimensional geometric structures that can be produced by three-beam interference patterns by varying angle and intensity of laser beams (Reprinted from Mücklich et al. 2006. With permission. Copyright Hanser.)

Two-dimensional and three-dimensional periodic patterns can be obtained by increasing the number of beams. The intensity distribution resulting from the super-position of four beams is given by (Kondo et al. 2001).

$$I(x, y) = 4I_0 \left[\cos\left(\frac{2\pi x}{l}\right) + \cos\left(\frac{2\pi y}{l}\right) \right]^2. \tag{11.3}$$

The superposition of four beams produces the interference pattern with two-dimensional modulated light field oscillating between zero and $16I_0$ and periodicity equal to $l\sqrt{2}$ (Kaganovskii et al. 2006). However, inference with multiple beams requires a complicated optical setup and its precise adjustment is often difficult.

Figure 11.2 presents the schematic of the possible one-dimensional and two-dimensional periodic geometric structures produced by three-beam interference patterns. As indicated in the figure periodic line or dot patterns can be produced by interference of the multiple beams.

11.3 Interferometry for Surface Processing of Materials

As mentioned previously, the three important elements of the interference arrangement are the laser, interferometer, and imaging surface of the material. In the context of surface processing of materials, each of these elements needs to be given careful consideration during the design of interferometer.

11.3.1 Laser and Materials Aspects

As mentioned earlier, the interference patterns are obtained from the beams which are coherent. Incoherent beams would not interfere to produce dark and bright fringes (due to intensity modulation in the resultant wave). Both the temporal and the spatial coherence of the beams need to be preserved to realize the interference pattern. Spatial coherence is related with the correlation between two points on the same wave front, whereas the temporal coherence is related with the correlation of similar points on different wave fronts. The spatial coherence of beams is greatly influenced by the presence of a number of longitudinal modes in the laser output (Engleman et al. 2005). Generally, loss of coherence occurs with the increasing number of longitudinal modes. Figure 11.3 presents the influence of a number of longitudinal modes on the visibility of the fringes in the interferometric experiment. As indicated in the figure, the visibility of the fringes decreases with path difference for multimode operation of laser. Thus, in multimode operations the allowable path difference is limited (Ready 1997). Temporal coherence is related with the spectral bandwidth of the source. Narrower bands result in longer coherence time. Coherence time (Δt) is expressed as the reciprocal of line width. The coherence length (Δx) is then given by the product of wave speed (c) and the coherence time (Δt). The coherence length is again influenced by the number of operational modes. For example, typical coherence length of multimode He–Ne laser is in the range of 20 cm, whereas the typical coherence length of single mode He–Ne laser is in the range of 100,000 cm (Ready 1997).

The type of laser source determines the interference pattern produced on the material surface. The most important laser parameters are the laser wavelength and the

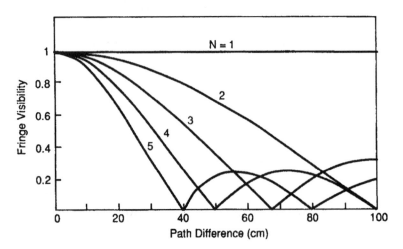

Fig. 11.3 Variation of fringe visibility with path difference for a laser operating with N longitudinal modes in an interferometric experiment. (Reprinted from Ready 1997. With permission. Copyright Elsevier.)

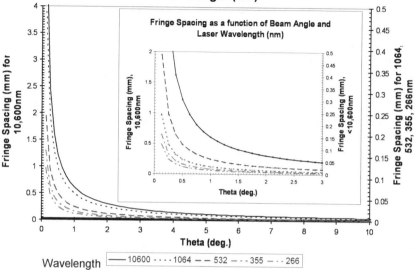

Fig. 11.4 Variation of theoretical fringe spacing with angle between the interfering beams for some common wavelengths of the lasers used in materials processing. (Reprinted from Engleman et al. 2005. With permission. Copyright The Minerals, Metals and Materials Society.)

angle between the interfering beams. These parameters determine the fringe spacing according to Eq. (11.2). Figure 11.4 presents the theoretical variation of fringe spacing with interference angle for some common wavelengths employed in laser materials processing. The figure indicates that for a given wavelength of the laser, shorter fringe spacing is produced with the beams interfering at large angle. The figure also indicates that shorter wavelengths (266, 355, 532, and 1,064 nm) produce fringe spacing which is proportionally smaller than that produced by lasers of longer wavelength (10.6 μm). The physical lower limit of the fringe spacing according to Eq. (11.2) is half the wavelength of the laser. Fringe spacing greatly influences the spatial resolution of the features on the surface of the material through the combination of effects such as physical, chemical, and metallurgical effects (Engleman et al. 2005).

In addition to the wavelength and the angle between the interfering beams, the other important laser parameter is the laser fluence (energy density). Laser fluence is determined by the laser power, the irradiated surface area, and the irradiation time. Laser fluence along with the thermophysical properties of materials determines the temperature distribution in the materials. The temperature distributions in the materials during laser surface processing are generally obtained by the solution of Fourier's equation of heat transfer. The Fourier's equation of heat transfer can be expressed as (Engleman et al. 2005):

$$\rho c_{p} \frac{\partial T}{\partial t} = q_{a} - q_{m} - q_{v} + \nabla \cdot (k \nabla T),$$ (11.4)

where $T = T(x,z,t)$ is the temperature at position(x,z) at time t; ρ, k, and c_p are the density, the thermal conductivity and the specific heat of the material respectively; and q_a, q_m, and q_v are the absorbed heat, the heat of melting, and the heat of vaporization respectively. The amount of heat absorbed by the material depends on the absorptivity of the material which is determined by various material and surface related factors such as surface roughness, surface contamination, angle of tilt, etc. The solution of the heat transfer equation gives the temperature distribution as a function of laser parameters and material properties. For a simplified case of one-dimensional conduction without convection and radiation effects, the solution of the heat transfer equation can be re-arranged to estimate the energy required to produce a single fringe of a particular surface feature size. This equation of energy can be expressed as (Engleman et al. 2005):

$$E = \frac{kT_m t_p 10^{-4}}{2\sqrt{\chi t_p}\, ierfc\left[\dfrac{z}{2\chi t_p}\right]}, \tag{11.5}$$

where E is energy fluence (J/cm^2), T_m is the melting point (°C), t_p is the pulse time (s), χ is the thermal diffusivity (m^2/s), and z is the surface feature size (m).

The region of the surface modified by melting, ablation, etc., determines the feature size (d_f) that can be created on the surface. To obtain the well-defined distinguishable periodic pattern, the feature size must be equal to or smaller than the fringe spacing (d_i). As the conductivity of the material increases, the heat is rapidly dissipated thus increasing the area modified by interference pattern. For low conductivity material, the thermal effects due to modulated intensity are limited to very narrow regions resulting in feature sizes smaller than the fringe spacing (Fig. 11.5). As the conductivity of the material increases, the feature size approaches the fringe spacing. Figure 11.5 also indicates the effect of increasing the angle of interference on the fringe spacing. Based on the heat transfer analysis discussed in Section 11.2, Table 11.1 provides the calculated amount of energies required to produce surface feature sizes equal to interference fringe spacings for a variety of materials irradiated with some common laser sources. The table, thus, provides the guidelines for the selection of appropriate laser processing parameters to achieve desired geometric structures in a given material by irradiating with laser interference pattern (Engleman et al. 2005).

11.3.2 Interferometer Design Aspects

Typical interferometer designs generally consist of a beam expanding telescope (BET), interferometer optics (beam splitters and a set of mirrors), and focusing optics. BET determines the size of the beam through the interferometer and thus determines the energy fluence at the sample surface. A beam of laser is then divided by a beam splitter into multiple beams which are subsequently superposed on the sample surface

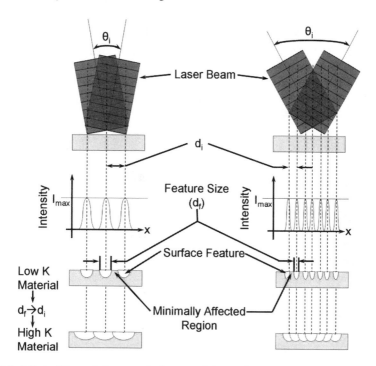

Fig. 11.5 Effect of the angle between the beams and the conductivity of the material on the interference spacing (d_i) and the feature size (d_f) obtained on the surface. (Reprinted from Engleman et al. 2005. With permission. Copyright The Minerals, Metals and Materials Society.)

using a set of mirrors. The contrast between the bright and dark fringes in the interference pattern is determined by the intensity distribution in the resultant wave. The optical path difference between the interfering waves is determined by the difference in the length of the interferometer arms. The optical path difference must be less than the coherence length to maintain the temporal coherence. The optical path difference also determines how well defined the pattern is. If one arm of the interferometer is shorter than the other, the beam from the shorter arm will arrive at the sample surface first thus initiating the surface modifications at the sample surface. For such a case, the interaction time between the beams will decrease. The interferometer design should be sufficiently flexible to allow the setting of any angle between the interfering beams with a minor adjustment of incidence angle and the movement of mirrors (Engleman et al. 2005). As described before, for a given wavelength of the laser beam, the spacing of the intensity distribution is determined by the angle between the interfering beams. The smaller the angle between the interfering beams, the larger is the spacing in the pattern. Thus, the upper limit of the spacing is determined by the smallest achievable angle with the interferometer optics. For larger spacing, special frameless optical elements could be designed to allow the smaller interference angle (Daniel 2006). Also, a set of focusing optics can be incorporated to adjust the energy fluence at the imaging surface of the sample.

Table 11.1 Calculated amount of energy required to produce a surface feature of given size (d_f) equal to interference fringe size (d_i) for a variety of materials using common laser sources. (Reprinted from Engleman et al. 2005. With permission. Copyright The Minerals, Metals and Materials Society.)

Material	Thermal conductivity K (W/mK)	Melting point (°C)	Q-Switched Nd:YAG, λ = 355 nm 10 ns pulse			Pulsed Nd:YAG λ = 1.064 μm 0.5 ms pulse			Pulsed CO_2 λ = 10.6 μm 10 μs pulse		
						Interference angle (θ)					
			5°	10°	30°	5°	10°	30°	5°	10°	30°
						Interference fringe size (μm)					
			4.07	2.04	0.69	12.21	6.11	2.06	121.51	60.97	20.48
			Energy required to produce surface feature of size equal to interference fringe size (J/cm2)								
Al	222	660	22.68	0.72	0.25	30.81	30.37	30.08	439.79	19.85	6.21
Ag	398	962	6.87	0.84	0.36	61.09	60.43	60.00	162.16	24.38	11.22
Cu	419	1,084	27.91	1.59	0.54	86.09	84.99	84.28	590.11	45.18	16.69
AISI1020	50.2	1,538	—	666.54	0.73	44.07	42.42	41.39	—	9.4 × 103	20.53
Ni	67	1,444	—	43.39	0.51	43.70	42.41	41.59	—	819.87	14.79
Si	170	1,410	23.81	1.00	0.31	47.84	47.19	46.77	481.09	27.96	9.47
SS304	16.2	1,427	—	—	2.78	24.88	23.20	22.19	—	—	62.96
Sn	67	232	20 × 102	0.63	0.05	5.16	5.04	4.97	19.8 × 103	14.701	1.33
Ti-6A-4V	7.3	1,688	—	—	61.00	15.09	13.50	12.60	—	—	905.77
Zn	105	420	43.9 × 102	1.64	0.12	14.43	14.12	13.92	43.6 × 103	38.7	3.69
Al2O3	33	2,050	—	23.2 × 103	1.2	47.0	44.9	43.5	—	22.7 × 104	32.2
HA	1.2	1,100	—	—	—	9.13	6.45	5.32	—	—	—
PVC	0.159	200	—	—	—	0.68	0.33	0.23	—	—	—
PMMA	0.199	140	—	—	—	0.54	0.27	0.19	—	—	—

11.4   Applications of Laser Interference Processing

This section briefly explains the various important applications of laser interference processing.

11.4.1   Crystallization and Structuring of Semiconductor Films

Recently, laser interference processing is attracting growing interests in the semiconductor industry. The applications which have received significant attentions include the laser-induced crystallization and the structuring of amorphous and nano-crystalline semiconductors. These applications are briefly discussed in the following sections.

When two or more beams are allowed to interfere on the surface of amorphous film, the modulation of intensity can induce the periodic crystallization patterns with alternating amorphous and polycrystalline lines (two-beam interference) or dots (three-beam or four-beam interference). The laser-induced crystallization involves ultrafast melting and solidification processes far from thermal equilibrium (Mulato et al. 2002). Laser-induced crystallization of amorphous semiconductors is of particular interest as it enables the fabrication of large area films for applications in flat panel displays and solar cells. The applications of interference patterns for producing periodic microcrystalline structures were first demonstrated for hydrogen-free amorphous silicon films using a pulsed dye laser (Heintze et al. 1994). Figure 11.6 presents the periodic line-like and dot-like crystalline structures produced by interference crystallization of amorphous silicon followed by selective plasma etching. As indicated in Fig. 11.6a, the sinusoidal intensity modulation in the two-beam interference results in the line grating of 400 nm wide square wave stripes separated by 340 nm wide trenches. The sharpness of the interface between the microcrystalline

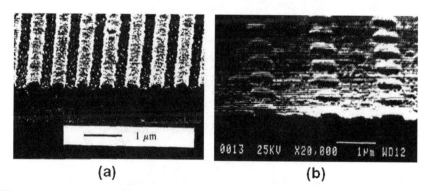

(a)   (b)

Fig. 11.6 (**a**) Crystalline line gratings and (**b**) dot gratings produced by laser interference crystallization and selective plasma etching. (Reprinted from Heintze et al. 1994. With permission. Copyright American Institute of Physics.)

and the amorphous region results from the well-defined threshold of laser crystallization of amorphous silicon (95 mJ/cm²). The periodic two-dimensional dot gratings can be produced by the interference of four beams such that each crystalline dot represents the crossing point of two superposed perpendicular line gratings (Fig. 11.6b). It is necessary to select the intensity of the beams such that crystallization is induced only at the interference maxima at the crossing points of two perpendicular line gratings. Microcrystalline dots with an average diameter of 700 nm and thickness of 200 nm have been produced using a combination of laser interference and selective plasma etching.

Similar laser interference crystallization studies have been conducted on the amorphous germanium films (Mulato et al. 1997; Mulato et al. 1998). Figure 11.7 presents the dot pattern of crystallized germanium with hexagonal lattice symmetry obtained by three-beam laser interference. The crystallinity of the dots can be confirmed using spatially resolved micro-Raman spectroscopy. Figure 11.8 presents the spatial variation (lateral resolution of 0.7 μm) of crystalline (300 cm⁻¹) and amorphous (~270 cm⁻¹) components of the Raman spectrum across the laser-crystallized dot. The figure indicates the highest crystalline contribution at the center of the dot and the highest amorphous contribution in between the dots (Mulato et al. 1997).

The amorphous silicon and germanium films grown by PECVD (plasma enhanced chemical vapor deposition) generally contain more than 10 at. % hydrogen. When such films are subjected to laser interference crystallization, explosive effusion of hydrogen takes place leading to the disruption of the film surface or the formation of free standing films. Recently, laser interference crystallization has been studied for hydrogen-free amorphous germanium–nitrogen (*a*-GeN) alloys to determine the role of nitrogen during phase transition. Figure 11.9 presents the

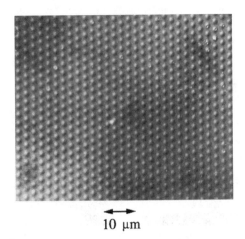

10 μm

Fig. 11.7 Laser interference crystallization of amorphous germanium showing the hexagonal lattice of crystallized germanium dots with a period of 2.6 μm obtained with three-beam interference. (Reprinted from Mulato et al. 1997. With permission. Copyright American Institute of Physics.)

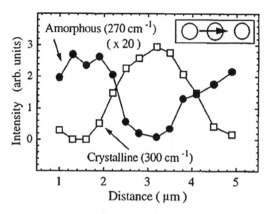

Fig. 11.8 Spatial variation of the crystalline (~300 cm⁻¹) and the amorphous (~270 cm⁻¹) components of the Raman spectrum across a laser-crystallized germanium dot. (Reprinted from Mulato et al. 1997. With permission. Copyright American Institute of Physics.)

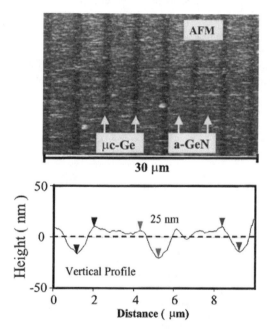

Fig. 11.9 AFM surface and vertical profiles of the amorphous GeN film irradiated with the two-beam interference pattern showing the periodic microcrystalline and amorphous lines. (Reprinted from Mulato et al. 2002. With permission. Copyright American Institute of Physics.)

surface profile and the vertical profile (measured with atomic force microscopy) of the periodic crystallization structure obtained with two interfering beams on the surface of *a*-GeN. The figure indicates the periodic darker lines corresponding to microcrystalline germanium and the clear lines corresponding to the unaffected amorphous GeN. The microcrystalline lines have the period of 4 μm and the width of 1 μm. Such surface interference structures with three-dimensional profiles, and different optical properties corresponding to the microcrystalline and amorphous regions obtained can be used as optical diffraction gratings. The vertical profile also shows that the crystallized portion of the film is around 25 nm lower than the amorphous region due to effusion of nitrogen similar to that of hydrogen in the case of amorphous silicon (a-Si:H) films. This can be confirmed by the characterization techniques such as infrared spectroscopy and Raman spectroscopy (Fig. 11.10). Figure 11.10a presents the infrared Ge–N stretching absorption band of GeN film before and after laser interaction. The difference in the strength of the absorption band indicates that the total number of Ge–N bonds has decreased after laser crystallization suggesting the effusion of nitrogen during crystallization. As mentioned before, the evidence of crystallization in the amorphous GeN films after laser interference processing can be obtained by Raman spectroscopy (Fig. 11.10b). The figure clearly indicates the absence of crystalline component corresponding to 300 cm^{-1} in the starting amorphous film. The peak appears in the laser-crystallized sample which can be compared with the reference crystalline germanium. The broadness of the peak in the laser-crystallized sample indicates that laser interference results in the formation of distribution of small crystallites instead of monocrystalline germanium film (Mulato et al. 2002).

For many thin film electronic applications, it is important to understand the grain-growth behavior during laser interference crystallization of amorphous or

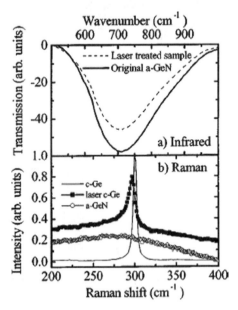

Fig. 11.10 (a) Infrared Ge–N stretching absorption bands, and (b) Raman spectra of the amorphous GeN films before and after irradiating with laser interference pattern. (Reprinted from Mulato et al. 2002. With permission. Copyright American Institute of Physics.)

nano-crystalline thin films. This is of particular importance where microcrystalliza-
tion facilitated by super lateral growth (SLG) is desired. As mentioned before the
laser-induced crystallization is associated with ultrafast melting and solidification.
The grains nucleate at the solid–liquid interface and grow toward the interference
maxima along the thermal gradient. The grains growing from either side of the
interference maxima meet at the center of the maxima and form a grain boundary.
The lateral grain growth under certain conditions is limited by the spontaneous
nucleation of smaller grains at the center of the energy maxima. Under these
conditions, the lateral grains cannot reach the center of the interference maxima.
This is shown in the AFM image (Fig. 11.11) obtained from the surface of amorphous silicon crystallized using symmetric two-beam interference pattern (by
frequency doubled Q-switched Nd:YAG laser with 532 nm wavelength). Asymmetric
laser interference crystallization, where the intensities of two laser beams are
different, can also be used to adjust and optimize the transient temperature profiles
and hence the grain-growth behavior (Rezek et al. 2000).

Similar studies on lateral grain-growth behavior during laser interference
crystallization of amorphous or nanocrystalline SiGe films, deposited on the quartz
substrates, have been conducted (Eisele et al. 2003). The crystallization experiments
were carried out with two distinct schemes: laser interference crystallization (LIC)
and scanning laser interference crystallization (SLIC). In LIC, the interference pattern is directly irradiated on the surface of the sample, whereas, in SLIC, the
interference pattern is shifted on the surface with a predefined stepwidth (Fig. 11.12).
Figure 11.13 presents the TEM images from the sections of laser-crystallized lines
of SiGe films crystallized at two different temperatures (25 °C and 740 °C). For the
case of laser-induced crystallization (LIC) at room temperature, the lateral grain
growth is limited due to spontaneous nucleation of smaller grains at the center of
the line. However, for the case of LIC at elevated temperature, the reduced cooling
rate results in reduced or no spontaneous nucleation. The spontaneous nucleation
can also be prevented by narrower lines obtained with three-beam interference. The
AFM image of SiGe film crystallized by three-beam interference pattern (with a
period 6 μm) using SLIC is presented in Fig. 11.14. As indicated in the figure, SLIC
results in longer grains (~2 μm).

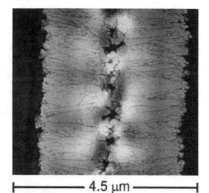

Fig. 11.11 AFM surface profile of crystallized
amorphous silicon film using symmetric two-beam
laser interference. (Reprinted from Rezek et al.
2000. With permission. Copyright Elsevier.)

⊢————— 4.5 μm —————⊣

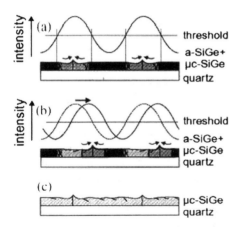

Fig. 11.12 Schematic of (**a**) laser interference crystallization (LIC), and (**b**) and (**c**) scanning laser interference crystallization (SLIC). (Reprinted from Eisele et al. 2003. With permission. Copyright Elsevier.)

11.4.2   Structuring of Monolayer and Multilayer Metallic Films

Recently, laser interference techniques have been applied for the long-range periodic patterning of thin film metallic surfaces. The interference technique offers a great potential for micromachining and micronanostructuring of thin films for applications in microelectronics and micromechanics. Extensive studies have been conducted on the interaction of laser interference patterns with monolayer and multilayer films. Various phenomena during such interactions include inhomogeneous melting, phase transformations, intermetallic reactions, etc.

11.4.2.1   Structuring of Monolayer Films

When a laser interference pattern with modulated energy distribution is irradiated on the surface of a film, the absorbed energy can cause the spatial heating, melting, and evaporation of the film depending on the energy thresholds for various effects. In most of the cases, laser structuring applications for thin films employ sufficiently high laser energies to induce the melting of the films. Due to the poor conductivity of the underlying substrates, most of the absorbed laser energy is confined in the thin film resulting in significantly longer melting duration time compared to the laser pulse time. The longer melting duration times give rise to physical processes such as hydrodynamic melt flow responsible for physical texturing of the surfaces. Figure 11.15 presents the typical periodic structures obtained by two-beam and four-beam interference patterns irradiated on monolayer (18 nm thick) gold films deposited on glass substrates. The formation of such periodic topographic periodic features is due to redistribution of molten film material in the "hot" and "cold" regions on the surface (Kaganovskii et al. 2006).

Film thickness plays an important role in influencing the formation of periodic surface structures during interference processing. For the case of very thin (thickness <17 nm) gold films on glass substrates, it was observed that beading (dewetting) of the molten film takes place in the hot regions followed by the motion of the beads toward cold regions. However, for thick films (thickness >17 nm),

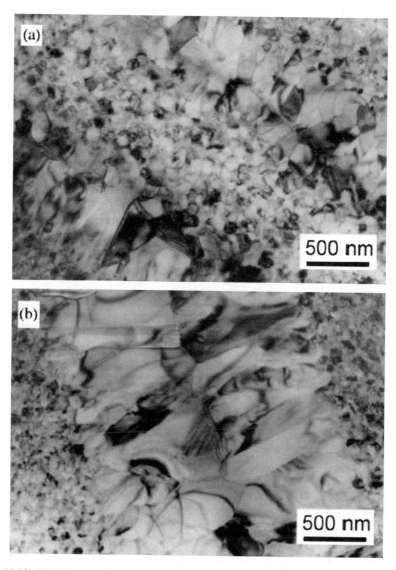

Fig. 11.13 TEM images of laser-crystallized lines of SiGe films crystallized at two different temperatures: (**a**) 25 °C and (**b**) 740 °C. (Reprinted from Eisele et al. 2003. With permission. Copyright Elsevier.)

complete hydrodynamic flow of molten film material (instead of dewetting) results in well-defined high and narrow periodic structure. Furthermore, the film thickness determines the threshold laser intensity (i.e., power density) required to induce the morphological changes and the fabrication of periodic structures. Figure 11.16 shows that for film thicknesses in the range of 5–15 nm, the threshold intensity decreases with film thickness; whereas, for film thicknesses above 15 nm, the threshold intensity increases with film thickness (Kaganovskii et al. 2006).

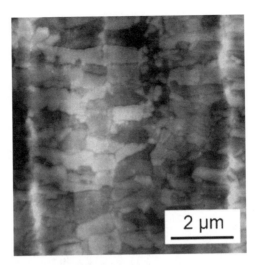

Fig. 11.14 AFM image of SiGe film crystallized using scanning laser interference crystallization (SLIC). Selective plasma etching was applied to visualize grain boundaries. (Reprinted from Eisele et al. 2003. With permission. Copyright Elsevier.)

Fig. 11.15 Periodic structures produced on 18 nm thick gold film by (**a**) two-beam laser interference (peak intensity, $I = 4I_0 = 1.3 \times 10^{11}$ W/m^2), and (**b**) four-beam laser interference (peak intensity, $I = 16I_0 = 1.5 \times 10^{11}$ W/m^2) using a single 7 ns pulse at 354 nm. (Reprinted from Kaganovskii et al. 2006. With permission. Copyright American Institute of Physics.)

11.4.2.2 Structuring of Multilayer Metallic Films

Most of the work in the area of laser interference processing of multilayer films is conducted by Prof. Mücklich and his research group at Germany. The combinations of a variety of metallic materials have been used to produce the bilayer and trilayer films on glass substrates which were subsequently irradiated with the laser

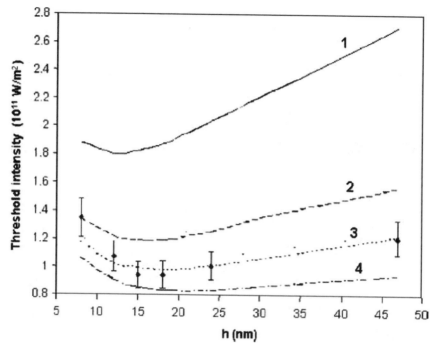

Fig. 11.16 Variation of calculated threshold intensity required to produce morphological changes and formation of periodic structure in the 18 nm gold film using laser interference processing. The curves labeled 1, 2, 3, and 4 correspond to the periodicity of 2, 3.5, 5, and 10 μm respectively. The experimental points shown in the figure were obtained for the periodicity of 5 μm. (Reprinted from Kaganovskii et al. 2006. With permission. Copyright American Institute of Physics.)

interference patterns. In contrast to monolayer films, the multilayer films present additional complexity due to difference in the thermophysical properties of the constituent metals and the correspondingly different responses to the laser irradiation.

For the multilayer films with high melting point material in the top layer, three distinct morphologies of the interference structures were observed depending on the laser energy fluence. The various systems studied for interference included Fe–Al–glass, Fe–Ni–glass, Ti–Al–glass, and Ti–Ni–glass. Above certain laser fluence, F_s, the absorbed laser energy is sufficient to cause the melting of the bottom layer which is composed of the low melting material. The melting of the bottom layer exerts the pressure on the unmelted top layer (composed of high melting point) resulting in the deformation of the top layer. The outward deformations of the top layer appear as a periodic pattern on the surface of the film. The mechanism is schematically presented in Fig. 11.17 where A represents the top layer of higher melting point material and B represents the bottom layer of lower melting point material. If the laser fluence is further increased beyond F_s, the melting of the B

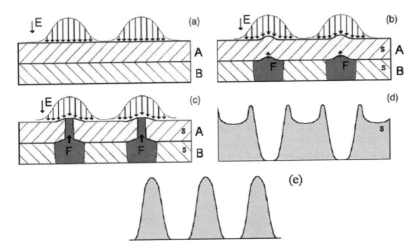

Fig. 11.17 Schematic of the mechanisms of formation of various surface morphologies during laser interference processing of two-layer films with higher melting point material (**A**) at the top layer and lower melting point material (**B**) at the lower layer: (**a**) irradiation of the surface with the modulated intensity distribution in interference pattern, (**b**) deformation of the upper layer induced by melting of the lower layer, (**c**) breaking of the top layer, (**d**) periodic pattern when material removal is initiated, and (**e**) periodic pattern at large value of laser fluence. (Reprinted from Lasagni and Mucklich 2005b. With permission. Copyright Elsevier.)

layer continues till the melting point of A is reached. Eventually, layer A breaks up resulting in the ejection of the material. This corresponds to the laser fluence, F_d at which the material removal is initiated. The removal of the material at the interference peak results in the depression between two consecutive peaks in the surface structure of the film. Further increase in the laser fluence beyond F_d, causes the increased material removal with the increasing depth of the depression at the interference maxima resulting in a well-defined periodic structure. These mechanisms have been confirmed by the experimental observation of the surface structures of bimetallic films irradiated with laser interference patterns at various fluences. Figure 11.18 shows the surface topographies and lateral profiles for Fe–Ni–glass system for which F_s and F_d corresponds to 151 and 201 mJ/cm^2 respectively (Lasagni and Mucklich 2005a, b).

Extensive thermal modeling efforts have been conducted to understand the melting behavior of the various layers in multilayer thin films composed of two differing constituent metals. These thermal modeling efforts were based on heat transfer equations similar to Eq. (11.4). Figure 11.19 presents one such modeling result based on finite element analysis for the multilayer Ni–Al film irradiated with laser interference pattern. Layer thicknesses of individual Al and Ni layers were 20 and 30.3 nm respectively and the film was irradiated with Q-switched Nd:YAG laser with a wavelength of 355 nm. The figure indicates that significant amount of aluminum melts in the top layers of the film causing the distortion of the nickel layers. The significant melting of corresponding nickel layers requires

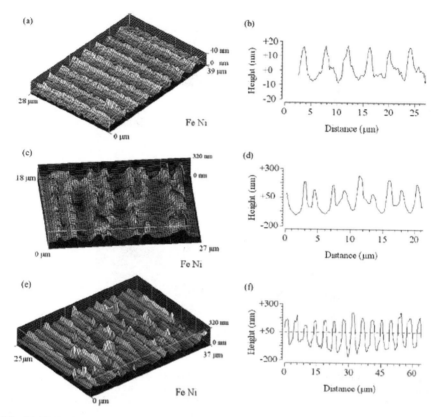

Fig. 11.18 Various surface topographies and vertical profiles of surface structures in Fe–Ni–glass films irradiated with laser interference patterns: (**a**), (**b**) 192 mJ/cm² (value between F_s and F_d); (**b**), (**c**) 210 mJ/cm² (value greater than F_d); and (**d**), (**e**) 215 mJ/cm². (Reprinted from Lasagni and Mucklich 2005a. With permission. Copyright Elsevier.)

higher laser fluence due to higher melting point of nickel than that of aluminum. Furthermore, periodic structuring of multilayer films with the laser interference pattern is associated with the changes in the stress and texture distribution depending on the thermal conditions prevalent during laser–material interactions (Daniel et al. 2004).

11.4.2.3 Phase–Microstructure Effects during Structuring of Films

In addition to the periodic physical topographical changes, the interaction of the laser interference pattern with the material often results in the metallurgical effects such as phase transformation, recrystalllization, intermetallic reactions, etc. Thus, periodic modulation of metallurgical microstructure (and dependent physico-mechanical properties) can be achieved (Daniel and Dahotre 2006). By combining

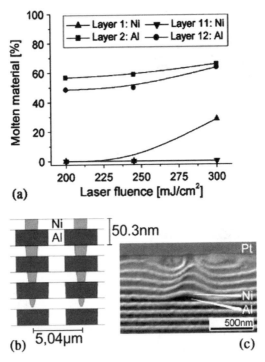

Fig. 11.19 (a) Variation of calculated fraction of molten material (Al and Ni) in various layers with laser fluence, (b) calculated cross section of multilayer film representing fractions of molten material (Al and Ni) in various layers (laser fluence of 300 mJ/cm²), (c) TEM micrograph showing individual Al and Ni layers after irradiation with laser interference pattern. (Reprinted from Daniel et al. 2004. With permission. Copyright Elsevier.)

the properties of the unaffected region and the laser interference irradiated region, surface composite film can be realized.

Sivakov et al. (2005) studied the laser interference-induced periodic phase transformations in iron oxide films due to chemical vapor deposited on silicon substrates. The periodic phase transformations from hematite to magnetite and magnetite to wustite have been reported based on detailed x-ray diffraction analysis before and after laser interference irradiation. The corresponding phase transformation equations are:

$$3\ Fe_2O_3 \rightarrow 2Fe_3O_4 + \tfrac{1}{2}O_2 \tag{11.6}$$

$$Fe_3O_4 \rightarrow 3FeO + \tfrac{1}{2}O_2 \tag{11.7}$$

It was proposed that reductive phase transformations from hematite to magnetite and magnetite to hematite in the regions of high energy are induced by spatially confined plasma density. The interaction of the laser with the film results in the generation of the plasma plume which prevents the interaction of oxygen with the

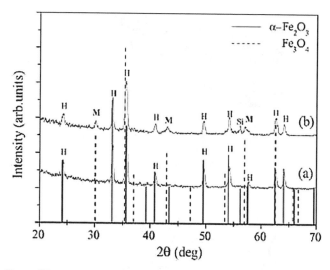

Fig. 11.20 X-ray diffraction patterns of hematite film (**a**) before and (**b**) after laser interference irradiation. H and M correspond to hematite (α-Fe$_2$O$_3$) and magnetite (Fe$_3$O$_4$) respectively. (Reprinted from Sivakov et al. 2005. With permission. Copyright Elsevier.)

CVD film and thus facilitates reductive transformations. The energy for such periodic transformations is supplied by the modulated intensity distributions in the incident interference pattern. Figure 11.20 presents the x-ray diffraction patterns of the CVD hematite film before and after laser interference irradiation. As indicated in the figure, the magnetite peak appears in the specimens of structured hematite films after interaction with laser interference pattern. The formation of periodic magnetic and nonmagnetic domains in the iron oxide films by laser interference processing offers several applications (Sivakov et al. 2005). Similarly, the indications of phase transformation from Cu$_6$Sn$_5$ to Cu$_3$Sn in hot-dipped tin CuSn-4 alloy (4 wt. % Sn) are irradiated with laser interference pattern (Daniel et al. 2003).

Irradiation of laser interference pattern can also initiate the periodic formation of intermetallic compounds in the homogenous matrix and thus realizing the composite surfaces with high strength of intermetallics and ductility of matrix material. This is demonstrated for the case of Ni–Al films deposited on Si wafers. Ni–Al films (900 nm thick) with stoichiometric ratio of 3:1 deposited by magnetron sputtering were modified by laser interference patterns. Based on x-ray diffraction, it was reported that Ni$_3$Al intermetallic is formed in the areas of laser interaction with the film. Furthermore, nanoindentation studies indicated that the formation of periodic intermetallic phases is associated with periodic modulation of mechanical properties. The indentation hardness in the range of 10 GPa is observed in the laser-modified area (where intermetallic reaction takes place) compared to average hardness of 4 GPa in the untreated areas (Fig. 11.21) (Liu et al. 2003).

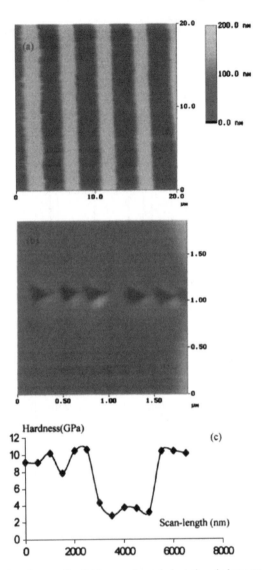

Fig. 11.21 (**a**) AFM surface profile, (**b**) image of nanoindentations in laser-treated region, and (**c**) hardness distribution across one interference period in a 900 nm Ni–Al film irradiated with laser interference pattern. (Reprinted from Liu et al. 2003. With permission. Copyright Elsevier.)

11.4.3   *Structuring of Biomaterials*

Recently, laser interference processing for modifying the surfaces of biomaterials is attracting significant research interests. It was suggested that the chemistry and topography of the biomaterials can be favorably modified by irradiating with laser interference pattern for enhanced cell–surface interaction and consequent attachment, spreading, and orientation of the cells on the surface. The interference techniques for modifying the surfaces of biomaterials are based on the selective ablation of the material at the interference maxima resulting in micropatterns consisting of well-defined ridges and grooves. Such micropatterns are expected to direct the cell growth in specific directions (contact guidance). The significant advantage of this technique compared with the random patterning is that the micropatterns on the surface of biomaterials can be efficiently controlled to the desired dimensions (periodicity, height, and width of lines or dots) by controlling the laser processing parameters. Also, a variety of biomaterials such as metal, ceramics, and polymers can be effectively modified (Li et al. 2003).

Most of the work recently reported on the studies of interference patterning of biomaterials is limited to few biopolymers. The important parameters of the laser interference patterned surfaces which are expected to have the influence on the cell adhesion, growth, and orientation are the contact angle, period dimension, morphology (lines or dots). Figure 11.22 presents the influence of laser fluence on the depth of micropattern and the contact angle in $100\,\mu m$ thick polycarbonate film irradiated with laser interference pattern. As indicated in the figure, the depth of micropattern increases and the contact angle decreases with the laser fluence. Thus, the topography and the wetting characteristics can be modified by laser interference patterning to promote the cell adhesion (Yu et al. 2005a, b).

Even though extensive studies are reported on the characterization of the interference structures obtained in various materials, very few studies have been reported on the interaction of cells with laser-modified surfaces. Figure 11.23 presents the results of one of the studies on the responses of HPF (human pulmonary fibroblast) cell to line structures and point structures obtained on the surface of polycarbonate (PC) films by two- or more-beam laser interference. The cells cultured on the structured surfaces were mostly spindle-like and bipolar. Also, as indicated in the light photographs, the cells cultured on the line patterns show directional growth parallel to lines, whereas the cells cultured on the point patterns showed mostly random orientations (Yu et al. 2005a).

To summarize, laser interference processing of advanced metallic, polymeric, and ceramic materials offers tremendous potential to be used in applications where periodic modulation of properties and topography is desired. The technology is still relatively new and presents various directions for research. To date, most of the studies dealt with the characterization of periodic patterns in various systems and the parametric influence of laser parameters on the morphology and topography of the interference structures generated on the material surfaces. The modulation of energy intensity in the interference pattern gives rise to complexity of thermal effects, such as inhomogeneous temperature distributions, temperature gradients, cooling rates, and thermal stresses. These thermal effects have a strong bearing on the fluid flow, solidification, microstructure development, thermal stresses, etc. A combination of modeling and experimental studies on the laser–material interaction during interference processing will provide further insights for the advancement of the process in the emerging applications.

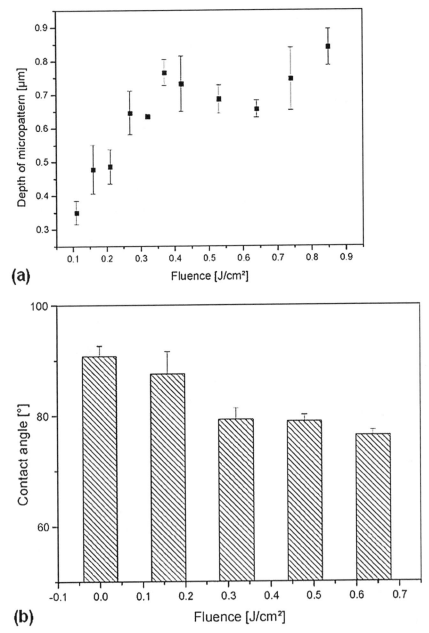

Fig. 11.22 Effect of laser fluence on (**a**) depth of line micropattern (period 5 μm) and (**b**) contact angle after laser interference irradiation with Q-switched Nd:YAG laser of wavelength 266 nm. (Reprinted from Yu et al. 2005a. With permission. Copyright American Chemical Society.)

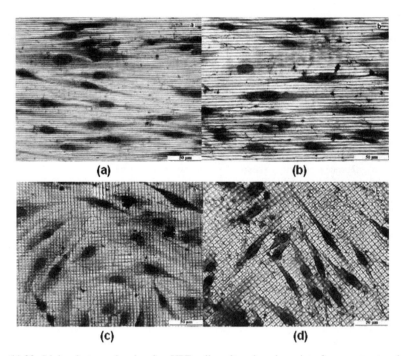

Fig. 11.23 Light photographs showing HPF cells cultured on laser interference-structured PC films: (**a**) line pattern with a period of 3 μm, (**b**) line pattern with a period of 9 μm, (**c**) point pattern with a period of 5 μm, and (**d**) point pattern with a period of 7 μm. All substrates were coated with collagen. (Reprinted from Yu et al. 2005a. With permission. Copyright American Chemical Society.)

References

Born M, Wolf E (1980) Principles of Optics. Pergamon Press, Oxford.

Daniel C (2006) Biomimetic structures for mechanical applications by interfering laser beams: more than solely holographic gratings. Journal of Materials Research 21:2098–2105.

Daniel C, Dahotre NB (2006) Phase-modulated hierarchical surface structures by interfering laser beams. Advanced Engineering Materials 8:925–932

Daniel C, Mucklich F, Liu Z (2003) Periodic micro-nano-structuring of metallic surfaces by interfering laser beams. Applied Surface Science 208–209:317–321.

Daniel C, Lasagni A, Mucklich F (2004) Stress and texture evolution of Ni/Al multi-film by laser interference irradiation. Surface and Coatings Technology 180–181:478–482.

Eisele C, Berger M, Nerding M, Strunk HP, Nebel CE, Stutzmann M (2003) Laser-crystallized microcrystalline SiGe alloys for thin film solar cells. Thin Solid Films 427:176–180.

Engleman PG, Kurella N, Samant A, Blue CA, Dahotre N (2005) The application of laser-induced multi-scale surface texturing. JOM 57:46–50.

Heintze M, Santos PV, Nebel CE, Stutzmann M (1994) Laser structuring of silicon thin films by interference crystallization. Applied Physics Letters 64:3148–3150.

Kaganovskii Y, Vladomirsky H, Rosenbluh M (2006) Periodic lines and holes produced in thin Au films by pulsed laser irradiation. Journal of Applied Physics 100:044317.

Kondo T, Matsuo S, Juodkazis S, Misawa H (2001) Femtosecond laser interference technique with diffractive beam splitter for fabrication of three-dimensional photonic crystals. Applied Physics Letters 79:725–727.

Lasagni A, Mucklich F (2005a) Structuring of metallic bi- and tri-nano-layer films by laser interference irradiation: control of the structure depth. Applied Surface Science 247:32–37.

Lasagni A, Mucklich F (2005b) Study of the multilayer metallic films topography modified by laser interference irradiation. Applied Surface Science 240:214–221.

Li P, Bakowsky U, Yu F, Loehbach C, Muecklich F, Lehr C (2003) Laser ablation patterning by interference induces directional cell growth. IEEE Transactions on Nanobioscience 2:138–145.

Liu Z, Meng X, Recktenwald T, Mucklich F (2003) Patterned intermetallic reaction of Ni_3Al by laser interference structuring. Materials Science and Engineering A 342:101–103.

Mücklich F, Lasagni A, Daniel C (2006) Laser interference metallurgy-using interference as a tool for micro/nano structuring. International Journal of Materials Research 97:1337–1344

Mulato M, Toet D, Aichmayr G, Santos PV, Chambouleyron IE (1997) Laser crystallization and structuring of amorphous germanium. Applied Physics Letters 70:3570–3572.

Mulato M, Toet D, Aichmayr G, Spangenberg A, Santos PV, Chambouleyron I (1998) Short-pulse laser crystallization and structuring of a-Ge. Journal of Non-crystalline Solids 227–230:930–933.

Mulato M, Zanatta AR, Toet D, Chambouleyron IE (2002) Optical diffraction gratings produced by laser interference structuring of amorphous germanium–nitrogen alloys. Applied Physics Letters 81:2731–2733.

Ready JF (1997) Industrial Applications of Lasers. Academic Press, San Diego.

Rezek B, Nebel CE, Stutzmann M (2000) Interference laser crystallization of microcrystalline silicon using asymmetric laser beam intensities. Journal of Non-Crystalline Solids 266–269:650–653.

Sivakov V, Petersen C, Daniel C, Shen H, Mucklich F, Mathur S (2005) Laser induced local and periodic phase transformation in iron oxide thin films obtained by chemical vapor deposition. Applied Surface Science 247:513–517.

Yu F, Mucklich F, Li P, Shen H, Mathur S, Lehr C, Bakowsky U (2005a) In vitro cell response to a polymer surface micropatterned by laser interference lithography. Biomacromolecules 6:1160–1167.

Yu F, Li P, Shen H, Mathur S, Lehr C, Bakowsky U, Mucklich F (2005b) Laser interference lithography as a new and efficient technique for micropatterning of biopolymer surface. Biomaterials 26:2307–2312.

Chapter 12
Laser Shock Processing

12.1 Introduction

Laser shock processing is a relatively novel process of surface modifications of materials. The surface modification is based on the generation of shock waves when the material is irradiated with laser radiation. Recently, significant interests have been attracted toward modifying the material surfaces to improve the fatigue and corrosion properties using laser shock processing. This chapter provides a very brief overview of the process principles, process parameters, and applications.

12.2 Fundamentals of Laser Shock Processing

When a material is irradiated with very high power (typically greater than 10^9 W/cm^2) and short-pulse (typically 1–50 ns) laser, the laser–material interactions at the very thin surface layer result in the generation of plasma (Fabbro et al. 1998). The volume of expansion of the plasma induces shock waves in the ablated target. There are two distinct modes in which laser shock processing can be carried out: direct ablation and confined ablation (Fig. 12.1).

Most of the early investigations were focused on the irradiation of the samples in vacuum such that the generated plasma expands freely. This regime of laser–material interactions with freely expanding plasma is generally referred as "direct ablation". The peak pressures attained in this regime ranged from 1 GPa to 1 TPa for the incident laser intensities in the range of 10^9–10^{15} W/cm^2. Also, the time duration of the applied pressure is approximately equal to the laser pulse duration due to rapid cooling of the plasma by its adiabatic expansion in vacuum.

Subsequently, it was observed that the magnitude of generated pressure can be substantially increased by confining the plasma during its expansion (Anderholm 1970). The confinement of the plasma is generally achieved by placing a transparent material (such as glass, water, etc.) in close contact with the material surface. This regime of laser–material interaction is generally referred as "confined ablation".

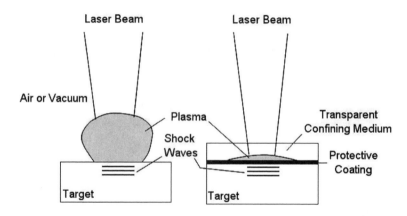

Fig. 12.1 Modes of laser shock processing: (a) direct ablation, and (b) confined ablation

In addition to the increased pressure, the confinement of plasma also delays its expansion resulting in longer shock waves (~2–3 times longer than that during free expansion) (Fabbro et al. 1990). In this confined ablation regime, thin protective absorbent films or coatings (such as aluminum and copper) are generally used to avoid the melting of material surfaces. In such systems, the irradiated laser beam vaporizes the coating and forms plasma. The coating thickness is such that the thermal effects are limited within the thickness of the coating. The confined plasma between the protective coating and the transparent material then induces shock waves in the target material (Montross et al. 2002). Very thick protective coatings may result in the attenuation of shock wave, whereas very thin coatings may result in the transfer of thermal effects to the target surface resulting in thermal damage.

The characteristics of the laser pulse and the stress profiles during direct (unconfined) and indirect (confined) ablation of 0.076 mm aluminum targets using Q-switched neodymium lasers are presented in Fig. 12.2. The stress was measured on the back surface of the target using quartz gauges. This introduces the delay time of 117 ns corresponding to wave transit time through the target. As indicated in the figure the stress reaches maximum and then decreases. The decrease in the stress corresponds to adiabatic cooling of the plasma. An important observation from these profiles is that the magnitude of maximum stress during confined ablation is almost an order of the magnitude greater than that during direct ablation. Furthermore, the peak stress reaches at longer time (~100 ns) in case of confined ablation when compared to that in direct ablation (~30 ns). The stress tail in case of confined ablation represents the larger fraction of the maximum stress (O'Keefe et al. 1973).

The stress pulse induced in the target by the plasma expansion propagates inside the target. Figure 12.3 presents the temporal pressure profiles at various depths during confined laser ablation of aluminum target. The figure indicates the long tail of each pressure pulse (decreasing pressure after maxima) characteristic of confined ablation. Most importantly, the peak pressure of the pressure pulse decrease with the increasing depth in the target (Fabbro et al. 1998).

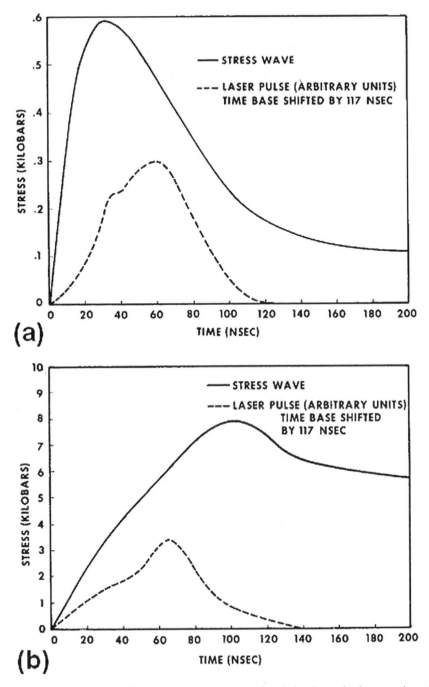

Fig. 12.2 Characteristics of laser pulse and stress profile during laser shock processing of aluminum target using Q-switched neodymium laser: (**a**) direct ablation with laser intensity of $0.92 \, GW/cm^2$, and (**b**) confined ablation with laser intensity of $1.3 \, GW/cm^2$. Laser pulse is shifted by 117 ns corresponding to wave transit time through target (Reprinted from O'Keefe et al. 1973. With permission. Copyright American Institute of Physics.)

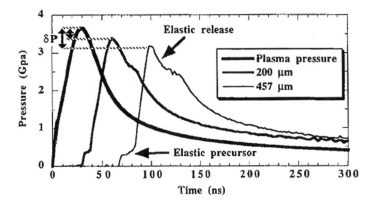

Fig. 12.3 Temporal pressure profiles at various depths during water-confined ablation of aluminum target with laser intensity of 5 GW/cm² and pulse duration (FWHM) of 25 ns (Reprinted from Fabbro et al. 1998. With permission. Copyright Laser Institute of America.)

12.3 Analysis of Laser Shock Processing Process

Recently excellent analysis of the laser shock processing with confined ablation has been provided by Fabbro et al. (1990). They described the confined ablation process as a three-step process. The first step corresponds to the irradiation of the target with the laser beam. During this laser on time (pulse duration), plasma generated by the laser–material interaction induces a shock wave which propagates into the target and the confining medium. The second step begins immediately after the laser is switched off. During this step, the plasma maintains the pressure which subsequently decreases due to adiabatic cooling. The target acquires an impulse momentum due to induced shock wave during these two steps. For a longer time (third step), the complete recombination of the plasma takes place and the additional momentum to the target is due to the "cannon-ball-like" expansion of the heated gas inside the interface (Fabbro et al. 1990).

Figure 12.4 presents the one-dimensional model geometry for the analysis of laser shock process. The geometry consists of a metallic target (1) and a transparent confining medium (2). When a laser beam (with parameters in the regime of shock processing) is irradiated at the interface, rapid generation and expansion of plasma creates pressure. This pressure induces shock waves in the target and the confining medium. The displacement of the interface between the two media is defined by the fluid motion behind the propagating shock waves (Fabbro et al. 1990).

If we consider u_i ($i = 1$ for metallic target and $i = 2$ for confining medium) as the fluid velocities behind the shock waves, then the length of interface L at time t can be expressed as (Fabbro et al. 1990):

$$L(t) = \int_0^t \left[u_1(t) + u_2(t) \right] dt. \tag{12.1}$$

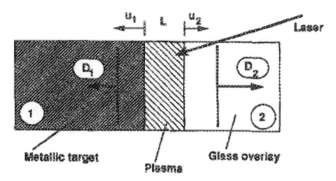

Fig. 12.4 Model geometry for confined ablation process. (Reprinted from Fabbro et al. 1990. With permission. Copyright American Institute of Physics.)

The pressure is given by the shock wave equation:

$$P = \rho_i D_i u_i = Z_i u_i,$$

(12.2)

where ρ_i, D_i, and Z_i are the density, the shock wave velocity, and the shock impedance respectively. In the case where both the media (target and confining media) are solids, the shock wave impedances can be taken as $Z_i = \rho_i D_i$ and are constant. Solving Eqs. (12.1) and (12.2), we obtain:

$$\frac{dL(t)}{dt} = \frac{2}{Z} P(t),$$

(12.3)

where $\dfrac{1}{Z} = \dfrac{1}{Z_1} + \dfrac{1}{Z_2}$.

The laser energy deposited at the interface, $I(t)dt$, is used to increase the internal energy of the plasma inside the interface, $E_i(t)$, and work as pressure forces $P(t)dL$ to open it. The energy balance can be expressed as:

$$I(t) = P(t)\frac{dL}{dt} + \frac{d\left[E_i(t)L\right]}{dt}.$$

(12.4)

A fraction α of internal energy is thermal energy $E_T(t)$. Remaining $(1 - \alpha)$ fraction is used for ionization of gas. If the plasma is considered as ideal gas, then the pressure can be expressed as:

$$P(t) = \frac{2}{3}E_T(t) = \frac{2}{3}\alpha E_i(t).$$

(12.5)

Eq. (12.4) becomes:

$$I(t) = P(t)\frac{dL(t)}{dt} + \frac{3}{2\alpha}\frac{d}{dt}\big[P(t)L(t)\big].$$ (12.6)

Assuming $I(t) = At^a$, the solution for the length of the interface can be obtained as $L(t) = Bt^b$. Relationships between various coefficients in these equations can be expressed as:

$$a = 2(b-1);$$

$$A = \left(\frac{Z}{2}\right)B^2 b\left[b + \left(\frac{3}{2\alpha}\right)(2b-1)\right].$$ (12.7)

For the case of constant laser energy I_0 and pulse duration τ, the coefficients can be obtained as: $a = 0$ and $A = I_0$. Also, If we assume that the length of the interface is zero at the start (i.e., $L(0) = 0$), then the constant pressure during the pulse duration can be obtained as (Fabbro et al. 1990):

$$P(\text{kbars}) = 0.10\left(\frac{\alpha}{2\alpha+3}\right)^{1/2} Z^{1/2}\left(g/cm^2s\right) \times I_0^{1/2}\left(GW/cm^2\right),$$ (12.8)

$$L(\tau)(\mu m) = 2\times10^4 P(\text{kbars})\tau(\text{ns}) \times Z^{-1}\left(g/cm^2s\right).$$ (12.9)

For direct ablation (ablation in vacuum), Phipps et al. (1988) have found that most of the experimental data for mechanical coupling coefficient, C_m, follows an empirical trend to within a factor of 2 over 7 orders of magnitude in the product $I\lambda\sqrt{\tau}$, where I is single-pulse laser intensity, λ is wavelength and τ is pulse width. Mechanical coupling coefficient is the measure of efficiency of ablation process in producing mechanical impulse during laser–material interaction. This empirical relationship can be expressed as:

$$C_m = \frac{P}{I} = b\left(I\lambda\sqrt{\tau}\right)^n,$$ (12.10)

where b is material dependent coefficient (5.6 for aluminum alloys and 8.5 for C–H material), and $n = -0.3\pm0.03$ for both type of materials (Phipps et al. 1988).

Based on the above analysis of direct and confined ablation processes, the impulse can be calculated. Impulse momentum and mechanical coupling coefficients during direct and confined ablation processes are plotted in Fig. 12.5. The figure indicates that confined ablation mode induces the impulse momentum

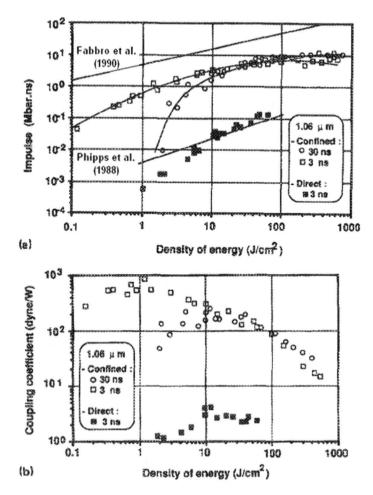

Fig. 12.5 Variation of (**a**) impulse momentum and (**b**) mechanical coupling coefficient with incident laser energy density for direct and confined ablation modes. The experimental data of impulse momentum is compared with the free ablation model (Phipps et al. 1988) and confined ablation model results. (Reprinted from Fabbro et al. 1990. With permission. Copyright American Institute of Physics.)

significantly greater than that during direct ablation. Hence confined ablation regimes are very attractive for materials processing. The figure also indicates that the impulse momentum for confined mode increases with laser energy density and then saturates above 90–100 J/cm². This seems to be due to internal breakdown of the transparent material such that energy is absorbed primarily inside the transparent material before reaching the glass–target interface. Thus dielectric breakdown of the confining medium limits the maximum peak pressure obtained by increasing laser energy density (Fabbro et al. 1990).

12.4   Processing Parameters

Plasma-induced pressure is an important parameter in the laser shock processing of materials. The plasma-induced pressure is influenced by a number of laser processing parameters. This section briefly explains the effect of various parameters on the plasma pressure generated during laser shock processing.

12.4.1   Effect of Laser Intensity

The effect of laser intensity on the peak pressure during water-confined and glass-confined ablation of metallic targets is presented in Fig. 12.6. In general, there are three distinct regimes of laser intensities corresponding to characteristic ablation behavior: low, intermediate, and high laser intensity regimes. For the case of water-confined ablation mode, these regimes approximately correspond to laser intensities $I < 0.5\,GW/cm^2$, $0.5\,GW/cm^2 < I < 30\,GW/cm^2$, and $I > 30\,GW/cm^2$. The figure indicates that there exists threshold laser intensity where the ablation of the material starts. This threshold intensity corresponds to about $0.5\,GW/cm^2$ for the case of water-confined ablation mode. In the intermediate intensity regime $0.5\,GW/cm^2 < I < 30\,GW/cm^2$, the pressure induced by plasma increases with laser power density and reaches maximum. Further increase in the laser intensity above $30\,GW/cm^2$ saturates the pressure induced by plasma (Devaux et al. 1993). The saturation of the pressure is due to dielectric breakdown of the confining medium.

The incident laser intensity is also expected to influence the duration of the pressure pulse, τ_p. Figure 12.7 presents the variation of pressure pulse duration (FWHM) with laser power density during water-confined ablation of metallic target. The figure clearly indicates that the pressure pulse duration decreases with increasing laser power density. The decrease in pressure pulse duration is more pronounced beyond 11–$13\,GW/cm^2$. This is due to effective dielectric breakdown of water. Thus the dielectric breakdown of material at high laser power densities causes the saturation of peak pressure and the shortening of the pressure pulse duration (Berthe et al. 1997).

12.4.2   Effect of Pulse Duration

Typical range of laser pulse durations for the shock processing regime is 1–$50\,ns$. The laser pulse duration significantly influences the pressure pulse duration, peak pressure, and breakdown thresholds. In direct ablation mode, the plasma-induced pressure closely follows the laser pulse profile with the pressure pulse duration of the order of pulse duration. In confined mode, the pressure pulse duration is longer

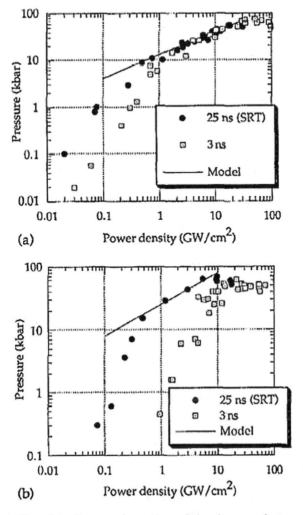

Fig. 12.6 Effect of laser intensity on maximum plasma-induced pressure for two pulse durations: (**a**) water confinement mode and (**b**) glass confinement mode. The figure also shows the analytical model predictions. (Reprinted from Devaux et al. 1993. With permission. Copyright American Institute of Physics.)

than the laser pulse duration. It is reported that typical pressure pulse durations are 3–4 times the laser pulse durations (Fabbro et al. 1998). Also, it is found that dielectric breakdown threshold of confining medium (fused silica) increases with decreasing pulse duration. Typical range of dielectric breakdown thresholds for 21–30 ps pulses is 47–70 GW/cm^2 compared to 20.5–25 GW/cm^2 for 1–12 ns pulses (Smith et al. 1977; Yokotani et al. 1989). The breakdown threshold determines the maximum achievable peak pressure. Hence, it is possible to achieve higher peak

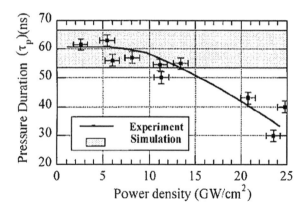

Fig. 12.7 Effect of laser power density on pressure pulse duration (FWHM) during water-confined ablation of metallic target. (Reprinted from Berthe et al. 1997. With permission. Copyright American Institute of Physics.)

pressures with shorter pulses. Peak pressures as high as 10 GPa can be achieved with pulse duration of 0.6 ns compared to around 5 GPa obtained with 25 ns pulse (Fabbro et al. 1998).

12.4.3   Effect of Temporal Pulse Shape

Temporal laser pulse shape also influences the pressure pulse shape. Figure 12.8 presents the two pulse shapes and the corresponding plasma-induced pressure pulses in confined ablation mode. In Fig. 12.8a, the pressure pulse shape is obtained with short-time-rise (SRT) pulse with a rise time of 10 ns and a pulse duration (FWHM) of 25 ns. Due to the short rise time of the SRT pulse, the breakdown of the confining medium occurs before the maxima in the laser pulse is reached. The figure clearly indicates that the peak pressure corresponding to the breakdown threshold is reached before the maxima in the laser pulse. Beyond the breakdown of the confining medium, there is no strong absorption of laser radiation at the interface. Furthermore, it is observed that SRT pulses result in significantly higher peak pressure compared to Gaussian pulse (Devaux et al. 1993).

12.4.4   Effect of Laser Wavelength

Wavelength of the laser influences the maximum peak pressure, duration of pressure pulse, and breakdown threshold of the confining medium during shock processing. Figure 12.9 presents variation of plasma-induced peak pressure with

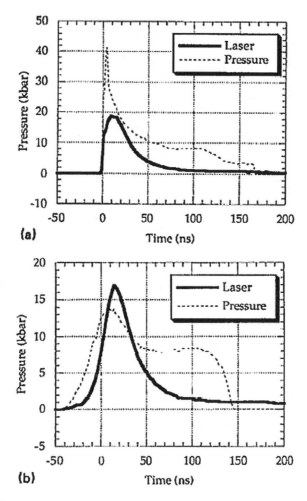

Fig. 12.8 Effect of temporal laser pulse shape on the plasma-induced pressure pulse shape during water-confined ablation: (**a**) short-rise-time (SRT) laser pulse and corresponding pressure pulse for power density of 8.8 GW/cm², (**b**) Gaussian pulse and corresponding pressure pulse for power density of 8.9 GW/cm². (Reprinted from Devaux et al. 1993. With permission. Copyright American Institute of Physics.)

laser power density for three different wavelengths. The data for 1.064 μm is taken from Berthe et al. (1997). For all the wavelengths, the figure indicates the typical relationship between the peak pressure and laser power density characterized by the saturation of plasma-induced pressure at high power densities. However, it seems that the dielectric breakdown threshold and the maximum peak pressure decreases with laser wavelength (Fabbro et al. 1998; Berthe et al. 1999). Furthermore, pressure duration decreases more sharply with laser power density for shorter wavelengths (Fig. 12.10) (Berthe et al. 1999).

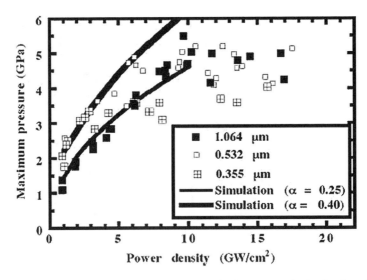

Fig. 12.9 Variation of plasma-induced peak pressure with laser power density for three different wavelengths. (Reprinted from Berthe et al. 1999. With permission. Copyright American Institute of Physics.)

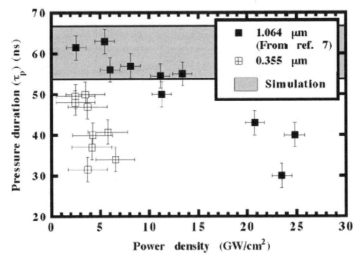

Fig. 12.10 Effect of laser power density on plasma-induced pressure duration (FWHM) for various laser wavelengths. (Reprinted from Berthe et al. 1999. With permission. Copyright American Institute of Physics.)

12.4.5 Effect of Spatial Energy Distribution

The plasma-induced pressure show significant inhomogeneities inside a laser impact. Figure 12.11 presents the variation of peak pressure along the focal spot radius. The figure indicates that the maximum intensity is located at a radius of

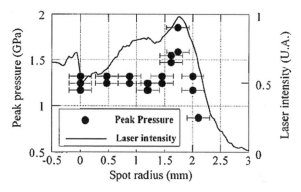

Fig. 12.11 Laser intensity profile and corresponding peak pressure profile inside the laser focal spot (laser intensity: 1.7 GW/cm², target: 100 μm thick aluminum foil, laser axis is located at 0 mm). (Reprinted from Berthe et al. 1997. With permission. Copyright American Institute of Physics.)

1.75 mm. The corresponding pressure is around 1.6 GPa. The pressure is almost constant (~1.25 GPa) within the ring of 1.5 mm. Beyond 1.75 mm, the pressure decreases below 1 GPa due to the radial release of waves (Berthe et al. 1997).

12.4.6 Effect of Coating Material

Plasma-induced pressure is an important parameter in laser shock processing of material. Significant efforts have been directed toward enhancing the plasma-induced pressure in the confined ablation mode. Earlier studies have indicated that the magnitude of the stress wave in the confined ablation mode can be increased by coating the target with thin metallic coating (Anderholm 1970). The pressure increased by the order of magnitude has been reported by irradiating the coated targets compared to bare targets. Figure 12.12 presents the variation of maximum stress in the coated aluminum target with laser fluence. The figure also presents the maximum stress obtained in bare targets with laser fluence of 91 J/cm². The figure indicates that significantly reduced laser fluence is sufficient to achieve the same magnitude of maximum stress in the coated targets compared to bare targets. The augmentation of maximum stress in the coated target is more pronounced at the lower laser fluences (O'Keefe and Skeen 1972).

The peak pressure during confined ablation is expected to be influenced by the physical properties of the target film. Figure 12.13 presents the variation of peak pressures generated in various target films confined by quartz or water with laser power density. For the laser power density less than 10^9 W/cm², the figure indicates that zinc targets covered with quartz overlays generate significantly higher peak pressure compared to aluminum targets covered with quartz overlay. This is due to

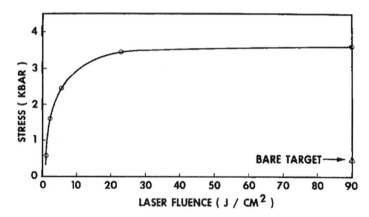

Fig. 12.12 Variation of maximum stress in coated aluminum targets with laser fluence (Reprinted from O'Keefe and Skeen 1972. With permission. Copyright American Institute of Physics.)

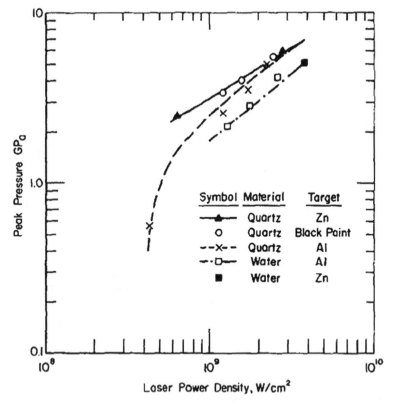

Fig. 12.13 Variation of plasma-induced peak pressures generated in various target films confined by quartz or water overlay with laser power density. (Reprinted from Fairand and Clauer 1979. With permission. Copyright American Institute of Physics.)

lower thermal conductivity of zinc compared to aluminum. Due to lower thermal conductivity of zinc, a smaller fraction of the absorbed laser energy thermally diffuses in the target resulting in higher peak temperatures and correspondingly higher peak pressures. The difference in the peak pressures in the two cases is not significant at higher laser power densities ($>4 \times 10^9$ W/cm²). At such higher laser power densities, other dominant mechanisms such as breakdown of confining medium limit the peak pressures (Fairand and Clauer 1979).

12.5   Mechanical Effects During Laser Shock Processing

As discussed in the previous sections, when a laser pulse (of sufficiently high intensity ~1–50 GW/cm² and short duration ~1–50 ns) is irradiated on the bare or confined target (with absorbent coating), it produces high temperature plasma which induces stress wave through the thickness of the target. The magnitude and duration of this stress wave for a given laser power density depend primarily on the extent of plasma confinement. One of the most important mechanical effects during laser shock processing is the generation of residual stresses. Figure 12.14 shows the schematic of the steps involved in the generation of residual stresses during laser shock processing. During laser–material interaction (laser on time), the rapid expansion of the plasma creates a uniaxial compression of the irradiated area along the direction of the shock wave and a dilation (tensile stretching) of the surface layers. At the end of the laser pulse time, the surrounding material responds to the volume change and introduces compressive stresses in the irradiated area (Peyre et al. 1996a).

Detailed analysis of the mechanical effects during laser shock processing is carried out by Ballard et al. (1991) by considering shock wave propagation into an elastic-perfectly plastic metal half space. This analytical elastic-perfectly plastic model assumed the shock waves to be longitudinal and planar, and the pressure pulse to be uniform on the impacted area. Furthermore, the material is considered to follow von Mises yield criteria in a biaxial condition. The model considered that

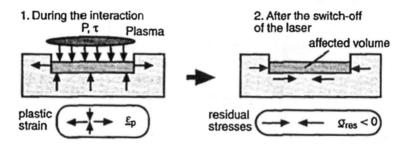

Fig. 12.14 Schematic of the steps in generation of compressive residual stresses during laser shock processing (Reprinted from Peyre et al. 1996a. With permission. Copyright Laser Institute of America.)

plastic deformation occurs when the peak pressure exceeds the Hugoniot limit (H_L) given by:

$$H_L = \left(1 + \frac{\lambda}{2\mu}\right)\sigma_y, \tag{12.11}$$

where σ_y is the compressive static yield stress, and λ and μ are the Lame constants. The plastic deformation increases between H_L to $2H_L$. Above $2H_L$, the plastic deformation is assumed to reach the maximum and above $2.52H_L$, and surface release waves occur. The plastically affected depth, L_p and surface residual stress, σ_{surf}, can be expressed as (Ballard et al. 1991):

$$L_p = \frac{C_{el}C_{pl}\tau}{\left(C_{el} - C_{pl}\right)H_L}\frac{P}{}, \tag{12.12}$$

$$\sigma_{surf} = \mu\varepsilon_p\left(\frac{1+v}{1-v}\right)\left[1 - \frac{4\sqrt{2}}{\pi}(1+v)\frac{L_p}{a}\right], \tag{12.13}$$

where C_{el} and C_{pl} are the elastic and plastic shock wave velocities, P is the shock wave pressure, ε_p is the plastic surface strain, μ is the pulse duration, a is the side of square impact. The above equations may be modified if the material consists of initial residual stress field.

A typical profile of the residual stresses introduced by confined laser shock processing is presented in Fig. 12.15. As indicated in the figure residual compressive stresses of high magnitude are introduced to greater depths of the sample. The compressive stresses are of great importance in improving the mechanical

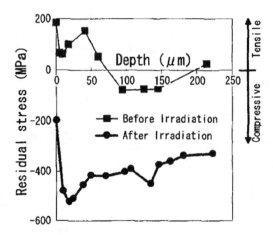

Fig. 12.15 Residual stress profile before and after water-confined laser shock processing of SUS304. (Reprinted from Sano et al. 1997. With permission. Copyright Elsevier.)

properties especially the fatigue resistance of the material. In case of direct irradiation, the tensile stresses may be introduced due to thermal effects and insufficient plasma pressure to cause plastic deformation (Sano et al. 1997). Hence, confined ablation mode is preferred for introducing the residual compressive stresses.

The magnitude of residual stresses and the depth of modification depend on a number of processing parameters which determine the profile of pressure pulse. These parameters may include laser power density, pulse duration, pulse shape, spot diameter, overlap between the spots, number of impacts, pulse densities, protective coating type, etc.

In general, the magnitude of residual compressive stress at the surface increases with increasing laser power density (Fig. 12.16a). This is a direct consequence of increasing plasma-induced pressure with increasing laser power density. As indicated in the previous sections, the plasma-induced peak pressure saturates due to breakdown of the confining medium. Thus, once breakdown threshold is reached, further increase in the laser power density saturates the maximum surface compressive stress and the depth of plastic deformation (Fabbro et al. 1998; Peyre et al. 1996a). Figure 12.16 also presents the effect of the number of impacts on the surface residual stress and the residual stress distribution in the depth of aluminum alloy sample. The figure indicates that for a given laser power density, a larger number of impacts gives enhanced magnitude of surface compressive stress and the depth of plastic deformation. Such as enhancement of the residual stresses with increased number of impacts seems to be due to Bauschinger-like effects where successive impacts causes enhanced deformation (Peyre et al. 1996a).

Residual stress distribution is also influenced by the pulse swept direction and the pulse density. Figure 12.17 presents the residual stress distribution in the depth of laser shock processed 6061-T6 aluminum alloy for two different pulse densities (pulses/area) in two directions. The figure indicates that compressive stress component perpendicular to laser swept direction, S_2, is significantly larger than that parallel to that direction, S_1. Furthermore, residual compressive stresses increase with increasing laser pulse density (Rubio-González et al. 2004). Residual compressive stress is also found to be improved by using protective coatings in the confined ablation modes (Peyre et al. 1998).

12.6 Microstructure Modification During Laser Shock Processing

Laser shock processing causes plastic deformation of the near surface material. Extensive studies have been conducted to analyze the microstructure changes after laser shock processing. One of the most important observations is the increase in the dislocation density after laser shock processing. Figure 12.18 presents the TEM bright field image of low carbon steel specimen shock processed with laser pulse energy of 111 J. The micrograph clearly indicates the high density of dislocations in a random tangles arrangement. The dislocation density in such microstructures

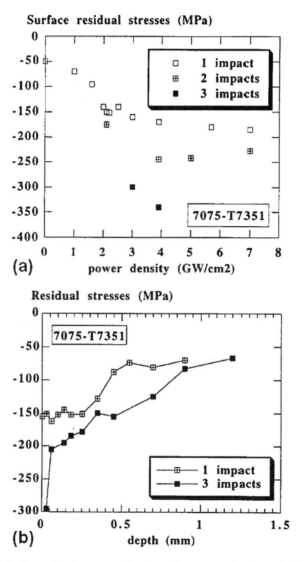

Fig. 12.16 (a) Surface residual stress as a function of laser power density, and (b) residual stress as a function of the depth from the surface for the various numbers of impacts of 7075 aluminum alloy. (Reprinted from Peyre et al. 1996b. With permission. Copyright Elsevier.)

was reported to be in the range of 2.6×10^{11} cm^{-2} (Chu et al. 1999). Similar disloca-tion structures are also observed in the case of aluminum alloys (Fairand et al. 1972). Laser shock processing may lead to the changes in the surface morphology due to undesirable thermal effects (such as melting) or the formation of deep inden-tations (Chu et al. 1999).

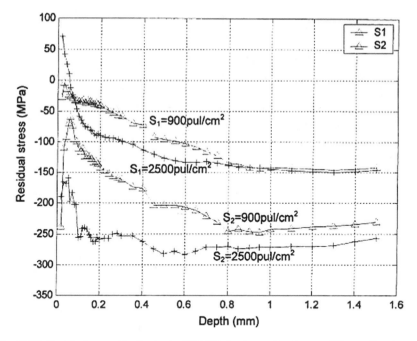

Fig. 12.17 Residual stress distribution in 6061-T6 aluminum alloy for two different direction and pulse densities. (Reprinted from Rubio-González et al. 2004. With permission. Copyright Elsevier.)

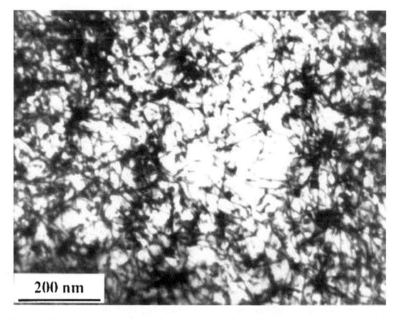

Fig. 12.18 TEM micrograph of laser shock processed low carbon steel showing the dense dislocation structure (Reprinted from Chu et al. 1999. 2004. With permission. Copyright Elsevier.)

12.7 Applications of Laser Shock Processing

Recently laser shock processing is attracting significant interests for improving the surface properties of materials. Three major areas where laser shock processing can be useful are the improvements in the fatigue resistance, crack corrosion resistance, and surface hardening.

Most of the studies on laser shock processing are probably motivated by the abilities of the process to improve the fatigue resistance. The improvement in the fatigue resistance is derived from the generation of surface compressive stresses

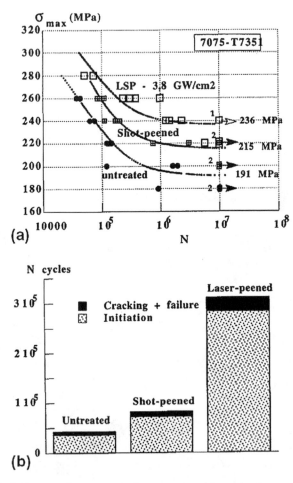

Fig. 12.19 (**a**) S–N curves for untreated, shot peened, and laser shock processed 7075 alloy, (**b**) crack initiation and propagation stages determined by a. c. potential drop method at the stress of 260 MPa. (Reprinted from Peyre et al. 1996b. With permission. Copyright Elsevier.)

during laser shock processing. Laser shock processing offers an attractive alternative to the conventional shot peening process of introducing compressive residual stresses for improvements in the fatigue performance. The process offers significant benefits over conventional shot peening process. Laser shock processing generally results in enhanced magnitude of residual stress and depth of plastic deformation. Also, the laser shock processing is expected to provide better surface finish compared to conventional shot peening. Furthermore, the distribution of stresses can be effectively controlled by optimizing the laser processing parameters. Figure 12.19a presents the typical S–N curves for untreated, shot peened, and laser shock peened aluminum 7075 alloy. The figure indicates that laser shock processing results in significant increase in the fatigue limit compared to shot peening. Furthermore, it has been reported that laser shock processing resulted in sevenfold increase in the crack initiation stage and threefold increase in the propagation stage for a stress of 260 MPa (Fig. 19b). The crack initiation was considered as the number of cycles corresponding to a change in potential down the root notch during a.c. potential drop method (Peyre et al. 1996b).

Laser shock processing is also reported to be useful for the improvements in the pitting corrosion resistance of 316 stainless steel. It has been suggested that increased pitting potentials after laser shock processing are due to large compressive stress and work hardening levels which play an important role in reducing pit formation on surface inclusions. The process is also suggested for the improvements in the stress corrosion cracking. One more area where laser shock processing is attractive is the surface hardening of the metallic materials. Significant improvements in the Vickers hardness have been reported for various alloys after laser shock processing (Peyre et al. 2000).

References

Anderholm NC (1970) Laser-generated stress waves. Applied Physics Letters 16:113–115

Ballard P, Fournier J, Fabbro R, Frelat J (1991) Residual stresses induced by laser shocks. Journal de Physique IV 1:487–494

Berthe L, Fabbro R, Peyre P, Tollier L, Bartnicki E (1997) Shock waves from a water-confined laser-generated plasma. Journal of Applied Physics 82:2826–2832

Berthe L, Fabbro R, Peyre P, Bartnicki E (1999) Wavelength dependent of laser shock-wave generation in the water-confinement regime. Journal of Applied Physics 85:7552–7555

Chu JP, Rigsbee JM, Banas G, Elsayed-Ali HE (1999) Laser-shock processing effects on surface microstructure and mechanical properties of low carbon steel. Materials Science and Engineering A 260:260–268

Devaux D, Fabbro R, Tollier L, Bartnicki E (1993) Generation of shock waves by laser-induced plasma in confined geometry. Journal of Applied Physics 74:2268–2273

Fabbro R, Fournier J, Ballard P, Devaux D, Virmont J (1990) Physical study of laser-produced plasma in confined geometry. Journal of Applied Physics 68:775–784

Fabbro R, Peyre P, Berthe L, Scherpereel X (1998) Physics and applications of laser-shock processing. Journal of Laser Applications 10:265–279

Fairand BP, Clauer AH (1979) Laser generation of high-amplitude stress waves in materials. Journal of Applied Physics 50:1497–1502

Fairand BP, Wilcox BA, Gallagher WJ, Williams DN (1972) Laser shock-induced microstructural and mechanical property changes in 7075 aluminum. Journal of Applied Physics 43:3893–3895

Montross CS, Wei T, Ye L, Clark G, Mai YW (2002) Laser shock processing and its effects on microstructure and properties of metal alloys: a review. International Journal of Fatigue 24:1021–1036

O'Keefe JD, Skeen CH (1972) Laser-induced stress-wave and impulse augmentation. Applied Physics Letters 21:464–466

O'Keefe JD, Skeen CH, York CM (1973) Laser-induced deformation modes in thin metal targets. Journal of Applied Physics 44:4622–4626

Peyre P, Fabbro R, Berthe L, Dubouchet C (1996a) Laser shock processing of materials, physical processes involved and examples of applications. Journal of Laser Applications 8:135–141

Peyre P, Fabbro R, Merrien P, Lieurade HP (1996b) Laser shock processing of aluminium alloys. Application to high cycle fatigue behaviour. Materials Science and Engineering A 210:102–113

Peyre P, Berthe L, Scherpereel X, Fabbro R (1998) Laser-shock processing of aluminum coated 55C1 steel in water-confinement regime, characterization and application to high-cycle fatigue behaviour. Journal of Materials Science 33:1421–1429

Peyre P, Scherpereel X, Berthe L, Carboni C, Fabbro R, Be' ranger G, Lemaitre C (2000) Surface modifications induced in 316L steel by laser peening and shot-peening. Influence on pitting corrosion resistance. Materials Science and Engineering A 280:294–302

Phipps CR, Turner TP, Harrison RF, York GW, Osborne WZ, Anderson GK, Corlis XF, Haynes LC, Steele HS, Spicochi KC (1988) Impulse coupling to targets in vacuum by KrF, HF, and CO_2 single-pulse lasers. Journal of Applied Physics 64:1083–1096

Rubio-González C, Ocana JL, Gomez-Rosas G, Molpeceres C, Paredes M, Banderas A, Porro J, Morales M (2004) Effect of laser shock processing on fatigue crack growth and fracture toughness of 6061-T6 aluminum alloy. Materials Science and Engineering A 386:291–295

Sano Y, Mukai N, Okazaki K, Obata M (1997) Residual stress improvement in metal surface by underwater laser irradiation. Nuclear Instruments and Methods in Physics Research B 121:432–436

Smith WL, Bechtel JH, Bloembergen N (1977) Picosecond laser-induced breakdown at 5321 and 3547 Å: Observation of frequency-dependent behavior. Physical Review B 15:4039–4055

Yokotani A, Sasaki T, Yoshida K, Nakai S (1989) Extremely high damage threshold of a new nonlinear crystal L-arginine phosphate and its deuterium compound. Applied Physics Letters 55:2692–2693

Chapter 13
Laser Dressing of Grinding Wheels

13.1 Introduction

Lasers are presently used in a number of noncontact manufacturing processes such as welding, drilling, cutting, shaping, scribing, etc. The applications of lasers in the machining of hard and brittle (difficult-to-machine) materials such as ceramics are well reported in literature and widely practiced in various industries such as aerospace, automotive, microelectronics, etc. The concept of laser machining can be further extended for novel applications such as dressing of grinding wheels. Dressing of grinding wheels refers to the resharpening operation designed to generate a specific topography on the cutting surface of the grinding wheel. This chapter briefly provides the background of the traditional approaches for dressing of grinding wheels and associated performance and quality issues. This is followed by a detailed explanation of the two major laser-based approaches namely, laser-assisted dressing and laser dressing which presents the potential for effective implementation in the manufacturing environment.

13.2 Grinding Process and Need of Wheel Dressing

Grinding is a machining process in which workpiece material removal is carried out by pressing an abrasive surface of the rotating grinding wheel against the workpiece surface with a force perpendicular to the contact zone (Lindsay 1997). Various commercial abrasive grinding wheel materials include conventional materials like aluminum oxide and silicon carbide, and other specialty materials like diamond and cubic boron nitride. The grinding process can be generally classified into two categories: stock removal grinding (rough grinding) and precision grinding (finish grinding). In stock removal grinding, the main consideration is the very high materials rates; whereas in precision machining, the main consideration is the high accuracy of the form and finish of the workpiece (Lin et al. 2002). In most of the grinding processes, the material removal is due to cutting, ploughing, and rubbing actions. Among the various parameters, the grinding performance of the grinding wheel is primarily determined by the topography of the grinding wheel surface.

N.B. Dahotre and S.P. Harimkar, *Laser Fabrication and Machining of Materials.*
© Springer 2008

This includes parameters like sharpness of the surface abrasive grains, grain size, surface porosity, grain density, etc. However, most of these important parameters undergo dynamic changes during grinding due to friction between the workpiece and the grinding wheel causing bond fracture, abrasive grain fracture, de-bonding, etc. Thus the grinding performance is greatly influenced by the changes in the grinding wheel surface during grinding. In general, a new commercial grinding wheel in an as-received condition consists of a large number of sharp, irregular abrasive grains projecting from the bond material. Each one of these grains act as a microcutting tool and give very high material removal rate in the initial stages of grinding. However, as the grinding continues, the abrasive grains undergo fracture and collision and eventually the surface becomes nearly flat during the final stages of grinding. This marks the condition where the grinding wheel can no longer perform the efficient material processes. Thus, the material removal rates decreases during grinding as a consequence of changes in the topography of the grinding wheel surface. The general changes taking place in the surface morphology of an individual abrasive grain on the surface of wheel during grinding are presented in Fig. 13.1. In addition, the schematic of the consequent variation in material removal rate during grinding is presented in Fig. 13.2 (Lin et al. 2002).

As explained, the efficiency of the grinding wheels in removal of material from the workpiece is greatly reduced due to gradual changes in surface condition of the wheels during grinding. Hence, impartment of the cutting ability to the grinding wheel surface is of prime importance to regain the grinding efficiency. This is conventionally carried out by single-point or multipoint diamond tools. The single-point diamond tool removes the worn-out surface from the grinding wheel surface and exposes the new irregular and sharp grains for cutting action. Also, the diamond tool may fracture the surface grains thus creating the new cutting edges. This process of resharpening the grinding wheel to provide the sharp cutting edges on the surface

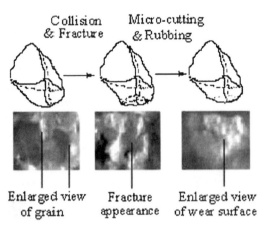

Fig. 13.1 Changes in the morphology of abrasive grain from sharp and irregular surface features during initial stages of grinding to flat and blunt surface features during final stages of grinding. (Reprinted from Lin et al. 2002. With permission. Copyright Elsevier.)

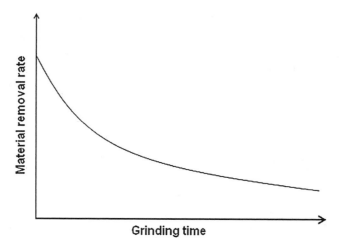

Fig. 13.2 Schematic of the variation of material removal rate with grinding time

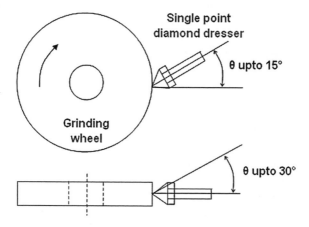

Fig. 13.3 Schematic of the single-point diamond (mechanical) dressing of grinding wheel

is often referred to as mechanical or diamond dressing (Fig. 13.3). The important parameters during dressing of grinding wheels are depth of dress and dressing lead. Even though mechanical dressing of the grinding wheels is widely used in the industrial practices, the process is associated with many limitations. Firstly, the process is a contact-type process and hence grinding operations needs to be stopped for every dressing cycle. This amounts to the loss of actual machine time and man time for machining thus decreasing the productivity. Furthermore, the diamond tool may undergo uneven wear during dressing operation leading to high consumable cost and no-uniform dressing results. Diamond dressing may also result in surface and subsurface defects (such as cracks) due to brittle nature of grinding wheels leading to premature fracture of the grinding wheels. It has also been reported that about 90% of the grinding wheel material is removed during dressing operation and not during actual grinding thus shortening the life of the grinding wheel (Buttery et al. 1979). To overcome all these limitations, the conventional diamond dressing process needs significant developments. Recently, significant interests have been attracted toward the development of laser-based technologies either to assist the conventional mechanical dressing operations or to use lasers for selectively modifying the wheel surfaces for efficient grinding performance.

13.3   Laser-Based Wheel Dressing Techniques

Dressing of grinding wheels using high-power lasers is demonstrated in various ways. When the grinding wheel surface is irradiated with a laser beam, the laser–material interactions at the surface of the wheel can cause heating and melting of the abrasive and/or bonding ingredients depending on the absorbed laser energy. Laser can also be used to assist the conventional mechanical dressing by locally heating the wheel surface ahead of the diamond dressing tool. This first approach is commonly referred to as laser-assisted dressing. In the second approach, the wheel is directly irradiated with the laser beam to generate sharp cutting edges either due to local melting / damage to the worn-out grits or dislodgement of the loaded chips. These two approaches are discussed in detail in the following sections.

13.3.1   Laser-Assisted Dressing

The concept of laser-assisted dressing was first demonstrated by the research group of Professor Shin at Purdue University (Zhang and Shin 2002). Laser-assisted dressing bears many similarities with the laser-assisted machining where the materials removal by traditional machining tool is facilitated by localized laser heating of the workpiece. In this method, a focused laser beam is used to locally heat the surface of the rotating grinding wheel ahead of the dressing tool, typically a single-point diamond dresser. Selection of suitable laser parameters such as energy density and heating time establishes the temperature distributions within the grinding wheel such that the temperature at the point of contact between the diamond tip and the wheel reaches sufficiently high to cause softening of the vitrified bond of the wheel. The subsequent removal of worn-out material from the wheel surface using a diamond dresser becomes easy. The main objectives behind such processes is to reduce the wear of dressing tool and improve the surface quality by locally changing the material removal mode of hard ceramics from brittle fracture to ductile flow by laser heating. The wear rate of the diamond dresser and hence the life can be well controlled by controlling the temperature at the tip of the diamond by adequately selecting the laser dressing conditions. The experimental setup for laser-assisted dressing and the corresponding thermal model are shown in Fig. 13.4 (Zhang and Shin 2002).

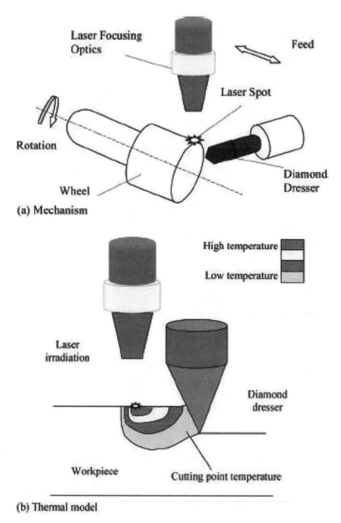

Fig. 13.4 Schematic of the laser-assisted dressing setup and associated thermal effects. (Reprinted from Zhang and Shin 2002. With permission. Copyright Elsevier.)

Extensive experiments have been carried out to investigate the feasibility of the laser-assisted dressing by characterizing and comparing the wheel profile, wear of diamond dresser, dressing force, etc. (Zhang and Shin 2002) during mechanical and laser-assisted dressing. Some of these results are summarized below.

13.3.1.1 Wheel Surface Condition

Surface conditions of the dressed grinding wheels have the great influence of the machining performance of the wheels. The two important characteristics of the surface of the dressed grinding wheels are the effective number of cutting edges and the porosity pockets. The dressing operation should generate/expose the large number of sharp and irregular grains with microcutting edges for cutting action. At the same time, the wheel surface should have enough porosity pockets for the collection (loading) of chips and flow of coolant. One of the most important abilities of laser-assisted dressing is that the subsequent re-dressings of the grinding wheel produces uniform wheel surface. This was demonstrated during the laser-assisted dressing of CBN grinding wheels. The CBN grinding wheel profiles showed the nearly same values of maximum and average peak-to-peak values (~80 and 65 µm, respectively) after each laser-assisted dressing operation (for two tests) indicating the uniformity of wheel profiles. In contrast, the mechanical dressing produces significantly different results for surface profiles after each test (for two tests). In general, the subsequent mechanical dressing operation produced surface profiles with significantly lower maximum and average peak-to-peak values indicating the inability of the diamond dresser to generate the uniform surface profile due to wear (Zhang and Shin 2002).

13.3.1.2   Wear of Diamond Dresser

One of the main objectives of the laser-assisted dressing is to reduce the wear of the diamond dresser by localized heating of the grinding wheel surface ahead of the dresser point. The effectiveness of this approach over the conventional mechanical dressing can be conveniently determined by comparing the wear of diamond dresser after each dressing operation. Figure 13.5 shows that wear of the diamond dresser during laser-assisted dressing of CBN grinding wheels is

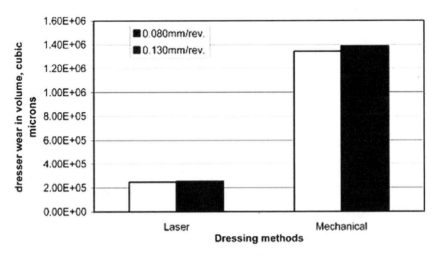

Fig. 13.5 Comparison of wear of diamond dressers during conventional mechanical and laser-assisted dressing of CBN grinding wheels. (Reprinted from Zhang and Shin 2002. With permission. Copyright Elsevier.)

significantly lower than that during conventional mechanical dressing. Such lower wear rates of diamond dresser during laser-assisted dressing may be due to ease of material removal at high temperature (Zhang and Shin 2002). Similar investigations have shown that the wear volume of the diamond dresser per unit material removal of the grinding wheel is significantly lower during laser-assisted dressing than during mechanical dressing (Zhang and Shin 2003). In all these studies the wear volume of diamond dresser was determined by measuring the diameter of the diamond dresser top.

13.3.1.3 Dressing Forces

The three important forces during machining of materials are primary cutting force (F_c), feed force (F_f), and thrust force (F_t). Localized heating of the grinding wheel ahead of the diamond dressing tool is expected to ease the material removal of the grinding wheel by lowering the machining (dressing) forces. Figure 13.6 presents variation in three dressing forces with the method of dressing (mechanical or laser-assisted) for two dressing conditions (feed of 0.08 and 0.13 mm/rev). The figure indicates that for each dressing condition, the forces during laser-assisted dressing of CBN grinding wheels are significantly less than that during mechanical dressing suggesting the effectiveness of laser-based approach for dressing of grinding wheels (Zhang and Shin 2002).

Thus, the experimental results on the dressing of CBN grinding wheels have shown that laser-assisted dressing and truing offers a number of advantages over conventional diamond dressing.

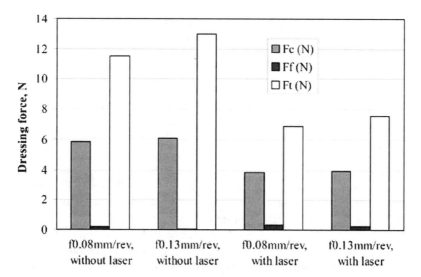

Fig. 13.6 Dressing forces during mechanical and laser-assisted dressing of CBN grinding wheels. (Reprinted from Zhang and Shin 2002. With permission. Copyright Elsevier.)

13.3.2   Laser Dressing

The previous section discussed the laser-assisted dressing of grinding wheels using mechanical dressing assisted by laser heating. Since lasers have been successfully used as machining tools in the past, significant attempts have been made to realize alternate purely laser dressing operation without any need for mechanical dressing tool. The primary motivation for such laser dressing operation is based on the ability of the high-power lasers to locally modify the surface of the grinding wheel to generate/expose the sharp cutting edges which can play an important role in efficient material removal during subsequent machining of the material. The feasibility of such laser dressing technique based on laser surface modification was first demonstrated by Babu et al. in 1989.

The experimental setup of laser dressing consisted of a grinding wheel mounted on a stepper motor shaft. A high power pulsed Nd:YAG laser was used to irradiate the cutting surface of rotating grinding wheel such that circular groove is produced on the surface. To dress the full cutting surface of the grinding wheel, the entire stepper motor assembly was moved after each revolution corresponding to the specified dressing lead. The laser dressing results in the generation of uniformly spaced circular grooves on the periphery of the grinding wheel. Laser causes selective damage to the abrasive or bond material of the grinding wheel thus exposing/generating the sharp grains for cutting action during machining (Babu et al. 1989).

13.3.2.1   Experimental Results of Laser Dressing

Extensive experimental work have been conducted on the laser dressing of grinding wheels by the research group of Professor Babu at the Indian Institute of Technology, Chennai (Babu et al. 1989; Babu and Radhakrishnan 1989, 1995; Phanindranath and Babu 1996). The summary of their reported results is presented in the following sections.

Surface Condition of Laser-Dressed Grinding Wheels

The interaction of laser pulses with the grinding wheel material during laser dressing depends on the laser parameters and the properties of the constituent phases (abrasive grains and bond) in the wheel material. For sufficiently high values of laser intensities and with the focused spot size relatively smaller than the abrasive grain size, the localized surface damage causes the formation of microcraters with resolidified material around it. When these microcraters are formed on the worn-out abrasive grains, a number of cutting edges are created. In addition, when these microcraters are formed on the bond material, the abrasive grains become loose and are removed during subsequent machining due to cracks initiated and propagated in the resolidified material around the crater. This exposes the sharp abrasive grains for cutting action. Thus laser dressing generates/exposes sharp abrasive grains for efficient removal of material during machining. The consequent changes in the topography of the grinding wheel depend on the dressing parameter such as dressing lead (feed rate). Figure 13.7 presents the topographic changes associated with the laser dressing of grinding wheel. The figure indicates that at high lead (0.8 mm/rev), the formation of craters is limited to very few grains/bond along the groove. However, at very small lead (0.35 mm/rev) the entire surface of the grinding wheel can be covered with craters (Babu et al. 1989). Furthermore, the surface topography of the laser-dressed grinding wheels is influenced by the laser pulse intensities (Fig. 13.8). Laser-dressed grinding wheel shows periodic surface features (grooves) at all the laser intensities (Fig. 13.8). As the laser intensity increases, the areas affected by the laser beam increases resulting in widening of the grooves. Thus careful selection of the laser dressing parameters can generate the surface topography of the grinding wheels for efficient removal of the material (Babu and Radhakrishnan 1989).

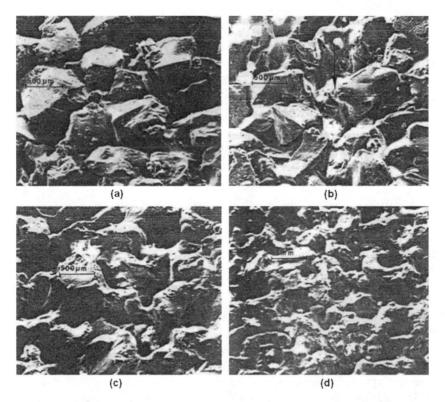

Fig. 13.7 Surface topography of Al_2O_3 grinding wheels: (**a**) before dressing, and (**b**), (**c**), and (**d**) after laser dressing with lead of 0.8, 0.5 and 0.35 mm/rev, respectively and laser intensity of 6.0×10^{10} W/m². (Reprinted from Babu et al. 1989. With permission. Copyright American Society of Mechanical Engineering.)

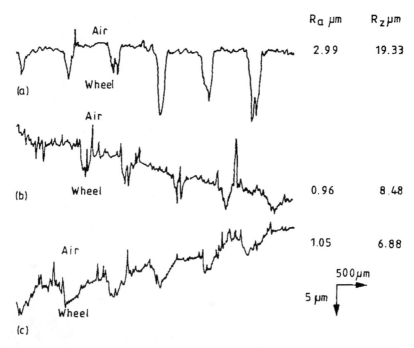

Fig. 13.8 Surface profiles of Al_2O_3 grinding wheels laser dressed at a feed rate of 0.8 mm/rev and various laser intensities: (**a**) 6.0×10^{10} W/m², (**b**) 6.0×10^{10} W/m², and (**c**) 6.0×10^{10} W/m². (Reprinted from Babu and Radhakrishnan 1989. With permission. Copyright American Society of Mechanical Engineering.)

Grinding Forces and Workpiece Finish

As explained earlier, the performance of laser-dressed grinding wheels can be explained based on the machining forces acting on the wheel and the surface finish of the machined workpieces. Figure 13.9 presents the comparison of the grinding forces (F_N and F_T) and workpiece surface roughness during machining of hardened carbon steel with grinding wheels dressed by conventional diamond dresser and lasers at various intensities. The figure clearly indicates that at intermediate laser intensity (6.0×10^{10} W/m²), the grinding forces and workpiece finish is comparable with that obtained with diamond-dressed grinding wheels. At lower (4.0×10^{10} W/m²) and higher (9.0×10^{10} W/m²) intensities, either the machining forces or the workpiece surface roughness values are significantly higher than that obtained with conventional dressing. Thus, selection of laser intensity during dressing of grinding wheels plays an important role in determining the machining performance of the dressed wheels. Such an understanding on the influence of laser intensities on the grinding forces and workpiece surface finish also facilitate the designing the machining process. For example, the laser-dressed grinding wheels at higher intensities offer more advantages for rough machining applications where high

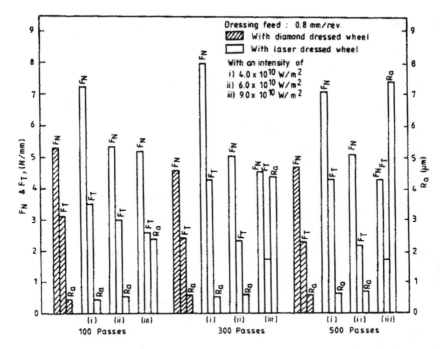

Fig. 13.9 Grinding forces and workpiece surface roughness during machining of carbon steel with diamond- and laser-dressed Al$_2$O$_3$ grinding wheels. (Reprinted from Babu and Radhakrishnan 1989. With permission. Copyright American Society of Mechanical Engineering.)

workpiece surface roughness is not critically important (Babu and Radhakrishnan 1989). The fluid media during grinding is expected to influence the wear behavior of the dressed grinding wheel. The effectiveness of the laser-dressed grinding wheels has been demonstrated further in such wet grinding conditions (Babu and Radhakrishnan 1995).

13.3.2.2 Analysis of Laser Dressing

Laser dressing is a complex process involving the interaction of axially moving laser beam (or grinding wheel) with the rotating composite grinding wheel material (consisting of abrasive grains, bond material, and porosity). In order to predict the evolution of surface topography during laser interaction with the grinding wheel, it is important to develop the theoretical model incorporating the complexity of the laser dressing process. Laser dressing is a thermal process. Hence, early attempts to model the laser dressing process involved the solution of the heat transfer equation. Following sequence was adopted to model the laser dressing process (Phanindranath and Babu 1996):

1. Modeling of grinding wheel structure: Grinding wheel structure consists of a distribution of abrasive grains and bond material. To develop the thermal model for predicting the thermal effects, it is necessary to define the grinding wheel structure in terms of the distribution of abrasive grit and bond material. Considering the random distribution of the grit spacing, the location of the grits and bond materials was obtained.
2. Estimation of grinding wheel properties: The properties of the constituent phases (grit and bond material) in the grinding wheel vary with the temperature during laser dressing. The average properties of the constituent phases within the temperature range of laser dressing were considered during thermal modeling.
3. Thermal model: Thermal model was developed considering the general three-dimensional heat conduction equation with a moving source of heat using finite difference formulation.

The thermal model gives the temperature distribution in the grinding wheel during laser dressing. By considering the solid–liquid and liquid–vapor boundaries corresponding to the melting and vaporization temperatures, respectively, the amount of ejected and resolidified material can be calculated. This facilitates the calculation of shape of the groove during laser dressing. Figure 13.10 presents the calculated shape of the groove along with resolidifed layer during laser dressing of Al_2O_3 grinding wheels. It was observed that the predicted widths of the grooves on the surface of the grinding wheel are in good agreement with the experimental results (Phanindranath and Babu 1996).

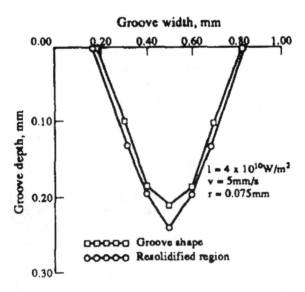

Fig. 13.10 Predicted shape of the groove formed on the surface of Al_2O_3 grinding wheel during laser dressing. (Reprinted from Phanindranath and Babu 1996. With permission. Copyright Elsevier.)

13.3.3 Recent Approaches of Laser Dressing

In the approaches for dressing the grinding wheels explained in the previous sections, a laser is used either to locally heat (laser-assisted dressing) or locally ablate the bond or abrasive grits (laser dressing) to expose the sharp abrasive grains. In laser-assisted dressing of grinding wheels, the topography of the dressed wheel is very similar to conventionally dressed grinding wheel. The primary objective of this approach is to reduce the wear of the diamond dresser. In laser dressing, the surface topography of the dressed grinding wheels consists of a number of craters caused by local damage (ablation) of the bond or abrasive grit material. Even though these approaches demonstrated the feasibility based on evaluation of dressing forces, wheel topography, and workpiece surface finish, these approaches have had limited success in realizing the full potential of noncontact laser dressing such as in-process dressing, improved productivity, low cost of production, and consistency in workpiece surface finish. Recently, significant research efforts have been conducted on the laser dressing of alumina grinding wheels to generate the highly well-defined, regular, and periodic topographic surface features having potential to play an important role in microscale material removal during grinding (Jackson et al. 2003; Khangar and Dahotre 2005; Khangar et al. 2005; Khangar et al. 2006; Harimkar and Dahotre 2006; Harimkar et al. 2006). Figure 13.11 presents the schematic of the process setup and surface morphology of a dressed wheel with this novel laser dressing approach and the conventional mechanical dressing approach. The characteristic surface features formed during laser dressing are associated with the evolution of morphological and crystallographic textures during rapid melting and solidification of abrasive material at the surface of the grinding wheel while laser dressing. The following sections briefly summarize the important aspects of this laser dressing process based on development of solidification microstructures (Harimkar and Dahotre 2006).

13.3.3.1 Morphological and Crystallographic Textures in Laser-Dressed Grinding Wheels

The morphological changes on the surface of the alumina grinding wheel after laser dressing are presented in Fig. 13.12. The rapid solidification associated with laser melting of alumina results in a wheel surface morphology is characterized by multifaceted polygonal surface grains with micropores between the faceted grains. Each multifaceted grain exhibits well-defined edges and vertex which can act as microcutting edges and points on the grinding wheel surface. Microporosity can assist in coolant flow and collection of finer chips of material removed during grinding. Additionally, since these surface features are on the micrometer scale, such a laser-engineered grinding wheel with surface microstructural features is expected to be well suited for microscale precision grinding (material removal) application. The highly faceted polygonal grains on the surface of alumina grinding

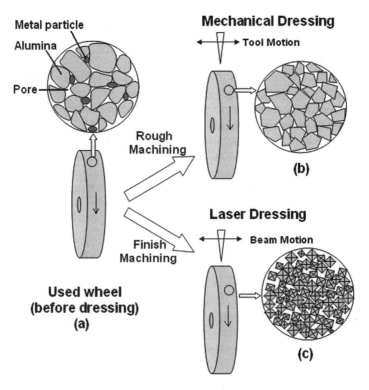

Fig. 13.11 Schematic of the setups and morphologies of the grinding wheel surface: (**a**) worn-out grinding wheel before dressing, (**b**) mechanical dressing, and (**c**) laser dressing (Reprinted from Harimkar and Dahotre 2006. With permission. Copyright American Ceramic Society.)

Fig. 13.12 Morphological changes associated with laser dressing of alumina grinding wheel: (**a**) before dressing, and (**b**) after laser dressing. (Reprinted from Harimkar et al. 2006. With permission. Copyright Institute of Physics.)

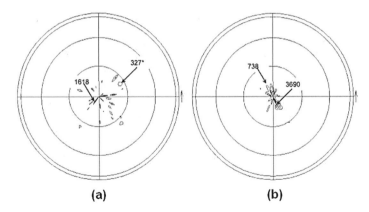

Fig. 13.13 (110) pole figure (**a**) before and (**b**) after laser dressing of alumina grinding wheels. (Reprinted from Khangar and Dahotre 2005. With permission. Copyright Elsevier.)

wheels during laser dressing are derived from the evolution of (110) crystallographic texture corresponding to the hexagonal alumina. This is clearly seen in the (110) pole figures obtained from the surface of grinding wheels before and after laser dressing (Fig. 13.13). Furthermore, the faceted surface grain size, the grain shape, the surface porosity, and the depth of melting are greatly influenced by laser processing parameters such as laser fluence. Thus, the solidification microstructures on a surface of the grinding wheel can be efficiently controlled by optimizing the laser processing parameters making the lasers reliable and efficient tools for dressing of grinding wheels for precision machining applications (Harimkar and Dahotre 2006).

13.3.3.2 Grinding Performance of Laser-Dressed Grinding Wheels

Early models of grinding were based on indentation fracture mechanics approach for the abrasive–workpiece interaction where the deformation and material removal mechanisms are assumed to be similar to the localized small-scale indentation events (Torrance and Badger 2000). The abrasive grains were idealized as pyramidal indenters and the material removal mechanisms were considered to be simultaneous ploughing and cutting of the material by the pyramidal indenter, where ploughing is the sideways flow of material round the indenter (Fig. 13.14). Since, the morphology of the laser-assisted rapid surface microstructured alumina grinding wheel with optimized laser processing parameters is expected to consist of multi-faceted grains closely resembling the pyramidal indenters, the idealized concept of material removal during grinding based on indentation fracture mechanics concept seems more appropriate in case of machining with laser-dressed grinding wheels. Figure 13.15 presents the illustration of this concept by comparing the idealized pyramidal indenter shape with the actual AFM picture of a faceted grain on the surface of laser dressed alumina grinding wheel.

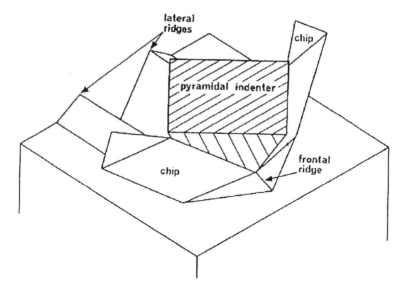

Fig. 13.14 Pyramidal indenter model for material removal during grinding of materials. (Reprinted from Torrance and Badger 2000. With permission. Copyright Elsevier.)

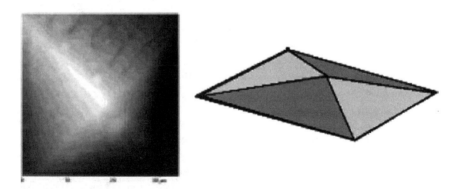

Fig. 13.15 AFM picture of faceted pyramidal grain on the laser-dressed alumina grinding wheel indicating its resemblance to an ideal pyramidal indenter (Harimkar and Dahotre 2007, unpublished data)

Extensive studies have been conducted to evaluate the performance of the laser-dressed grinding wheels in terms of the topography of the grinding wheels after laser dressing and the surface finish of the ground workpieces. Figure 13.16 presents the surface roughness profiles of the laser-dressed alumina grinding wheels dressed using three different laser powers (500, 750, and 1,000 W). The figure indicates that the laser-dressed grinding wheels are characterized by microlevel peaks and values suitable for precision machining of the material. In general, the alumina grinding wheels dressed at higher laser powers are smoother than those

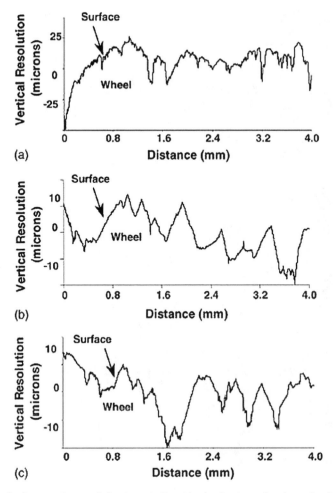

Fig. 13.16 Surface roughness of alumina grinding wheels after laser dressing using laser powers of (**a**) 500 W, (**b**) 750 W, and (**c**) 1,000 W. (Reprinted from Khangar and Dahotre 2005. With permission. Copyright Elsevier.)

dressed at lower laser powers. This is primarily due to increased extent of melting on the surface of the grinding wheel (Khangar and Dahotre 2005).

The surface finish of the workpieces ground using laser-dressed grinding wheels is mainly by the laser power used for laser dressing. Figure 13.17 presents comparison of the surface roughness profiles of the steel (AISI 1010) workpieces ground using laser-dressed grinding wheels (using three different laser powers) and undressed grinding wheels. The corresponding data of the surface roughness parameters is presented in Table 13.1. As indicated in the figure and table, the surface roughness of the workpieces ground using undressed grinding wheel is significantly higher than those ground using laser-dressed grinding wheels. The

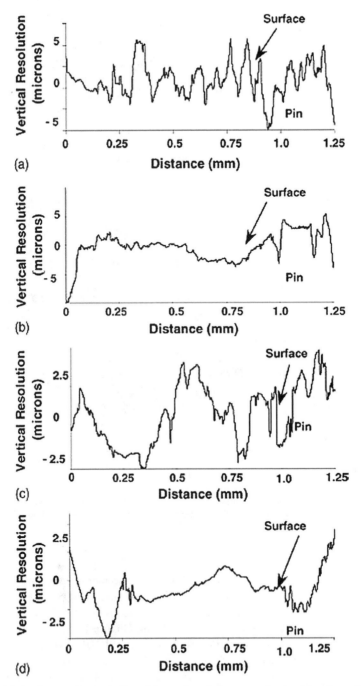

Fig. 13.17 Surface roughness profiles of the steel workpiece ground using (**a**) undressed and (**b**), (**c**), and (**d**) laser-dressed grinding wheels with laser powers of 500, 750, and 1,000 W, respectively. (Reprinted from Khangar and Dahotre 2005. With permission. Copyright Elsevier.)

Table 13.1 Surface roughness parameters of the AISI 1010 plain carbon steel workpiece used for grinding test on laser-dressed and undressed grinding wheel surface. (Reprinted from Khangar and Dahotre 2005. With permission. Copyright Elsevier.)

Laser power (W)	Ra (μm)	Rz (μm)	Rmax (μm)	RPc (1.0, −1.0) (cm⁻¹)
1,000	0.557 (±0.015)	3.1033	5.38	27
750	0.429 (±0.11)	2.1566	3.7816	11
500	0.467 (±0.185)	2.673	6.673	10
Undressed sample	1.859(±0.21)	10.237	14.2	112

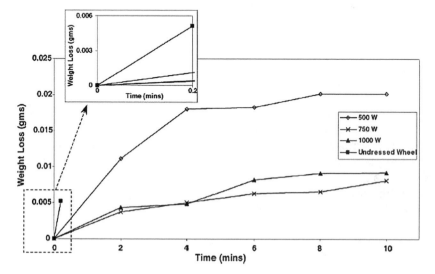

Fig. 13.18 Cumulative weight loss of steel workpieces during grinding tests using undressed and laser-dressed grinding wheels. (Reprinted from Khangar and Dahotre 2005. With permission. Copyright Elsevier.)

observed differences in the surface roughness of the workpieces ground using laser-dressed and undressed grinding wheels comes from the differences in the material removal characteristics during corresponding grinding processes. Figure 13.18 presents the cumulative weight loss of AISI 1010 steel during various grinding tests. The figure indicates that the weight loss of the workpiece is significantly lower in case of the grinding tests using laser-dressed grinding wheels. This suggests the potential of laser dressing in the precision machining of materials.

One of the most significant advantages of the laser dressing is that the dressed grinding wheel undergoes minimal wheel material loss (wheel wear) during grinding tests. This result in the very large values of grinding efficiency (grinding ration) defined as the weight loss in workpiece per unit weight loss of the grinding wheel material during grinding test (Table 13.2). Thus, the undressed or conventionally dressed grinding wheels are more suitable for the rough machining; whereas the laser-dressed grinding wheels are expected to be suitable for precision machining applications (Khangar and Dahotre 2005).

Table 13.2 Grinding ratio for grinding at varying laser power. (Reprinted from Khangar and Dahotre 2005. With permission. Copyright Elsevier.)

Sample	Undressed wheel	Laser-dressed wheel		
		500 W	750 W	1,000 W
Time (min)	0.2	10	10	10
Cumulative weight loss in grinding wheel G (g/unit contact area)	0.0444	0.0257	0.0203	0.0091
Cumulative weight loss in steel workpiece P (g/unit contact area)	0.00515 (in 0.2 min)	0.02017	0.0055	0.00922
Ratio of weight loss in steel workpiece to weight loss in grinding wheel: P/G	0.0115991	0.784825	0.270936	1.013187

13.3.3.3 Thermal Analysis of Laser Dressing

In order to get insight into the thermal effects during laser dressing of grinding wheels, significant modeling efforts have been carried out (Harimkar et al. 2006). Simplified thermal model considers the one-dimensional heat transfer in the grinding wheel to predict the temperature distribution and the depth of melting. Such an approach is considered to be reasonable based on the uniformity of the microstructure in the laser-dressed grinding wheel surface. The governing heat transfer equation is given by (Carslaw and Jaeger 1967):

$$\frac{\partial T(x,t)}{\partial t} = \alpha \frac{d^2 T(x,t)}{dx^2}, \tag{13.1}$$

where α is the thermal diffusivity of the material and is equal to $k / \rho C_p$; k is the thermal conductivity of the material; ρ is the density of the material; and C_p is the specific heat. The initial condition of $T = T_0 = 298\,K$ was applied at time $t = 0$.

If L is the length of the sample, then the convection occurring at $x = L$ is given by:

$$-k \frac{\partial T(L,t)}{\partial x} = h(T(L,t) - T_0), \tag{13.2}$$

where h is the convective heat transfer coefficient.

At $x = 0$, the heat balance equation can be expressed as:

$$-k \frac{\partial T(0,t)}{\partial x} = \delta A I - \varepsilon \sigma (T(0,t)^4 - T_0^4),$$

$$\delta = 1 \quad 0 \le t \le t_p$$

$$\delta = 0 \quad t > t_p \tag{13.3}$$

where t_p is the irradiation time; I is the laser power intensity (power per unit area) of the incident beam; A is the absorptivity of alumina; ε is the emissivity of alumina for thermal radiation; and σ is the Stefan Boltzmann constant = 5.67×10^{-8} W/m²K⁴. The heat transfer equations were solved using finite element approach to predict the temperature distribution in the alumina grinding wheel material during laser dressing. The properties of the alumina grinding wheels were appropriately adjusted to consider the effect of porosity. Figure 13.19 presents the temporal evolution of the depth of melting during laser irradiation of grinding wheel. As indicated in the figure, the depth of melting increases with time (heating cycle), reaches maximum, and the decreases with time (cooling cycle). This corresponds to the advance (heating cycle) and recession (cooling cycle) of the melting front during melting and resolidification. The comparison of the calculated maximum depth of melting with the experimental values is presented in Fig. 13.20. As indicated in the figure, the experimental values of maximum melt depth exceed the calculated values. This is most probably due to inflow of molten alumina material into the underlying pores during laser dressing (Harimkar et al. 2006). Such an analysis of the microstructure evolution during laser dressing process is expected to facilitate the design of process for precision machining of materials.

Thus, lasers have demonstrated the feasibility of being used as flexible tools in the dressing of grinding wheels to generate/expose the sharp cutting abrasive grains for machining action. Extensive studies on the surface topography of the laser-dressed grinding wheels and the surface finish of the ground workpieces indicated that the laser dressing approaches are well suited for the precision machining applications. The proposed laser dressing approaches are expected to offer significant advantages over the conventional diamond dressing such as noncontact process, increased productivity, and rapid processing. To facilitate the utilization of this approach in the actual industrial practices it is necessary to necessitate further studies on the laser interaction with the grinding wheels in the actual industrial setups, and the efficiency of these approaches in generating/exposing the sharp abrasive grains with a large number of effective cutting edges.

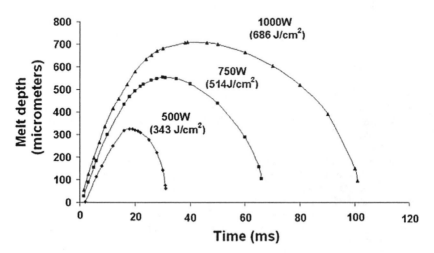

Fig. 13.19 Temporal evolution of depth of melting during laser irradiation of alumina grinding wheel. The wheel was irradiated at various laser powers with irradiation time of 14.4 ms. (Reprinted from Harimkar et al. 2006. With permission. Copyright Institute of Physics.)

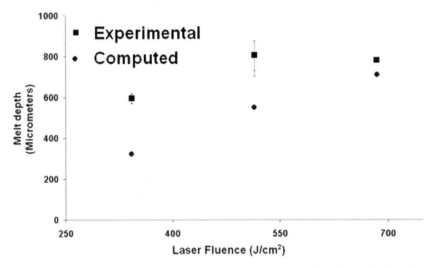

Fig. 13.20 Comparison of calculated and observed maximum depth of melting during laser irradiation of alumina grinding wheel. (Reprinted from Harimkar et al. 2006. With permission. Copyright Institute of Physics.)

References

Babu NR, Radhakrishnan V, Murti YV (1989) Investigations on laser dressing of grinding wheels-Part I: preliminary study. Transactions of the ASME 111:244–252.

Babu NR, Radhakrishnan V (1989) Investigations on laser dressing of grinding wheels-Part II: grinding performance of a laser dressed aluminum oxide wheel. Transactions of the ASME 111:253–261.

Babu NR, Radhakrishnan V (1995) Influence of dressing feed on the performance of laser dressed Al_2O_3 wheel in wet grinding. International Journal of Machine Tools and Manufacture 35:661–671.

Buttery TC, Statham A, Percival JB, Hamed MS (1979) Some effects of dressing on grinding performance. Wear 55:195–219.

Carslaw HS, Jaeger JC (1967) Conduction of Heat in Solids. Oxford University Press, Oxford.

Harimkar SP, Samant AN, Khangar AA, Dahotre NB (2006) Prediction of solidification microstructures during laser dressing of alumina-based grinding wheel material. Journal of Physics D: Applied Physics 39:1642–1649.

Harimkar SP, Dahotre NB (2006) Evolution of surface morphology in laser-dressed alumina grinding wheel material. International Journal of Applied Ceramic Technology 3:375–381.

Jackson MJ, Robinson GM, Dahotre NB, Khangar AA, Moss R (2003) Laser dressing of vitrified aluminium oxide grinding wheels. British Ceramic Transactions 102:237–245.

Khangar A, Dahotre NB, Jackson MJ, Robinson GM (2006) Laser dressing of alumina grinding wheels. Journal of Materials Engineering and Performance 15:178–181.

Khangar A, Dahotre NB (2005) Morphological modifications in laser-dressed alumina grinding wheel material for micro-scale grinding. Journal of Materials Processing Technology 170:1–10.

Khangar AA, Kenik EA, Dahotre NB (2005) Microstructure and microtexture in laser laser-dressed alumina grinding wheel material. Ceramic International 31:621–629.

Lindsay RP (1997) Principles of grinding. ASM Handbook 16:421–429.

Lin B, Li ZC, Xu YS, Hu J (2002) Theoretical generalization and research on the mechanism of the unsteady-state grinding technique. Journal of Materials Processing Technology 129:71–75.

Phanindranath V, Babu NR (1996) A theoretical model for prediction of groove geometry on laser dressed grinding wheel surface. International Journal of Machine Tools and Manufacture 36:1–16.

Torrance AA, Badger JA (2000) The relation between the traverse dressing of vitrified grinding wheels and their performance. International Journal of Machine Tools and Manufacture 40:1787–1811.

Zhang C, Shin YC (2002) A novel laser-assisted truing and dressing technique for vitrified CBN wheels. International Journal of Machine Tools and Manufacture 42:825–835.

Zhang C, Shin YC (2003) Wear of diamond dresser in laser-assisted truing and dressing of vitrified CBN wheels. International Journal of Machine Tools and Manufacture 43:41–49.

Chapter 14
Laser Processing in Medicine and Surgery

14.1 Introduction

Since the invention of the first pulsed ruby laser in 1960 by Maiman, extensive investigations have been conducted to examine and utilize the properties of lasers for applications in medicine and surgery. These investigations have been primarily motivated by the ability of the lasers to locally damage the biological tissue. The earliest clinical studies were focused on the pulsed laser-induced photocoagulation of the tissue in the treatment of eye diseases (Kapany et al. 1963; Campbell et al. 1963). These studies eventually led to the successful application of lasers in the routine ophthalmic procedures. With the improvements in the laser systems and the enhanced understanding of the laser–tissue interactions, the applications of lasers have been extended toward many medical specialties such ophthalmology, dermatology, dentistry, otolaryngology, angioplasty, etc. This chapter begins with a brief discussion of the laser–tissue interaction followed by a review of laser applications in medicine and surgery.

14.2 Laser–Tissue Interactions

When the laser light, which is an electromagnetic radiation, is incident on the surface of the tissue, various effects include reflection, refraction, absorption, scattering, and transmission (Fig. 14.1). The extent of these effects primarily depends on the characteristic of the radiation and the properties of the target tissue (Niemz 2004). Due to differing optical properties of various biological tissues (eye, skin, tooth, etc.), some of these effects may be more dominant over the other effects. One of the most important effects during laser–tissue interaction is the penetration of laser energy in the tissue by absorption or scattering. If the laser–tissue interaction is considered as the propagation of photons through a randomly distributed absorption and scattering centers, the attenuation of laser intensity occurs inside the tissue. This is generally expressed according to the Beer-Lambert Law:

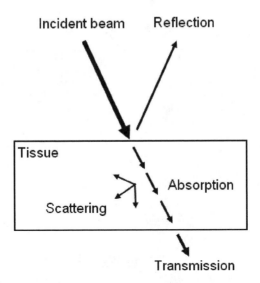

Fig. 14.1 Schematic of various effects (reflection, absorption, scattering, and transmission) during laser interaction with tissue

$$E(z) = (1 - r_s) E_0 e^{-\mu_t z} \qquad (14.1)$$

where E_0 is the incident energy, $E(z)$ is the attenuated energy at depth z, r_s is the Fresnel specular reflection, and μ_t is the attenuation coefficient (Welch and Gardner 2002). Assuming, the absorption and scattering as disjoint events, the attenuation coefficient can be expressed as the sum of absorption coefficient (μ_a) and scattering coefficient (μ_s):

$$\mu_t = \mu_a + \mu_s \qquad (14.2)$$

The laser–tissue interactions are significantly influenced by the composition of the tissue. Biological tissues are primarily composed of water (~70%) and organic material (~25%). The organic material forms the various cellular and extracellular structures and includes proteins, lipids, nucleic acids (DNA and RNA), etc. The molecular or submolecular regions which absorb the laser radiation in the organic material are referred to as chromophores. Figure 14.2 presents the absorption properties of principal chromophores in tissues (Vogel and Venugopalan 2003). As indicated in the figure, water shows the strong absorption in the deep ultraviolet (UV) region corresponding to a wavelength around 170 nm. However, over the range of wavelength in UV region, the absorption coefficient of water at room temperature is negligible. Hence, water is not generally considered as a significant chromophore for UV radiation. The absorption characteristics of water to UV radiation may be enhanced at higher temperatures. After water, the most significant component of the tissue is the organic matter, which greatly determines the overall

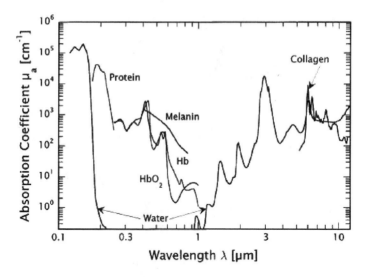

Fig. 14.2 Absorption properties of important tissue chromophores in the spectral range 0.1–12 μm. (Reprinted from Vogel and Venugopalan 2003. With permission. Copyright American Chemical Society.)

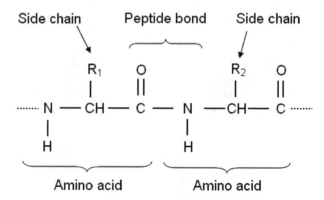

Fig. 14.3 Molecular structure of protein consisting of amino acid chains linked by peptide bonds

absorption characteristics of the tissue. The organic matter of the tissue such as proteins, lipids, etc., is the dominant tissue chromophores, which strongly absorb the UV radiation. Figure 14.3 presents the molecular structure of a protein. As shown in the figure, the proteins are composed of amino acid chains linked by peptide bonds. The important types of chromophores in the proteins are thus the amino acid chains and the peptide bonds. The various amino acids (20 standard amino acids) show different absorption properties and exhibit various different wavelengths at which the strong absorption bands are centered. In general, amino acids in the proteins exhibit strong absorption in the UV radiation. The peptide bonds also exhibit the significant absorption at the far UV radiation. The absorption of radiation by peptide bonds fall significantly with increasing wavelength and reaches a minimum at around 240 nm. The absorption of radiation by other organic biological matters (nucleic acids, lipids, carbohydrates, etc.) is not as important as that of proteins in determining the overall absorption characteristics of tissues (Pettit 2002).

The absorption of laser radiation by tissue results in laser-induced photothermal effects such as tissue heating and vaporization (thermal ablation). In addition to the thermal effects, the interaction of laser with tissue may be associated with various photochemical and photodisruptive processes. These laser–tissue interactions depend on properties of the tissue and the laser parameters. One of the most interesting characteristic of such interactions is that there exist distinct regimes of laser parameters (laser power density and exposure time) where these interactions dominate. Figure 14.4 presents the famous laser–tissue interaction map showing the regimes of laser power densities and exposure times for each mechanism. The areas are plotted on the constant energy density lines corresponding to $1 \, J/cm^2$–$1,000 \, J/cm^2$. This range of laser energy density is most relevant in medical application of

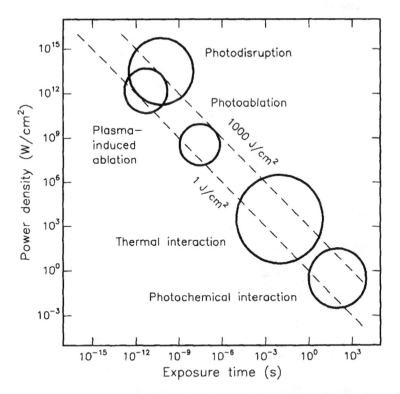

Fig. 14.4 Laser–tissue interaction map showing the approximate regimes of various interactions in medical applications. (Reprinted from Niemz 2004. With permission. Copyright Springer.)

lasers. As indicated in the figure, the exposure times for various mechanisms can be expressed in the following ranges (Niemz 2004):

Photothermal interactions: 1 μs to 1 min
Photoablation: 1 ns to 1 μs
Photodisruption: <1 ns

The following sections briefly explain these laser–tissue interactions.

14.2.1 Photothermal Interactions

In photothermal interactions, the light energy absorbed by the tissue is converted into heat energy resulting in temperature rise. Depending on the magnitude and rate of temperature rise, the laser interaction with tissue can involve effects such as photocoagulation, melting, or photovaporization (thermal ablation) of tissue. If the temperature rise is very small (<10 °C) and occurs over longer times (>1 s), the laser–tissue interactions are usually associated with the cell damage or death without structural changes. Temperature rise in the range 10–20 °C may cause thermal denaturation of the tissue. Thermal denaturation refers to the changes in the three-dimensional molecular structure due to temperature rise (Krauss and Puliafito 1995). During denaturation, the temperature rise is high enough to increase the kinetic energy of the molecules such that they overcome the bonds responsible for stabilizing the protein structure. This is a rate process and often explained based on Arrhenius equation. When the temperature of the tissue exceeds around 100 °C during laser irradiation, the tissue may get carbonized. Carbonization of tissue during laser surgery is generally considered detrimental because it reduces visibility during surgery.

Most significant photothermal interactions in the context of laser surgery are photocoagulation, melting, and photovaporization. The irradiation of tissue with the laser light initiates the coagulation when the temperature reaches above 60 °C. Photocoagulation has been extensively used in the ophthalmology for treating various eye diseases. The laser–tissue interactions may eventually heat the tissue beyond its melting or boiling point resulting in tissue removal by melt expulsion or photovaporization. Depending on the laser parameters and the tissue properties, various thermal effects such as coagulation, carbonization, and photocoagulation often occur together. Such effects can be observed during the laser irradiation of the human cornea (Fig. 14.5). Figure 14.6 presents the schematic of the various thermal effects taking place during laser–tissue interactions. The extent of these thermal effects is influenced by the temperature distribution in the tissue (Niemz 2004).

The temperature rise in the tissue during thermal interactions depends primarily on the properties of the tissue (composition and morphology, absorption coefficient, etc.) and laser parameters (wavelength, fluence, pulse time, etc.). The temperature (T) in the tissue at any time (t) can be calculated from the solution of heat transfer equation of the form (expressed in cylindrical coordinates):

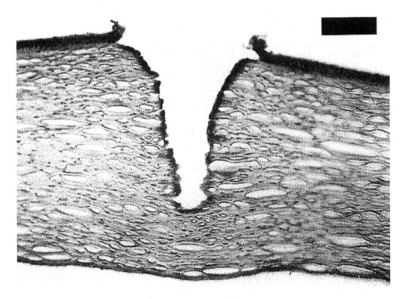

Fig. 14.5 Photocoagulation of the human cornea using Er:YAG laser (laser parameters: 120 pulses, 90 μs pulse duration, 5 mJ pulse energy, and 1 Hz repetition rate; bar length corresponds to 100 μm). (Reprinted Ref. Niemz 2004. With permission. Copyright Springer.)

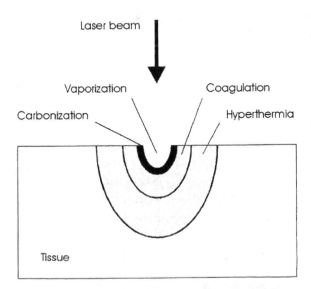

Fig. 14.6 Schematic of various thermal effects during laser–tissue interactions. (Reprinted from Niemz 2004. With permission. Copyright Springer.)

$$\rho c_{\mathrm{p}} \frac{\partial T}{\partial t} = k \left(\frac{\partial}{\partial r} \frac{1}{r} \frac{\partial T}{\partial r} + \frac{\partial^2 T}{\partial^2 z} \right) + Q, \tag{14.3}$$

where ρ is the density, c_{p} is the specific heat, k is the thermal conductivity, and Q is the heat source term. The above equation neglects the heat transfer due to perfusion and metabolic heat generation. The heat source term Q in the above equation corresponds to the heat generation by absorption of light and can be expressed in terms of fluence rate (ϕ) and tissue absorption coefficient (μ_{a}):

$$Q(r) = \mu_{\mathrm{a}}(r)\phi(r) \tag{14.4}$$

The solution of the above heat transfer equation requires the assumption of appropriate insulating, convective, and/or radiative boundary condition. The general scheme for modeling the temperature distribution and corresponding tissue damage during laser–tissue interaction is presented in Fig. 14.7. The three important steps are the estimation of input parameters (tissue properties and laser parameters), heat transfer analysis, and reaction rate analysis (Welch 1984).

To reliably use the thermal models for predicting the temperature during laser–tissue interaction, the calculated results needs be confirmed by the experimental data. Extensive investigations have been conducted by Torres et al. (1993) to verify various models for predicting the thermal response of tissue to laser irradiation. Figure 14.8 presents the comparison of the experimental and calculated temperature during irradiation of human aorta using argon laser. As indicated in the figure, the experimental and calculated results of tissue temperature show pronounced differences. To improve the accuracy of such thermal models, careful considerations needs to be given to the definition of physical problem, and the selection of the tissue properties and the laser parameters. For example, most of the models consider the constant absorption coefficient and negligible scattering. In such cases the penetration depth (δ) is given by the reciprocal of the absorption coefficient (μ_{a}). However, in case of a weakly absorbing

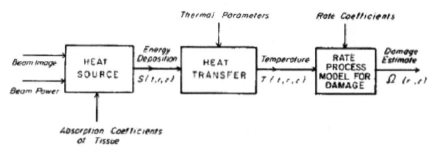

Fig. 14.7 Block diagram of steps required for modeling the thermal damage during laser irradiation of tissue. (Reprinted from Welch 1984. With permission. Copyright Institute of Electrical and Electronics Engineers.)

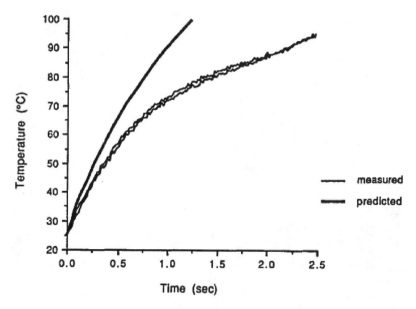

Fig. 14.8 Comparison of measured and calculated temperatures during irradiation of aortic wall with a 2 W of laser power and spot size of 2 mm. (Reprinted from Torres et al. 1993. With permission. Copyright Optical Society of America.)

tissue, the light scattering may play an important role. The attenuation coefficient (μ_t) in cases is a sum of absorption coefficient (μ_a) and scattering coefficient (μ_s). In such cases, the penetration depth is smaller than the reciprocal of absorption coefficient, μ_a (Jacques 1993). In addition, the thermo-physical properties of the tissue such as thermal conductivity, specific heat, and density change with the water content in the tissue. As the temperature is raised during laser–tissue interaction, the water gets vaporized resulting in a decrease in thermal conductivity and specific heat. Hence, the thermal models which take into account the temperature variation of the thermo-physical properties and the phase change corresponding to vaporization of water at 100°C are expected to improve the accuracy of temperature predictions.

14.2.2 Photochemical Interactions (Photoablation)

Photochemical interactions are also referred to as photochemical decompositions and photoablation. The earliest reports of the photoablation of organic materials using excimer lasers were published in 1982 (Kuwamura et al. 1982; Srinivasan and Mayne-Banton 1982). These studies demonstrated that when intense pulses of excimer lasers are irradiated on the synthetic polymers, submicron features can be etched on the surface without any noticeable thermal damage. Such precise

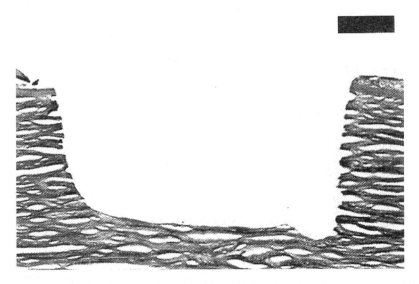

Fig. 14.9 Cross section of a cornea photoablated using ArF excimer laser (laser parameters: 14 ns pulse duration, and 180 mJ/cm² energy density; bar length corresponds to 100 µm). (Reprinted from Niemz 2004. Copyright, Springer. With permission.)

photoetching of polymers without thermal damage have been attributed to the high absorption coefficients for 193 nm radiations and high efficiency for bond breaking. These studies, which were primarily directed toward materials processing, were soon extended for the applications in ophthalmology where precise removal of corneal tissue is desired (Trokel et al. 1983). Figure 14.9 presents the cross section of a cornea photoablated using ArF excimer laser. The figure clearly indicates that photoablation results in precise removal of material with no or minimum thermal damage to the surrounding tissue.

In photoablation, the energy of the incident photon causes the direct bond breaking of the molecular chains in the organic materials resulting in material removal by molecular fragmentation without significant thermal damage. This suggests that for the ablation process, the photon energy must be greater than the bond energy. The UV radiation with wavelengths in the range 193–355 nm corresponds to the photon energies in the range 6.4–3.5 eV. This range of photon energies exceeds the dissociation energies (3.0–6.4 eV) of many molecular bonds (C–N, C–O, C=C, etc.) resulting in efficient ablation with UV radiation (Vogel and Venugopalan 2003). However, it has been observed that ablation also takes place when the photon energy is less than the dissociation of energy of molecular bond. This is the case for far UV radiation with longer wavelengths (and hence correspondingly smaller photon energies). Such an observation is due to the multiphoton mechanism for laser absorption. In multiphoton mechanism, even though the energy associated with each photon is less than the dissociation energy of the bond, bond breaking is achieved by simultaneous absorption of two or photons.

The laser–tissue interaction during ablation is complex and involves interplay between the photothermal (vibrational heating) and photochemical (bond breaking) processes. One of the important considerations during the laser–tissue interaction studies is the thermal relaxation time (τ). Thermal relaxation time is related with the dissipation of heat during laser pulse irradiation and is expressed as (Thompson et al. 2002):

$$\tau = \frac{d^2}{4\alpha},$$ (14.5)

where d is the absorption depth and α is the thermal diffusivity. For longer pulses (with pulse time longer than the thermal relaxation time), the absorbed energy will be dissipated in the surrounding material by thermal processes. To facilitate the photoablation of material with minimum thermal damage, the pulse time must shorter than thermal relaxation times. For such short pulses (pulse times in the range of microseconds), the laser energy is confined to a very thin depth with minimum thermal dissipation. Thus efficient ablation of the tissue during laser–tissue interactions necessitates laser operation at shorter wavelengths with microsecond pulses.

The ablation process is generally explained on the basis of "blow-off" model which assumes that the ablation process takes place when the laser energy exceeds the characteristic threshold (Srinivasan and Mayne-Banton 1982; Andrew et al. 1983). Ablation threshold represents the minimum energy required to remove the material by ablation. Figure 14.10 presents the schematic distribution of absorbed energy in the tissue irradiated with incident laser energy, E_0. Above the ablation threshold energy ($\mu_a E_{th}$), the material removal is facilitated by bond breaking; whereas below ablation threshold energy, thermal effects such as heating takes place. Absorption properties of tissue and incident laser parameters determine the location at which the absorbed energy reaches the ablation threshold thus determining the depth of ablation. The depth of ablation according to this model is given by:

$$\delta = \frac{1}{\mu_a} \ln\left(\frac{E_0}{E_{th}}\right)$$ (14.6)

Above ablation threshold, the depth of ablation predicted based on "blow-off" model increases continuously with the laser energy (Fig. 14.10b). However, actual ablation depths depends on the number of other effects such as plasma shielding and radiation-induced changes in the material absorption coefficient, which tend to deviate the actual ablation depth from that predicted by simple "blow-off" model. Figure 14.11 presents the variation of ablation depths of corneal tissue with the laser energy for two wavelengths of excimer lasers. As indicated in the figure, the ablation threshold corresponding to ArF laser ($\lambda = 193$ nm) irradiation is less than that corresponding to KrF laser ($\lambda = 248$ nm) irradiation. This is due to the stronger absorption of radiation by corneal tissue at 193 nm (Pettit 2002).

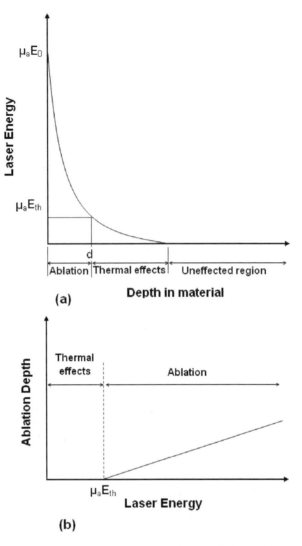

Fig 14.10 Blow-off model of laser ablation: (**a**) distribution of absorbed laser intensity in the depth of material, and (**b**) variation of depth of ablation with laser energy

14.2.3   Photodisruptive Interactions

The photodisruptive laser–tissue interactions are based on the "laser-induced breakdown" process. The laser-induced breakdown process generally involves nonlinear absorption of light or cascade ionization resulting in the generation of plasma. Unlike, photothermal processes, the optical breakdown process allow the deposition of energy in the transparent materials. Plasma is generally regarded as

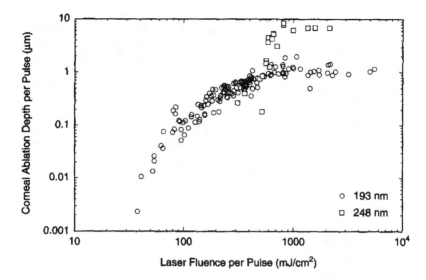

Fig. 14.11 Variation of ablation depth per pulse with incident laser fluence during laser ablation of cornea using two different wavelengths (Reprinted from Pettit 2002. With permission. Copyright CRC Press.)

an ionized gas formed by the ionization of molecules on the target surface and exhibits the maximum temperature exceeding 10^4 K.

There are two mechanisms of laser-induced breakdown resulting in plasma formation. These are cascade ionization (or avalanche ionization) and multiphoton absorption. The cascade ionization considers the presence of free electrons often termed "seed" electrons in the focal volume at the beginning of the laser pulse. These seed electrons absorb the laser energy by inverse bremsstrahlung absorption process. When the energy acquired by the free electrons exceeds the ionization potential of the molecules, ionization of molecules is initiated by collision. The ionization of molecules generates the new free electrons, which continue to absorb the photon energy and ionize the molecules resulting in "avalanche" breakdown. Thus the breakdown process basically consists of electron generation and cascade ionization. Plasma generation by optical breakdown is a threshold phenomenon. Ablation threshold is considered to be reached when the free electron density in the plasma corresponds to about $10^{18}/cm^3$. Experimentally, ablation threshold can be identified by the observation of luminescent plasma or the cavitation bubble formation in the liquid. In multiphoton absorption mechanism each electron is independently ionized thus requiring no seed electrons, collision, or particle–particle interactions (Kennedy 1995). Once the plasma is formed at the focal region corresponding to threshold intensity, further increase in the pulse energy causes the growth of plasma toward the incoming laser beam.

The rapid expansion of the plasma laser-induced breakdown generates the shock waves. In addition, when the application site is surrounded by water, plasma formation is accompanied by cavitation. These mechanical effects can cause the disruption

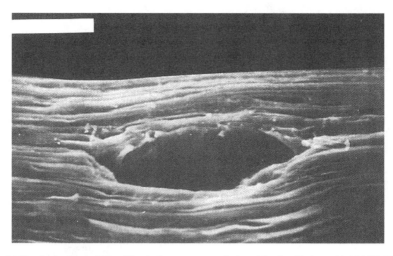

Fig. 14.12 Cavitation bubble effect in human cornea induced by irradiation with Nd:YLF laser (laser parameters: 1 pulse, 30 ps pulse duration, and 1 mJ pulse energy; bar length corresponds to 30 μm). (Reprinted from Niemz 2004. With permission. Copyright Springer.)

of tissue adjacent to the optical breakdown site. The photodisruption of the tissue using laser is a well established surgical technique used in various medical applications such as intraocular surgery and iridotomy (Vogel et al. 1990). Figure 14.12 presents the cavitation effects in human cornea induced by a single pulse of Nd: YLF laser (Niemz 2004). Short-pulse Nd:YAG lasers, and Q-switched lasers with pulse length in the range of nanoseconds and pulse energy in the range of millijoules are most popular in the photodisruptive clinical procedures.

14.3 Laser Applications in Medicine and Surgery

Lasers are currently used as therapeutic and diagnostic tools in a number of medical and surgical specialties. The applications of lasers in medicine and surgery are broad and are based on distinct laser–tissue interactions. The following sections briefly discuss the major medical applications of lasers.

14.3.1 Lasers in Ophthalmology

Historically, the earliest reported and most widespread applications of lasers in medicine have been in ophthalmology. Pulsed Nd:YAG laser was first introduced to induce the photocoagulation of tissue in the treatment of eye diseases (Kapany et al.

1963; Campbell et al. 1963). With the enhanced understanding of the laser–tissue interactions, and the development of argon, krypton, tunable dye, and diode lasers, lasers are now used in the treatment of a variety of eye diseases ranging from diabetic retinopathy to glaucoma. These applications are based on various laser–tissue interactions such as photocoagulation, photoablation, and photodisruption. Laser applications in ophthalmology have been extensively reviewed by Krauss and Puliafito (1995). Following subsections present the brief summary of these applications.

14.3.1.1 Diabetic Retinopathy

Diabetic retinopathy is an eye disease associated with adverse effects of diabetes on circulatory system of retina. The two phases of the disease are background diabetic retinopathy and proliferative diabetic retinopathy. These phases cause the formation of hemorrhages (red, dot-blot configurations at the venous ends of capillaries) and development of new fragile vessels (neovascularization), respectively (Harney 2006). These new vessels hemorrhage easily leading to decreased vision. Under serious conditions where the abnormal growth of vessels occurs, detachment of retina may take place (Fig. 14.13a). Lasers have established the effectiveness in the treatment of diabetic retinopathy by causing destruction in the peripheral retina. This laser surgery is often referred to as panretinal photocoagulation and involves placing of laser burns on the retina (Fig. 14.13b). Panretinal photocoagulation causes the blood vessels to shrink and disappear thus reducing the chances of hemorrhage and retinal detachment. The destruction of peripheral retina (panretinal photocoagulation) also reduces the stimulus for new blood vessel growth (neovascularization). Common lasers used in diabetic retinopathy include argon (wavelength: 488–514 nm) and krypton (647 nm).

14.3.1.2 Retinal Vein Occlusion

Retinal vein occlusion is an eye condition causing loss of vision due to retinal vascular disease. The two forms of retinal vein occlusion include branch retinal vein occlusion (BRVO) and central retinal vein occlusion (CRVO). Retinal vein occlusion involves either blockage of veins (BRVO) or closure of the final retinal vein (CRVO). This is associated with the hemorrhaging or leakage of fluid on the retina. Some of the complications associated with retinal vein occlusion (especially BRVO) can be appropriately treated with laser treatment. These are based on photocoagulation for sealing the leaking capillaries and stabilizing the hemorrhage (Fig. 14) (Krauss and Puliafito 1995). Similar lasers which cause the photocoagulation during diabetic retinopathy can be used for treatment of retinal vein occlusion.

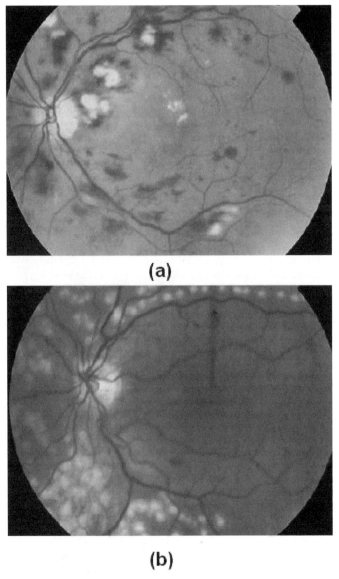

Fig. 14.13 (a) Diabetic retinopathy and (b) panretinal photocoagulation (Reprinted from Harney 2006. With permission. Copyright Elsevier.)

14.3.1.3 Age-Related Macular Degeneration

Macular degeneration is one of the leading causes of vision loss and blindness in the United States. It is associated with the damage to the middle portion of the retina called macula which is responsible for focusing the central vision in the eye. The loss of central vision deteriorates the abilities (reading, driving, etc.)

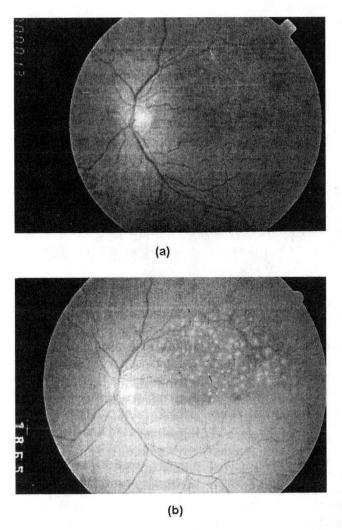

(a)

(b)

Fig. 14.14 (a) Branched retinal vein occlusion, and (b) grid photocoagulation with argon green laser. (Reprinted from Krauss and Puliafito 1995. With permission. Copyright Wiley.)

which require fine vision. There are two forms of macular degeneration: wet or dry. Wet macular degeneration is generally associated with the neovascularization where new vessels are formed to supply blood to the oxygen-deprived regions of the retina. These new fragile blood vessels cause hemorrhage and swelling. Dry macular degeneration is relatively less severe. Even though the restoration of vision is difficult, laser photocoagulation can help in preventing further vision damage. Macular degeneration may be treated using photodynamic therapy (PDT).

14.3.1.4 Glaucoma

Glaucoma is an eye disease primarily associated with the generation of intraocular pressures (IOPs). Under certain conditions of eyes, the outflow of the aqueous humor through the trabecular meshwork and Schlemm's canal into the episcleral space is obstructed resulting in high IOPs (>22 mm Hg). Such high IOP causes the distortion of the eye structure leading to loss of vision. Laser surgeries can be used for treating glaucoma. Lasers can create small holes in the iris to provide alternate pathways for the outflow of aqueous humor (angle closure glaucoma). Such pathways help in reducing the IOP. The laser–tissue interactions during creation of holes are the result of strong absorption of laser radiation (corresponding to argon, krypton, and dye lasers) by the pigmentation in the iris. If the pigmentation is not sufficient to strongly absorb the radiation, Q-switched laser may offer disruptive interactions to create holes in the iris. Alternatively, photocoagulation of trabecular meshwork may be used to stretch the structures to facilitate the outflow of the aqueous humor.

14.3.1.5 Corneal Surgery

Recently, there has been an increasing interest toward the possible use of lasers in corneal surgery. Cornea strongly absorbs the mid-infrared and infrared radiation allowing the precise cutting of the corneal tissues during surgical procedures. Some of the possible applications of laser in corneal surgery are laser photothermal keratoplasty, cataract surgery, etc.

Lasers were proposed to heat the corneal tissue to adjust the shape of the cornea and thus the refractive power of the eye. However, the process is still under development and major issues include the minimization of mechanical and thermal damage to the corneal tissue during laser interactions. Lasers can also play an important role in cataract surgery. Cataract is associated with the opacification of the transparent crystalline lens resulting in blurred vision. Lasers may be used to create small incisions through which the cataractous lens can be removed while retaining the lens capsule intact. The empty capsule can then be filled with biocompatible synthetic gel to create the compatible intraocular lens in situ. In future, this process may allow restoration of the accommodation which is the ability of the lens to change the curvatures to focus the objects at different distances (Thompson et al. 2002).

14.3.1.6 Ocular Tumor Treatment

Lasers also offer a great potential to be used in the treatment of ocular tumors such as retinoblastoma, choroidal melanomas, eyelid tumors, etc. These malignant tumors tend to increase in size and exert pressures on the eye resulting in loss of vision. Lasers can also be used in the treatment of eyelid tumors. Argon lasers have

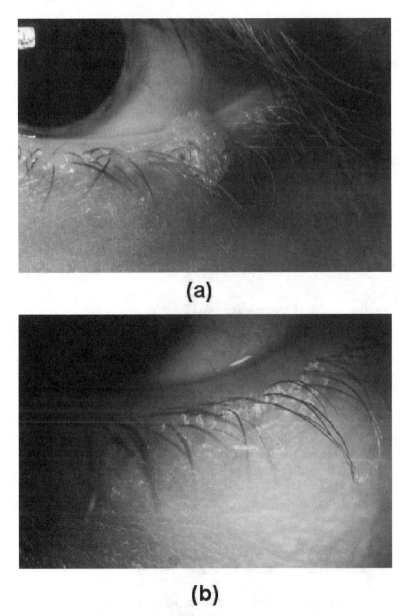

Fig. 14.15 Laser treatment of eyelid tumors: (**a**) before, and (**b**) 3 months after argon laser therapy. (Reprinted from Wohlrab et al. 1998. With permission. Copyright Elsevier.)

been reported to cause the phocoagulation of tissue during treatment of ocular tumors. Due to the match of the wavelength of argon laser with the absorption spectra of oxyhemoglobin, the occlusion of blood vessels occurs during laser treatment. Figure 14.15 presents the conditions of the eyelid tumors before and after laser treatment.

14.3.2 Lasers in Dermatology

Lasers are increasingly used in dermatology for treating a variety of skin problems. These applications are based on the ability of the lasers of specific wavelengths to cause localized thermal changes (phototherapy) while limiting the thermal damage to the surrounding tissue. A variety of lasers such as pulsed dye, Q-switched ruby, Nd:YAG, alexandrite, CO_2, etc. can be used for depending on the specific condition of the skin to be treated. Major applications of lasers in dermatology are explained in following subsections.

14.3.2.1 Vascular Lesions

Vascular lesions are skin disorders characterized by the appearance of abnormal blood vessels under the surface of the skin. Various types of vascular lesions include port-wine stains, telangiectasias, cherry angiomas, spider veins, etc. Particularly, the argon laser treatment of port-wine stains is reported to be highly effective (Fig. 14.16) (Bernstein and Brown 2005). The selective laser-induced damage causes the sealing and disappearance of abnormal vessels. The laser treatment of vascular lesions is based on the selective absorption of laser radiation by hemoglobin in the vessels. Ideally, the effective laser treatment of vascular lesions requires the lasers with wavelengths matching the absorption peaks of hemoglobin (corresponding to 418, 542, and 577 nm). Even though the exact matching of the laser wavelength with absorption peaks in hemoglobin are difficult, significant absorption of argon laser (wavelength: 488–514 nm), pulsed dye laser (577 nm), and frequency-doubled Nd:YAG laser (532 nm) can be used for the treatment of vascular lesions. Important consideration must be given to the pulse duration. Theoretically, the pulse length must be shorter than the thermal relaxation time to avoid the significant thermal damage to the surrounding tissue. Pulse lengths in the range of microseconds to milliseconds have been used for the treatment of lesions.

14.3.2.2 Removal of Pigmented Lesions or Tattoos

Lasers are finding increased utilization in the removal of pigmented lesions such nevi, melasma, tattoos, etc. The selection of lasers for the treatment of these pigmented lesions and tattoos depends on the composition of the pigments which determine the absorption wavelength of laser. In general, the important chromophores in these skin disorders are melanine and tattoo ink. These chromophres absorb laser radiation selectively and break down into smaller particles. This minimizes and often causes the disappearance of the tattoos (Fig. 14.17). Common lasers for removal of pigmented lesions and tattoos include Q-switched ruby, Nd:YAG, and alexandrite lasers.

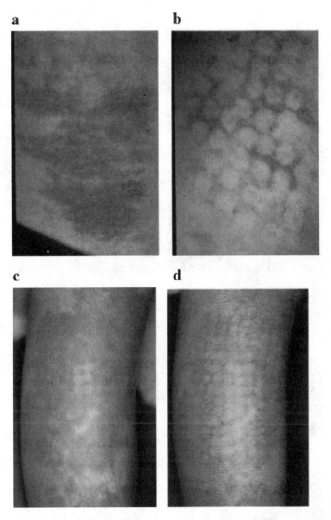

Fig. 14.16 Laser treatment of port-wine stains: (**a**), (**c**) before, and (**b**), (**d**) after laser treatment (laser parameters: 585 nm pulsed dye laser and 1.5 ms pulse duration). (Reprinted from Bernstein and Brown 2005. With permission. Copyright Wiley.)

14.3.2.3 Hair Removal

Recently, considerable interests have been directed toward laser-assisted removal of unwanted hair follicles by selective photothermolysis. The treatment is based on the selective targeting the melanin in the hair follicles using pulsed ruby laser. Long-pulse lasers are believed to be more efficient for hair removal applications due to better conduction of heat and thermal damage to the hair follicles. Lasers treatment for unwanted hair removal is particularly attractive due to the possibility of

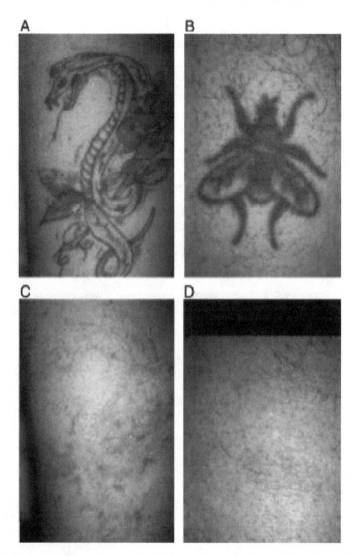

Fig. 14.17 Laser treatment of tattoos: (**a**), (**b**) before laser treatment, and (**c**), (**d**) after laser treatment using Nd:YAG and alexandrite lasers, respectively. (Reprinted from Bernstein 2006. With permission. Copyright Elsevier.)

permanent hair removal and usefulness with a variety of hair colors. Extended or permanent hair removal generally requires direct insertion methods and several sessions. In direct insertion method, a laser beam is delivered through an optical needle (~130 μm diameter) into each hair follicle. Figure 14.18 presents the results of such laser removal using diode-pumped Nd:YAG laser (Hashimoto et al. 2003).

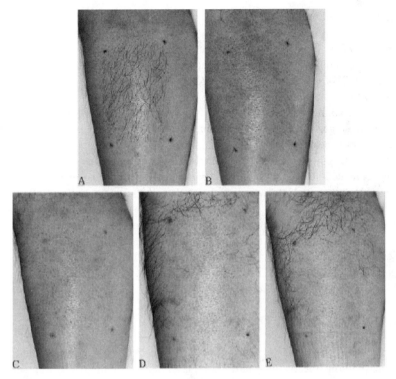

Fig. 14.18 Laser removal of hairs: (**a**) rectangular area selected for hair removal (surrounding hairs are shaved), (**b**) immediately after hair removal, (**c**) 7 days after hair removal (this area was again laser treated five times), (**d**) 13 months after first hair removal, and (**e**) 22.5 months after first hair removal. (Reprinted from Hashimoto et al. 2003. With permission. Copyright American Academy of Dermatology.)

14.3.2.4 Laser Skin Resurfacing

Lasers can also find applications in the resurfacing of skin where areas of damaged or wrinkled skin are removed layer by layer. Laser resurfacing is believed to be particularly useful for minimizing facial wrinkles (rhytides) and scars. The treatment is based on controlled vaporization of thin skin layers up to controlled depths by scanning a laser beam. The selection of laser parameters is critical to control the thermal effects during resurfacing. Laser treatment is expected to offer significant advantages such as precise resurfacing with little or no bleeding and postoperative discomfort. Lasers commonly used for resurfacing of skin include CO_2 and Er:YAG lasers. The laser treatment for resurfacing of skin is still a novel technique and needs extensive assessment before introduction into routine surgical procedures.

14.3.2.5 Photodynamic Therapy

Recently, PDT is found to be highly effective in the treatment of actinic keratoses, basal cell carcinomas, Bowen's disease, superficial squamous cell carcinomas, etc. (Sieron et al. 2006). PDT involves the injection of photosensitizer or the photosensitizing agent into the bloodstream of the patient. The photosentizer is then allowed to distribute into all the soft tissues of the body. Due to the nature of the tumor cells, the photosensitizer gets selectively concentrated into the tumors; whereas it is almost cleared from the healthy cells after a certain period (~48–72 days). When the laser of specific wavelength is irradiated on the malignant tissues, the photosensitizers in the cells get activated and produce singlet oxygen (active form of oxygen) or free radicals. The photochemical interactions of the tissue with the singlet oxygen and radicals cause the destruction of the malignant tissues. Dye lasers are the most commonly used lasers in PDT. While, the commonly used photosensitizers include hematoporphyrin derivative, photofrin, etc., Fig. 14.19 presents the results of one of the research studies on the treatment of Bowen's disease using 5-aminolaevulinic acid (ALA) in topical PDT (Ibbotson et al. 2004).

14.3.2.6 Laser Tissue Welding/Soldering

In dermatology, lasers have been proposed to selectively weld or solder the skin incisions without mechanical sutures. In laser tissue welding, the light energy

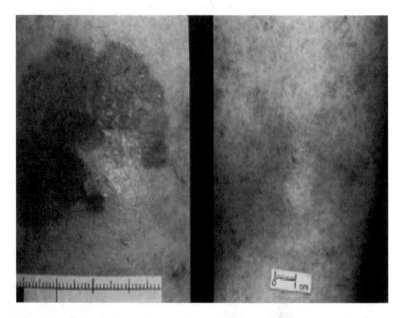

Fig. 14.19 Treatment of Bowen's disease using PDT: (**a**) before, and (**b**) after single-laser treatment. (Reprinted from Ibbotson et al. 2004. With permission. Copyright Elsevier.)

absorbed by the tissue is rapidly converted into heat energy. The resulting thermal effects facilitate the adhesion of tissue edges. Laser tissue welding involves the application of adhesive proteins (like albumin) and wavelength-specific dye absorbers (like indocyanine green) in combination with the irradiation with deep penetrating lasers. The application of dye solders improves the strength of the bonded tissue. Even though, a variety of lasers have been reported to be useful in laser tissue welding experiments, diode lasers operating in the wavelength range of 780–830 nm are becoming increasingly popular due to their compact size, easy use, and low cost (McNally et al. 1999; Menovsky et al. 1999). Figure 14.20 compares the results of laser tissue welding (using 980 nm diode laser) with the mechanical suture closure in rats during the recovery period. As indicated in the figure, the incisions were successfully closed by laser tissue welding. To exploit the full potential of laser tissue welding the surgical procedures, careful consideration must be given to minimize the thermal damage to the surrounding tissue (Gulsoy et al. 2006).

14.3.3 Lasers in Otolaryngology–Head and Neck Surgery

Otolaryngology–head and neck surgery is a medical specialty which deals with the medical and surgical treatment of ears, nose, throat, and related structures of head and neck. Lasers are increasingly used in otolaryngology. Some of the applications include endoscopic laser excision of oropharyngial and laryngopharyngial carcinomas, laser stapedotomy (surgical procedure for making a small hole in the footplate to improve hearing), laser-assisted uvulopalatoplasty (treatment for snoring), etc. Lasers offer significant advantages such as less bleeding and pain than the conventional surgical procedures. Various lasers such as CO_2, Nd:YAG, argon, and KTP lasers have been used in the surgical procedures in otolaryngology–head and surgery.

14.3.4 Lasers in Angioplasty

Laser angioplasty involves the opening or widening of the coronary arteries blocked or narrowed by plaque. In laser angioplasty, an excimer laser beam delivered through the optical fiber is inserted into the artery using flexible catheters and positioned at the site of plaque. The plaque is then vaporized by delivering the controlled amount of laser energy, thus, opening and widening the artery for normal blood flow (Fig. 14.21). Lasers can also be used in combination with the conventional balloon angioplasty. The main risk factor is the piecing of the hole in the arterial wall.

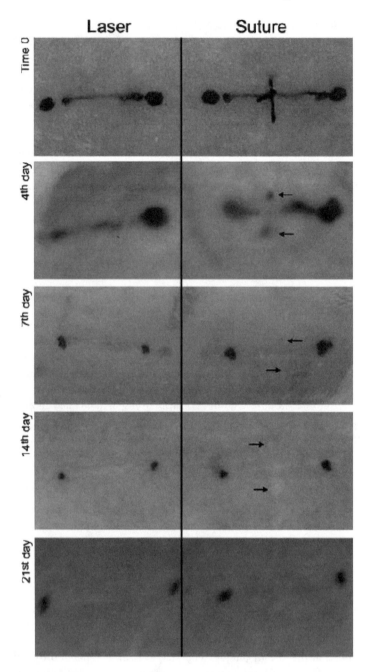

Fig. 14.20 Closure of skin incisions: laser tissue welding and mechanical suture closing of 1 cm long incision during recovery period in rat model. (Reprinted from Gulsoy et al. 2006. With permission. Copyright Springer.)

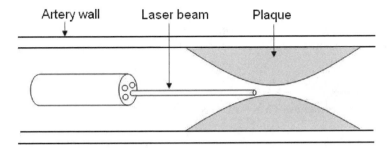

Fig. 14.21 Schematic of laser angioplasty for removing the arterial plaque

14.3.5 Lasers in Osteotomy

Lasers can also be used for surgical sectioning of the bones (osteotomy). They offer significant advantages over conventional mechanical bone-cutting methods, such as sawing and drilling, such as precision, aseptic nature, and minimal thermal damage to the surrounding bone tissue. The method is based on vaporization of the bone which is accompanied by the denaturation of proteins resulting in carbon residue. The important consideration during laser osteotomy is the healing response of the bone.

14.3.6 Lasers in Dentistry

14.3.6.1 Caries Removal and Cavity Preparation

Treatment of caries was the first suggested application of lasers in dentistry. Early dental application involved the use of ruby, Nd:YAG, and CO_2 lasers for the removal of caries. However, low ablation rates and extensive thermal damage to the surrounding tissues made them unsuitable for the treatment. Recently, Er:YAG lasers (wavelength: 2.94 nm) have attracted considerable interests in this application based on thermally induced mechanical ablation of mineralized tooth tissues with minimum thermal damage/defects (Keller et al. 1998).

14.3.6.2 Surface Modifications of Root Surfaces

Laser surface modifications of dentine and cementum have been suggested for restorative procedures. Dentine and cementum are primarily composed of carbonated hydroxyapatite mineral which has absorption bands in mid-infrared region. Laser surface modifications using Er:YAG lasers cause effective removal of calculus for root etching and creation of biocompatible surfaces for cell or tissue reattachment. The treatments of major consideration in such applications are the minimization of thermal damage to the pulp tissue and undesired removal of sound root structure. Lasers can also be used for surface modifications of dental implants.

14.3.6.3 Treatment of Oral Soft Tissues

Lasers have also been suggested for the soft tissue periodontal surgery. This includes the application in frenectomy (removal of lingual frenulum under the tongue), gingivectomy (surgical removal of gingival), etc.

References

Andrew JE, Dyer PE, Forster D, Key PH (1983) Direct etching of polymeric materials using a XeCl laser. Applied Physics Letters 43:717–719.

Bernstein EF, Brown DB (2005) Efficacy of 1.5 millisecond pulse-duration, 585 nm, pulsed dye laser for treating port-wine stains. Lasers in Surgery and Medicine 36:141–146.

Bernstein EF (2006) Laser treatment of tattoos. Clinics in Dermatology 24:43–55.

Campbell CJ, Noyori KS, Rittler MC, Koester CJ (1963) Intraocular temperature changes produced by laser coagulation. Acta Ophthalmologica S76:22–31.

Gulsoy M, Dereli Z, Tabakoglu HO, Bozkulak O (2006) Closure of skin incisions by 980-nm diode laser welding. Lasers in Medical Science 21:5–10.

Harney F (2006) Diabetic retinopathy. Medicine 34:95–98.

Hashimoto K, Kogure M, Irwin TL, Tezuka K, Osawa T, Kato K, Ebisawa T (2003) Permanent hair removal with a diode-pumped Nd:YAG laser: a pilot study using the direct insertion method. Journal of the American Academy of Dermatology 49:1071–1080.

Ibbotson SH, Moseley H, Brancaleon L, Padgett M, O'Dwyer M, Woods JA, Lesar A, Goodman C, Ferguson J (2004) Photodynamic therapy in dermatology: Dundee clinical and research experience. Photodiagnosis and Photodynamic Therapy 1:211–223.

Jacques S (1993) Role of tissue optics and pulse duration on tissue effects during high-power laser irradiation. Applied Optics 32:2447–2454.

Kapany NS, Peppers NA, Zweng HC, Flocks M (1963) Retinal photocoagulation by lasers. Nature 199:146–149.

Keller U, Hibst R, Guertsen W, Schilke R, Heidemann D, Klaiber B, Raab WH (1998) Erbium: YAG laser application in caries therapy. Evaluation of patient perception and acceptance. Journal of Dentistry 26:649–656.

Kennedy PK (1995) A first-order model for computation of laser-induced breakdown thresholds in ocular and aqueous media: part I-theory. IEEE Journal of Quantum Electronics 31:2241–2249.

Krauss JM, Puliafito CA (1995) Lasers in ophthalmology. Lasers in Surgery and Medicine 17:102–159.

Kuwamura YK, Toyoda K, Namba S (1982) Effective deep ultraviolet photoetching of polymethyl methacrylate by an excimer laser. Applied Physics Letters 40:374–375.

Maiman TH (1960) Stimulated optical radiation in ruby. Nature 187:493–497.

McNally KM, Sorg BS, Welch AJ, Dawes JM, Owen ER (1999) Photothermal effects of laser tissue soldering. Physics in Medicine and Biology 44:983–1002.

Menovsky T, Beek JF, Van Gemert MJ (1999) Commentary: Photothermal effects of laser tissue soldering. Physics in Medicine and Biology 44 (doi:10.1088/0031–9155/44/4/027).

Niemz MH (2004) Laser-Tissue Interactions. Springer, Heidelberg.

Pettit GH (2002) The physics of ultraviolet laser ablation. In: Waynant RW (ed) Lasers in Medicine, CRC Press, Boca Raton, FL, pp. 109–133.

Sieron A, Kawczyk-Krupka A, Adamek M, Cebula W, Zieleznik W, Niepsuj K, Niepsuj G, Pietrusa A, Szygula M, Biniszkiewicz T, Mazur S, Malyszek J, Romanczyk A, Ledwon A, Frankiewicz A, Zybura A, Koczy E, Birkner B (2006) Photodynamic diagnosis (PDD) and photodynamic therapy (PDT) in dermatology: "How we do it". Photodiagnosis and Photodynamic Therapy 3:132–133.

Srinivasan R, Mayne-Banton V (1982) Self-developing photoetching of poly(ethylene terephthalate) films by far-ultraviolet excimer laser radiation. Applied Physics Letters 41:576–578.

Thompson KP, Ren QS, Parel JM (2002) Therapeutic and diagnostic application of lasers in ophthalmology. In: Waynant RW (ed.) Lasers in Medicine, CRC Press, Boca Raton, FL, pp. 211–245.

Torres JH, Motamedi M, Pearce JA, Welch AJ (1993) Experimental evaluation of mathematical models for predicting the thermal response of tissue to laser irradiation. Applied Optics 32:597–606.

Trokel SL, Srinivasan R, Braren B (1983) Excimer laser surgery of the cornea. American Journal of Ophthalmology 96:710–715.

Vogel A, Schweiger P, Frieser A, Asiyo MN, Bringruber R (1990) Intraocular Nd:YAG laser surgery: laser-tissue interaction, damage range, and reduction of collateral effects. IEEE Journal of Quantum Electronics 26:2240–2260.

Vogel A, Venugopalan V (2003) Mechanisms of pulsed laser ablation of biological tissues. Chemical Review 103:577–644.

Welch AJ (1984) The thermal response of laser irradiated tissue. IEEE Journal of Quantum Electronics 20:1471–1481.

Welch AJ, Gardner C (2002) Optical and thermal response of tissue to laser radiation. In: Waynant RW (ed.) Lasers in Medicine, CRC Press, Boca Raton, FL, pp. 27–45.

Wohlrab TM, Rohrbach JM, Erb C, Schlote T, Knorr M, Thiel HJ (1998) Argon laser therapy of benign tumors of the eyelid. American Journal of Ophthalmology 125:693–697.

Index

A

Ablation
  damage, 257, 260
  non-thermal, 248
  rate, 63, 64, 250–253, 547
  thermal, 61, 249, 525, 526
  threshold, 62–64, 250–253, 258,
    531, 533
Absorption
  coefficient, 35, 36, 63, 104, 250, 251, 294,
    421–424, 523, 526
  depth, 62, 250, 251, 531
  multi-photon, 39, 59, 60, 249, 533
Absorptivity, 34, 36, 41, 55, 60, 131, 132, 148,
  194, 196
Adhesive bonding, 74, 75
Age related macular degeneration, 536–537
Amorphization, 258
Amorphous film, 459, 462
Amplification, 3, 6, 9–13
Angioplasty, 522, 545, 547
Annealing, 39, 258
Arc augmented laser cutting, 200–202
Assist gas, 85, 98, 115, 117, 120–124,
  129–134, 144–146, 152, 154, 183
Auxiliary inert gas jets, 193, 196
Avalanche breakdown, 59, 533

B

Ballistic particle manufacture (BPM), 357
Beam divergence, 15, 16, 31, 128, 241
Beam interference solidification (BIS), 356
Beam polarization, 186, 187, 227, 228,
  230, 242
Beam splitter, 268, 269, 452, 456
Beat expanding telescope (BET), 456
Beer-Lambert law, 35, 421, 424, 522
Bend correction, 80, 341

Bending
  angle, 91, 291, 296, 298–301, 304,
    307, 308
  rate, 323, 325, 326
Biomaterials, 472, 473
Blow-off, 62, 63, 531, 532
Boltzmann distribution law, 6
Bond breaking, 61, 62, 248, 249, 264, 273,
  530, 531
Break out station (BOS), 380
Brightness, 17
Brittle temperature range (BTR), 430
Buckling, 73, 293, 294, 296–300, 308, 309
Build style, 365, 367
Build up time, 354
Bulging, 297

C

CAD model, 353, 354, 356, 359, 360, 370,
  372, 373, 379, 402
Carries, 162, 378
Cascade ionization, 59, 532, 533
Casting, 69–71, 74, 75, 79–81, 86, 88, 90, 91,
  95, 367
CO_2 laser, 14, 23, 25–27, 36, 37, 39, 98, 99,
  144, 157, 159, 160, 167
Coherence
  length, 454, 457
  spatial, 454
  temporal, 454, 457
  time, 454
Collimation, 15, 16
Compressive stress, 170, 225, 297, 299, 346,
  347, 491–493, 496, 497
Computer tomography, 353
Concept modelers, 359
Confined ablation, 477–484, 486, 487, 489,
  490, 493

Controlled fracture, 145, 146, 167–171, 185, 186, 198
Cooling rate, 48, 50, 393, 404, 418, 438, 444, 463, 473
Corneal surgery, 538
Corrosion resistance, 75, 496, 497
Coupling coefficient, 482, 483
Crack susceptibility, 430
Cracking, 79, 90, 91, 117, 137, 180, 276, 406, 429, 430
Cross-hatch, 231
Crystallization, 70, 250, 260, 459–464, 466
Curing, 356, 360, 361, 363–365, 367, 368, 375
Cutting
   arc augmented, 200–202
   assist gas, 145, 152, 183, 190, 192, 193, 200
   ceramics, 198
   composites, 200, 202
   controlled fracture, 145, 167–170, 198
   dross, 154, 171, 176, 178
   edges, 142, 172, 176, 189, 191, 196, 197, 200, 500, 502
   energy balance, 165, 166
   evaporative, 145–152, 158, 163, 165, 167, 198
   exothermic reaction, 158, 160
   focus, 178, 186, 187
   forces, 152, 155, 158, 160
   fusion, 145, 146, 152–158, 163–166, 176, 178
   glasses, 198–199
   groove, 147–152, 165, 167–169
   heat affected zone, 179–182, 235, 280
   heat transfer, 147, 148, 152, 168, 169, 174
   kerf width, 145, 154, 157, 160, 161, 163, 165, 181–184, 187, 191
   laminates, 202–203
   mass balance, 154, 160, 161, 163, 164
   materials, 193–199
   mild steel, 146, 154, 157–160, 163, 164, 172–176, 179, 190, 193, 195, 199–201
   molten layer, 158, 160–165, 174, 175
   momentum balance, 155–156
   nozzle, 152, 155, 176, 178, 187–191
   oxygen-assisted, 158–166, 174, 179, 191, 196, 201
   penetration velocity, 146
   polymers, 146, 197–198, 200
   quality, 145, 146, 163, 167, 171–183, 186–191, 193, 196–198, 200–203
   reactive, 145, 146, 152, 158–166
   roughness, 176, 181–184, 186

   speed, 144, 157, 160, 161, 163–167, 170–192, 194–197, 199–203, 217, 221, 222
   stainless steel, 158–160, 176–178, 184, 191, 193–195, 199, 203, 280
   striations, 171–175
   underwater, 201–202

D
Deformation zone, 76, 225
Degree of ionization, 59
Density gradient field, 132, 188
Dentistry, 522, 547
Depth
   dress, 501
   kerf, 235, 239
   melting, 43–48, 418, 513, 518–520
   penetration, 150, 238, 251, 253, 364, 417, 419, 426, 434, 528, 529
   vaporization, 45, 46, 235
Dermatology, 522, 540, 543, 544
Diabetic retinopathy, 535, 536
Diamond dressing, 501, 502, 505, 519
Dicing, 167, 276, 277
Dielectric breakdown, 483–485, 487
Diffraction, 11, 15, 18, 101, 225, 266, 270, 278, 279, 356, 462, 470, 471
Direct
   ablation, 87, 137, 477–479, 482–484
   writing, 265–266
Directed light fabrication (DLF), 403, 404
Dislocation density, 325, 493
Drawing, 71–73, 97, 136, 291, 361, 384
Dressing
   forces, 505
   lead, 501, 506, 507
Drilling
   advances, 137–140
   analysis, 103–117
   aspect ratio, 84, 117, 119
   assist gas, 98, 115, 124, 129–133
   cooling holes, 97, 134–136
   diamonds, 136
   energy balance, 98–102, 104, 111
   focusing conditions, 128–129
   gas pressure, 131–134
   melt expulsion, 99–103
   microdrilling, 136–137
   multiple reflections, 115, 116
   nozzle design, 129, 134
   percussion, 85, 98, 99, 120, 122, 134, 138, 272
   quality, 117–124, 134, 138–140

recast layer, 99, 100, 115–117, 122
single-pulse, 132
spatter, 100, 117, 120–123, 137, 138
taper, 117, 120, 121, 125, 137
temperature distribution, 103, 104, 107
temperature, 99–101, 104–111, 113, 117, 118, 131, 134, 139
trepanning, 98, 134
velocity, 101–107, 111–113, 118
Dross, 85, 100, 152, 154, 171, 176–179, 186, 190, 191, 193, 194, 196, 197

E

Edge effects, 323, 328–330
Electromagnetic radiation, 3–6, 34, 35, 522
Electron-hole pair, 263
Electrosetting (ES), 357
Engraving, 277
Etch rate, 257, 262–265
Etching
dry, 260–261
wet, 260–262
Evaporative cutting, 198
Excimer laser, 27, 137, 249, 252, 253, 256, 261–264, 267
Exothermic reaction, 130, 132, 146, 156, 158, 160
Extrusion, 70–73, 80, 86, 91, 291, 357, 371

F

Faceted grains, 511, 513
Fatigue resistance, 180, 493, 496
Feature size, 456, 457
Flexible straightening, 343–344
Focus, 18, 115, 128–130, 178, 186, 187, 231, 242, 255, 266, 270, 403, 429, 435, 436, 538
Forging, 69–73, 80, 86, 91, 291
Forming
advances, 344–349
analysis, 299–312
applications, 341–344
bending, 341
buckling, 296–299, 308–312
composites, 349
edge effects, 329, 330
microstructure, 337–341
process control, 333–337
scanning strategies, 80, 329–333
temperature gradient, 294–296, 300–308
tubes, 344–347
upsetting, 299

Fourier Equation, 455
Fracture toughness, 338, 340
Free expansion, 292, 295, 297, 478
Frequency multiplication, 22
Fringe spacing, 455, 456
Fuel injector nozzle, 137, 283, 284
Fused deposition modeling (FDM), 118, 356–357
Fusion cutting, 145, 146, 152, 154, 157, 158, 164–166, 176, 178

G

Gain coefficient, 11
Gas
pressure, 129, 131–134, 178, 179, 187, 188, 190, 191
type, 129, 190
Gaussian beam, 16, 18–20, 51, 138, 153, 165, 186, 272, 418
Glaucoma, 535, 538
Grinding
forces, 508, 509
performance, 499, 500, 502, 513
wheel, 77, 78, 209, 243, 499–520
Groove depth, 150, 165, 240, 241, 253, 255, 277

H

Hair removal, 541–543
Heat affected zone, 74, 83, 85, 91, 117, 136–138, 179–182, 235, 280, 337, 338, 424, 431–432, 441
Heat corralling, 331
Heat transfer equation, 147, 149, 168, 236, 312, 415, 416, 456, 468, 509, 518, 526, 528
Heating, 34, 37, 38, 40–45, 52–54, 62, 74, 80, 86, 103, 106, 137
Holding time, 312, 313, 316, 317, 325
Holographic interference solidification (HIS), 356
Hot isostatic pressing (HIP), 395, 396

I

Implantable prostheses, 407
Impulse momentum, 480, 482, 483
Incubation pulses, 254–256
Ink jet nozzle, 357
Integrity, 222

Interaction
  ablation, 61–64
  absorption, 34–39
  heating, 40–43
  melting, 43–45
  plasma, 59–61
  recoil pressure, 55–59
  reflection, 34, 35
  scattering, 34
  vapor expansion, 55–59
  vaporization, 45–50, 55, 56, 58, 59, 61
Interference, 188, 190, 258, 265, 268–270,
      356, 451–475
Interferometer, 269, 452, 453, 456, 457
Inverse bremsstrahlung, 59, 421, 533
Ionization
  avalanche, 59, 60, 533
  cascade, 59, 532, 533

J

Joining, 74, 75, 79, 83, 86, 91, 95, 365, 412,
      438, 440
Joint configurations, 437–438

K

Kerf width, 145, 154, 157, 160, 161, 163, 165,
      181–184, 187, 191, 242, 243, 276
Keyhole, 83, 412–414, 417–424, 427–429,
      434, 435, 443

L

Laminated object manufacturing (LOM), 80,
      82, 358, 397–402
Laser
  casting, 79–81
  CO_2, 14, 23, 25–27, 36, 39, 99, 144, 157,
      159, 160, 167–169
  cutting, 84, 85, 144–203, 237, 280, 281,
      358, 397, 400, 402, 423
  dicing, 276–277
  dressing, 194, 499–520
  drilling, 48, 84, 85, 97–140, 271, 273,
      284, 423
  engineered net shaping (LENS), 80, 82,
      358, 402–405, 407
  engraving, 277–280
  excimer, 27, 137, 249, 252, 253, 256,
      261–264, 267, 272, 273
  forming, 80–82, 91, 291–349
  interference crystallization, 459, 460,
      462–464, 466
  joining, 83–84, 91
  liquid dye, 23, 31–32
  machining, 18, 24, 25, 34, 84–85, 88, 91,
      207–244, 499
  marking, 277–280
  micromachining, 247–286
  milling, 84, 226, 227, 231, 234
  moving source, 147, 415
  Nd:YAG, 24–25, 36, 98, 99, 122, 123, 132,
      134, 140, 167–169
  operation mechanism, 6–13
  point source, 415
  scribing, 167, 276–277
  semiconductor, 16, 23, 27–31
  shaping, 80–82, 451
  Ti:sapphire, 24, 250, 258, 260, 261
  turning, 234, 243
  welding, 46, 83, 343, 412–444
Laser-assisted dressing, 502–506, 511
Laser-assisted machining (LAM), 84,
      207–225, 502
Laser-supported
  absorption wave (LSAW), 61
  combustion wave (LSCW), 39, 61
  detonation wave (LSDW), 39, 61
Layer pitch, 365
Layered manufacturing, 353
Line energy, 317, 318
Liquid
  dye laser, 23, 31–32
  phase sintering (LPS), 383, 384
  temperature polymerization (LTP), 356

M

Machining
  abrasive jet, 78, 94
  chemical, 78, 94
  electric discharge, 94, 97
  electrochemical, 78, 94
  intersecting beams, 234, 242
  laser, 84–85, 91, 225
  non-traditional, 78–79
  plasma arc, 94
  single beam, 233
  three-dimensional, 215, 225, 233, 243–244
  traditional, 75–78
  ultrasonic, 78, 94
Magnetic resonance imaging, 353, 355
Manufacturing
  adhesive bonding, 74, 75
  brazing, 74
  casting, 69–71, 91
  ceramics, 79, 85, 86, 88

cold working, 70
composites, 78, 79, 85, 86, 88
defects, 91
drawing, 71, 73
drilling, 76–77
economics, 92–95
extrusion, 71–73, 91
forging, 69, 70, 72, 73, 91
hot working, 70
machining, 70, 75, 91
material removal, 70, 75, 76, 78, 79
metals, 80, 85
milling, 76, 77
molding, 74, 91
planning, 77
plastics, 86–87
powder metallurgy, 73–74, 91
primary, 69–70
quality, 90–92
rolling, 70, 72, 91
secondary, 69–70
selection, 85–95
shaping, 77, 80
sheet metal forming, 71–73, 80, 91
soldering, 74, 83
turning, 76, 77
welding, 74, 83, 91
Marking, 64, 95, 266, 277–279
Mask projection, 265, 267–268, 282
Mass balance, 154, 160, 161, 163, 164
Material removal rate, 58, 79, 115, 129, 152,
    209, 219, 229–232, 234, 243, 500, 501
Mechanical dressing, 501, 502, 504–506,
    511, 512
Melt
    ejection fraction, 101, 102
    expulsion, 57, 58, 99–103, 111, 145, 526
Melting, 37, 40, 43–46, 48, 73, 79, 80, 83
Micro
    columns, 260
    craters, 506
    drilling, 84, 136–137, 271, 273
    fabrication, 88, 89, 247, 344
    machining, 136, 137, 247, 249, 251, 253,
        257, 260
    stereolithography, 377–379
Micromachining
    ablation, 247–260, 265, 270, 271, 281–283
    ablation rate, 250–253
    applications, 136, 137, 251, 266, 271–286
    dicing, 276, 277
    direct writing, 265, 266
    engraving, 277
    etch depth, 255, 256, 262

etch rate, 257, 262–265
etching, 247, 260–265, 270, 283
femtosecond laser, 250, 259–261
fuel injector, 283–285
incubation, 254–257
inkjet nozzle, 272, 273
interference, 258, 265, 268–270, 451
marking, 266, 277–279
mask projection, 265, 267, 268, 282
mechanisms, 247–265
microchannels, 281, 282
microvia, 271, 272
photodissociation, 262, 264
resistor trimming, 273–275
scribing, 266, 276, 278
stents, 280, 281
stripping of wire insulation, 285, 286
techniques, 265–271
thin films, 282, 283
Microstructure, 48, 179, 325, 326, 337, 338,
    340, 365, 367, 385, 387
Microvia, 271, 272
Mode locking, 20, 22
Molding, 71, 72, 74, 86, 91, 407
Momentum balance, 155, 156, 162
Monochromaticity, 13–15
Morphology, 64, 140, 220, 222, 258, 260, 274,
    382, 383, 404, 434, 473, 494, 500, 511,
    513, 526
Moving source, 53–55, 110, 147, 153, 162,
    235, 236, 415, 417, 510
Multiple reflections, 115–117, 152, 165, 413,
    421, 434

N
Nd:YAG laser, 8, 20, 23–25, 36, 37, 98, 99,
    122, 123, 132, 134
Neural networks, 322, 324
Non-traditional machining
    chemical, 78, 79
    electrical, 78, 79
    mechanical, 78, 79
    thermal, 78, 79

N
Normal pulsing, 20, 21, 126, 127, 138
Nozzle design, 85, 129, 134, 187, 188, 190

O
Ophthalmology, 526, 530, 534–535
Osteotomy, 547

Otolaryngology, 522, 545
Out-of-plane bends, 292, 293, 296, 330, 332,
        344, 346, 349
Overlapping multiple grooves, 225, 226,
        230–232
Oxygen assisted cutting, 160, 191, 196

P
Paper lamination technology (PLT), 399
Path difference, 454, 457
Patterning, 64, 267, 271, 282, 283, 464, 473
Peak pressure, 477, 478, 483–489, 491–493
Penetration depth, 150, 238, 251, 253, 364,
        417, 419, 426, 434, 528, 529
Percussion drilling, 98, 99, 120, 122, 134,
        138, 272
Periodicity, 269, 452, 453, 467
Phase transformation, 40, 43, 91, 147, 152,
        156, 464, 469–471
Photoablation, 37, 61, 273, 526, 529–532, 535
Photochemical, 37, 248, 249, 261, 525, 531, 544
Photocoagulation, 526, 527, 534, 535, 537, 538
Photodisruption, 526, 534, 535
Photodissociation, 262, 264
Photodynamic therapy, 537, 544
Photolithography, 359
Photopolymerization
        cationic initiated, 362, 363
        radical initiated, 362
Photothermal, 61, 261, 264, 526–529, 531, 538
Pigmented lesion, 540–541
Plasma
        absorption coefficient, 63, 421, 422
        coupling, 60, 61,
        shielding, 61, 63, 531
Point source, 53–55, 110, 166, 236, 415, 416
Polarization, 9, 34, 186, 187, 227–229, 242
Population inversion, 6–8, 24, 27, 29
Porosity, 71, 91, 382, 389, 392, 401, 427–429,
        439, 443, 500, 513, 519
Post-processing, 354–355, 361, 368–370, 386,
        407
Powder metallurgy, 73–74, 91
Preload, 347–348
Process control
        feedback, 334–336
        predictive, 334, 335
Process planning
        orientation, 371–372
        path planning, 375
        slicing, 373–375
        support structures, 372–373
Protein, 523–526, 545, 547

Pulse
        duration, 21, 63–64, 113, 126, 137, 253
        multi-pulse, 52, 53
        rectangular, 52, 53
        repetition rate, 20–22, 25, 63, 64, 253, 274
        shape, 21, 52–53, 124, 126–127, 183, 486,
                487, 493
        triangular, 52, 53, 115
Pumping, 6, 8, 9, 22, 24, 25, 31

Q
Q-switching, 20–22
Quantum theory, 4–6

R
Radiance, 17
Rapid
        manufacturing, 292, 343, 407
        solidification, 404, 438, 439, 511
        tooling, 367, 407
Rapid prototyping
        applications, 405–408
        ballistic particle modeling (BPM), 357
        beam interference solidification (BIS), 356
        classification, 355–359
        concept modelers, 359
        electrosetting (ES), 357
        fused deposition modeling (FDM),
                356–357
        holographic interference solidification
                (HIS), 356
        laminated object manufacturing (LOM),
                82, 358, 397
        laser engineered net shaping (LENS), 80,
                82, 358, 402–405
        liquid thermal polymerization (LTP), 356
        post-processing, 354, 361
        selective laser sintering (SLS), 80, 82, 357,
                379
        shape melting (SM), 357
        solid foil polymerization (SFP), 358
        stereolithography, 80, 354, 356, 359
        three dimensional printing (3DP), 358
Rate of densification, 383, 393
Reaction sintering, 219, 224, 401
Reactive cutting, 145, 146, 152, 158–160, 163
Recoil pressure, 55–59, 115, 139, 413, 423
Recrystallization, 70, 250, 260
Reflectivity, 9, 36, 37, 114, 194, 424, 434, 452
Refractive index, 36, 132, 188, 271
Residual stress, 91, 209, 225, 276, 298, 299,
        305, 340, 492, 493, 495

Resin, 80, 87, 198, 360–363, 365, 368, 370
Resistor trimming, 273–275
Resonant cavity
    mirror configurations, 12, 13
    stability, 12, 13
Retinal vein occlusion, 535–536
Reverse engineering, 353
Ripple, 258
Rolling, 70–72, 86, 91
Runaway instability, 423

S

Saha equation, 59
Scan pitch, 365
Scanning
    near-field optical microscopy
        (SNOM), 270
    speed, 229–230, 277, 307, 313, 329, 382
Scanning strategy
    alternating, 330, 331
    circumferential, 345, 346
    dashed line, 330, 331
    single direction, 330, 331
Scattering, 11, 253, 522, 523, 528, 529
Scribing, 109, 167, 198, 266, 276–277
Selective laser melting (SLM), 382, 383
Selective laser sintering (SLS), 80, 357,
        379, 391
    ceramics, 381–388
    metals, 381–388
    polymers, 388–389
Semiconductors
    film, 459–464
    laser, 23, 27–31
Sequential pulse delivery pattern control
        (SPDPC), 121, 123, 127, 138, 139
Shape melting (SM), 357
Sheet metal forming, 71–73, 80, 91, 292
Shielding gas, 427, 429, 435–437, 439
Shock
    processing, 477–497
    wave, 103, 139, 478, 480, 481, 491, 492,
        533
Shot peening, 496, 497
Shrinkage, 71, 74, 88, 91, 362, 371, 380, 389,
        399, 429, 440
Skin resurfacing, 543
Soldering, 74, 83, 544–545
Solid
    foil polymerization (SFP), 358
    freeform fabrication, 353
    ground curing (SGC), 360, 361
    state sintering (SSS), 381–383

Specific cutting energy, 215–219
Spot size, 18, 128, 240, 317, 339, 390, 427,
        506, 529
Standoff, 134, 155, 187, 190
Stefan problem, 114
Stereolithography
    microstereolithography, 377–379
    orientation, 370–372
    path planning, 370, 375
    resins, 363, 368, 376
    support structure, 370, 372–373
Stimulated emission, 6, 8–9, 12, 14, 29
Strain hardening, 91, 325, 338
Striations, 85, 91, 171–175, 181, 182
Stripping, 285–286
Structuring, 459–469
Super lateral growth, 463
Superposition, 268, 269, 451–453
Supersonic nozzle, 188, 242–243
Surface condition, 36, 196, 209, 222, 437,
        500, 503–504, 506
Surface finish, 71, 90–92, 207, 222–225,
        242, 244, 370, 372, 374, 380, 389,
        511, 514, 519
Surface roughness, 91, 93, 181, 183,
        209, 222, 224, 373, 456, 509,
        514–517
Surgery, 407, 522–548
Synchronized overlay scanning, 268

T

Tailor welded blanks, 444
Tattoo, 540, 542
Temperature
    distribution, 37, 40, 52, 54, 55, 103,
        107, 168
    gradient, 48, 80, 101, 109, 213, 292
Temperature gradient mechanism
        (TGM), 294–296, 300–308, 313,
        332, 349
Temporal modes, 20
Texture, 371, 469, 511, 513
Thermal
    contraction, 295
    degradation, 250
    diffusivity, 40, 55, 62, 110, 119, 235, 250,
        302, 415, 417
    effects, 40–55, 325, 378, 456, 473
    expansion, 80, 111, 297–299, 305, 310,
        311, 329
    relaxation time, 62, 249, 350,
        531, 540
Thin films, 61, 282–283, 452, 464, 468

Three-dimensional machining
   analysis, 209–215
   ceramics, 207, 224, 243
   cutting forces, 209, 215–218
   heat flow, 235, 236
   heat transfer, 209, 236, 415
   integrity, 222–225
   intersecting laser beams, 233–243
   laser assisted machining (LAM), 207–215
   laser machining (LM), 225–243
   metal matrix composites, 223, 440
   overlapping multiple grooves, 226, 232
   single laser beam, 225–233
   specific energy, 215–219
   surface finish, 222–225
   temperature, 209, 213, 215, 220, 236, 301
   tool wear, 220–222
   turning, 207, 208, 243
Three dimensional printing, 358
Threshold intensity
   melting, 40, 43
   plasma, 59–61
Ti:sapphire, 24, 250, 258, 260, 261
Tissue
   absorption coefficient, 528
   welding, 544–546
Tool wear, 93, 94, 97, 98, 217, 220–222, 225
Transverse modes, 18–20
Trepanning, 85, 98, 134, 272
Tumor, 538–539, 544

U
Underwater
   cutting, 201–202
   drilling, 139, 140
Upsetting, 293, 294, 299, 300

V
Vapor expansion, 55–59
Vaporization, 37, 45–50, 55, 56, 59, 61, 85,
   99, 100
Vascular lesion, 540

W
Wave theory, 6
Wear rate, 97, 502, 505
Weld pool, 412, 413, 426, 434, 438, 439
Welding
   analysis, 423
   arc, 443
   arc augmented, 443
   ceramics, 439
   composites, 440
   conduction, 412
   cracking, 429
   deep penetration, 83, 412, 413, 418–420,
     428
   dissimilar metals, 442
   dual beam, 443, 444
   focusing conditions, 435
   heat affected zone, 424
   heat transfer, 415, 424
   joint configurations, 437, 438
   materials, 438, 442, 444
   multi beam, 443
   penetration, 83, 412, 413, 418–420, 428
   polymers, 440
   porosity, 427–429, 432, 439, 440, 443
   quality, 412, 427, 429, 439, 442–444
   shielding gas, 427, 429, 435–437, 439
   speed, 83, 414, 422–429, 431, 434, 435,
     437, 439, 440, 443
   tailor welded blanks, 444